THE CARE OF CONGENITAL HAND ANOMALIES

ADRIAN E. FLATT
M.D., M.Chir., F.R.C.S., F.A.C.S.

Professor of Anatomy and of Orthopaedic Surgery,
University of Iowa, Iowa City, Iowa;
Hunterian Professor, Royal College of Surgeons of England;
Civilian Consultant to U.S. Air Force in Hand Surgery;
Past President, American Society for Surgery of the Hand;
Past President, Midwestern Association of Plastic Surgeons

WITH 745 ILLUSTRATIONS

THE C. V. MOSBY COMPANY
SAINT LOUIS 1977

Printed in the United States of America

Distributed in Great Britain by Henry Kimpton, London

The C. V. Mosby Company
11830 Westline Industrial Drive, St. Louis, Missouri 63141

Library of Congress Cataloging in Publication Data

Flatt, Adrian E
The care of congenital hand anomalies.

Bibliography: p.
Includes index.
1. Hand—Abnormalities. 2. Hand—Surgery.
I. Title. [DNLM: 1. Hand—Abnormalities.
WE830 F586cb]
RD778.F5 617'.397 77-5932
ISBN 0-8016-1586-0

CB/CB/B 9 8 7 6 5 4 3 2 1

PREFACE

This book is offered as a guide to those who would treat the congenitally malformed hand. It is a personal report that describes one man's surgical approach and philosophy of care in dealing with these problems. I have summarized the experience gained in trying to help over 1,400 patients with malformed hands. Many of this large group of patients have been followed through to adulthood, and some have brought their own malformed children for treatment.

I have not altered my style of writing from that in my previous books, and I carry the blame for anything that may offend. I repeat my belief that it is no sin to occasionally split an infinitive. Nor do I apologize for occasional repetition; I have used it to make the subject read better and to avoid too many infuriating references to earlier pages.

I have chosen to use simple English words wherever practical. Much of the terminology in this field is derived from Greek or Latin roots, which have little or no current value because of the absence of a classical education in the schools of today. A glossary of the more common terms used in this field is included to aid those unfamiliar with the Greek and Latin roots, and in addition, I have included summaries of the more common syndromes associated with congenital malformations of the hand.

This book is not written for geneticists. They may enjoy the pictures, but I do not fully understand their field and have not attempted anything other than genetic generalizations in relation to incidence. It is my hope that those responsible for the early care of these children, such as obstetricians, pediatricians, and family practitioners, may also benefit from the large number of illustrations showing the hands before and after surgery.

This book is not intended to be an encyclopedia. I have been selective in the conditions described and have not given equal prominence to all diagnoses. In general, the longer the chapter, the more common the condition. I have tried to be practical in my advice, but no doubt each chapter will contain statements with which some experienced surgeon could reasonably disagree. This does not disturb me. I would, however, be concerned if it could be shown that my advice was harmful to a child. I have tried to follow the precepts of my former chief, the late Sir Reginald Watson-Jones, who once told me, "In all my writings on orthopaedic, fracture, and traumatic surgery I have tried to emphasize the operative procedures that are safe in the hands of surgeons with average experience rather than to describe more adventurous procedures which may be safe in the hands of highly specialized surgeons though with others may be fraught with danger."

I believe that this type of hand surgery should be performed by highly specialized surgeons and many of the operations described are not to be done by the tyro. I have deliberately included some of the more adventurous procedures because to exclude them would leave an important gap in the experience I am reporting. Surgical abilities differ and ultimately it is the conscience of the individual surgeon that will decide if he is to perform a certain procedure.

I have benefited from seeing the long-term results of surgery done in childhood and have applied this knowledge in the surgical plans I suggest. I have not attempted to describe every operation; standard procedures such as skin grafting, osteotomy, and others have not been included. I have, however, tried to include those practical details that I have found make an operation easier or an end result better.

I do not discuss the fitting of prosthetic replacements for absence of parts. This is an entirely specialized field that is well covered by surgeons working in special centers scattered throughout the country. There would be no purpose in repeating or copying their work in a book of this size.

Much of the material in these pages has appeared in papers written with my Fellows training in Hand Surgery. I am grateful to them for their help over the years, and although it may seem invidious, I especially thank Drs. William Engber and Thomas Gillespie because they have read—and corrected—the whole manuscript.

Throughout the many years that I have been engaged in the care of the congenitally anomalous hand, my secretary, Mrs. Joyce Roller, has been invaluable in tracing family histories, maintaining patient records, and in encouraging me to complete the manuscript. My research secretary, Mrs. Margaret Washburn, produced the final typed copy with incredible speed and accuracy. Without their help this book would never have seen the light of day.

Adrian E. Flatt

CONTENTS

SECTION FOUR

THE HAND

SECTION ONE

THE MALFORMED HAND

CHAPTER 1

THE ROLE OF RECONSTRUCTIVE SURGERY

Infants are expected to be born perfect in every respect; the emotional impact when parents first see their abnormal child is overwhelming. Although ultimately most parents adopt a realistic and practical attitude toward their child's malformation, they should be allowed to go through a period of self-recrimination, remorse, and indignation that this tragedy should have happened to them.

In fact, it is not the parents but the child who is really the victim and who will need strong and consistent loving support from the parents if the abnormality is to be accepted and emotional stability attained. The mother will be the mainstay in the child's support, and it is she who needs the most emotional support in the early postpartum period. There are conflicting reports about the degree of emotional disturbance suffered by the mothers of malformed children, but it must be an extremely rare individual who would not benefit from comforting and counseling during the early months after delivery. The principal factor in the parents' acceptance of the situation is the attitude of the attending physicians and the ancillary staff. Unrealistic prognoses are cruel, but so much can be done for the child with a malformed hand that sympathetic optimism is completely justified.

The best and most practical support that can be given is a detailed and factual explanation that surgery can, in most cases, improve deformity and increase function. Pictures are better than words, and visits from children who have had reconstructive surgery are far better than pictures. I have for many years asked parents if, should the occasion arise, they would be prepared to show the results of their child's surgery to others who have produced similar children. I have never been refused such a request. There is fellowship in adversity, and this sharing of information by the parents seems to be of great comfort to all concerned.

Congenital malformations are some of the most difficult problems confronting the hand surgeon. Anatomical abnormalities, unpredictable growth potential, and lack of practical cooperation by the patient make it extremely difficult to devise a proper reconstructive plan and define the timing of the chosen procedures. It is not reasonable to expect the obstetrician who delivered the child or the pediatrician who is responsible for neonatal care to formulate such a plan. But it is reasonable for these physicians to adopt an understanding and sympathetic approach and for a full physical examination to be carried out to exclude the presence of any other important, or even life-threatening, associated anomalies.

Probably the greatest mistake that can be made in the management of these malformations is to treat the child like a miniature adult with a similar deformity produced by disease or trauma. The infant or toddler does not know that others are different, and this state of innocence lasts for several years. For all malformed children some degree of withdrawal is reasonable; its extent is largely dependent upon how much attention has been directed to the malformation by the family and playmates. Most parents accept the fact that their child is "different" and realize that the child should not lead a life of isolation. The handicap must be acknowledged and the child must learn acceptance. The upper limbs are an integral element of one's body image and are exposed in the most elementary of social contacts. For children they are the tools used to explore and communicate with their expanding world.

Congenitally malformed children make every effort to keep up with their peers and frequently develop successful substitution patterns that provide the essential functions of the malformed part. There is nothing more cruel than the sometimes innocent questions and comments of young playmates. Once the malformed children have been made self-conscious they try to hide their deformity, a situation that cannot go on forever. Because of this and because some school teachers inadvertently undo the parents' early teaching by oversolicitous attention to the deformity, my reconstructive plans are always designed to be completed by the time the child is of school age. If severe deformity dictates dependence on others, then the child's functional frustration and emotional anger will appear early and may lead to rejection and hiding of the malformed limb. Some individuals cannot reconcile themselves to their obvious deformity and develop antisocial behavior that may progress to frank delinquency (Fig. 1-1).

The objective of reconstructive surgery is to provide a hand normal in function and appearance. The former is often nearly possible but the latter is frequently impossible. In planning the maximum restoration of function for these congenitally deformed hands one is confronted with such a huge variety of permutations that one has to work from general principles and meld them into a plan for each individual.

The unique feature of the congenitally anomalous hand is that it represents a relatively static situation in which the only changes are caused by growth or surgery. The former is slow and gradual and the latter is controllable. In addition, it is rare to find neurological abnormalities such as spasticity, paralysis, or lack of sensibility. There is therefore no great urgency in carrying

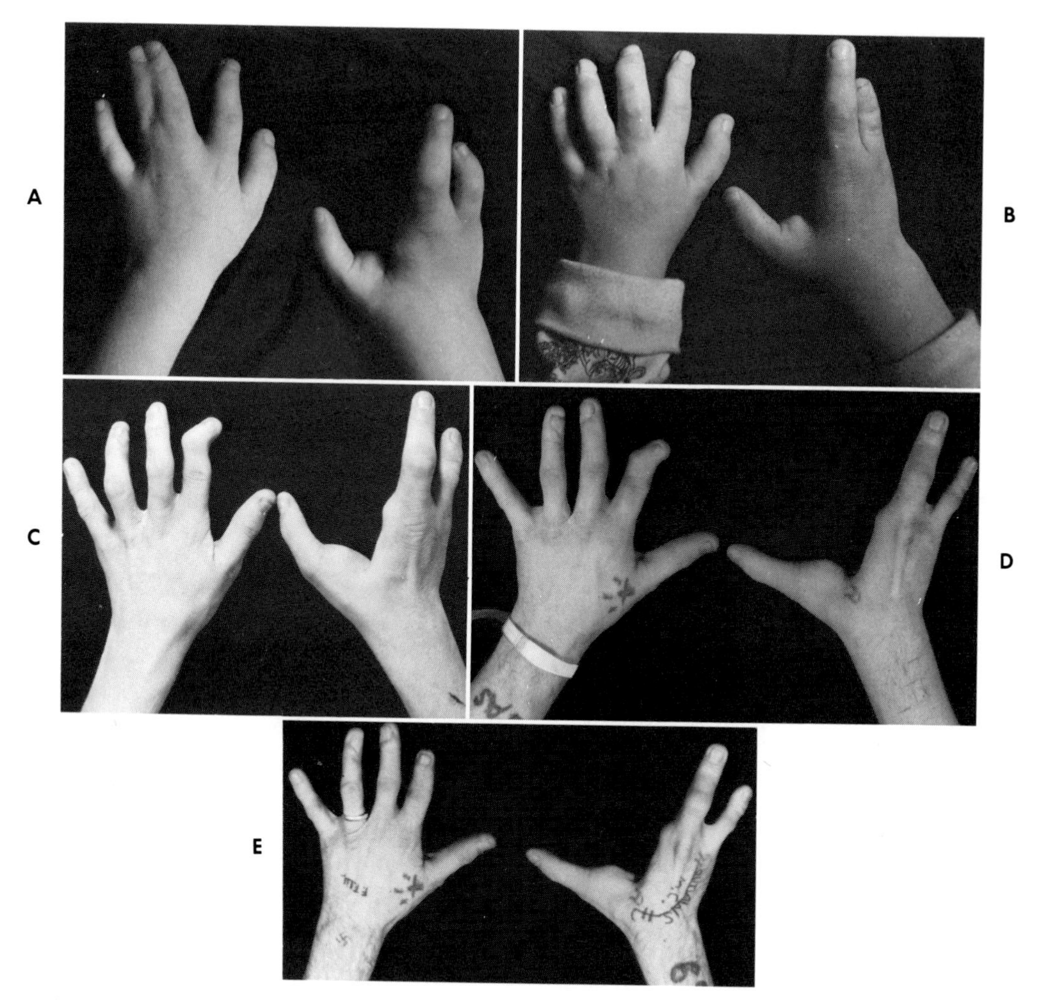

Fig. 1-1. *Deformity and delinquency.* This 30-year-old man was first seen at the age of 4 years with bilateral hand deformities. Surgery provided additional function, but as he grew his behavior became steadily worse. He is currently a long-term inmate of a state correctional school and has agreed to allow these illustrations to be published. **A,** before surgery (age 4 years); **B;** after release of the syndactyly of the left hand; **C,** age 16 years; **D,** age 23 years; and **E,** age 27 years.

out an operation. Some conditions will need surgical correction in the child's first 6 months of life, most should be operated upon by the child's first birthday, and a few can be safely postponed beyond that date.

However, some conditions need to be recognized at the outset because without correction the deformity will increase. Congenital clubhand deformities are a good example of this in that they are much more amenable to later surgical treatment if corrective plaster splinting is applied in the newborn nursery. Also, an infant with a ballooned finger may need almost immediate surgical relief of the constriction ring if the circulation to the digit is to be saved.

An extra small finger held on by a narrow skin pedicle can be removed or a narrow skin bridge joining two fingers in children with acrosyndactyly can be divided by tightly tying a black silk suture around the pedicle or bridge. Complex syndactyly of the border digits, in which the tips of the thumb and index or ring and small digits are joined, will cause a significant flexion contracture in the longer digit; therefore, the sooner these digits can be separated, the less likely is it that a permanent flexion contracture will result. Certainly the operation should be delayed a few months until the infant is thriving, and then only such surgery as is necessary to allow unimpeded growth of the digital skeleton need be done.

It is around 1 year of age that the hand becomes a fully integrated, useful tool to the child. Grasp and pinch between thumb and fingers, although present for some time, is accurately controlled and frequently used at about this age. Waving "bye-bye" and other signs used in nonverbal communication develop around this time. Accuracy of prehension, refinement of coordination, and increase in strength all continue to develop until by the age of 3 years all basic control is established. Gesell has stressed that emotional control and ability to cooperate also develop around this age. A little over 10 years ago Edgerton, Snyder, and Webb studied six children who had had thumb reconstruction operations. They concluded that these children accepted their hands as normal in direct proportion to the degree of pinch and grasp provided by the thumb reconstruction. While pointing out that the series was too small to be statistically significant, they concluded that the best adjustment to pollicization of the index could be expected if the operation was carried out between 3½ and 5 years of age. I believe that results of more recent work justify the lowering of this age considerably. Cortical control of placing the upper limb in space and of strong grasp are well developed by 1 year of age, and most hand surgeons experienced in this field consider this as the target age by which pollicization should have been done. However, a number of very experienced surgeons operate at a significantly earlier age.

In the early discussions with the parents it must be stressed that although conventional appearance of the hand is desirable, it is not always attainable and that the primary objective is the provision of useful function. I also explain to the parents that because of the dynamic nature of the developing hand, all but the simplest malformations must be regularly followed until epiphyseal growth has ceased.

The three major goals of reconstructive surgery are to provide the ability to control the placing of the hand in space, good skin cover with adequate sensibility, and satisfactory power grasp and precision handling. It is not always possible to supply all features to both hands, and I believe that the principles of redistribution of functioning parts, which are frequently used in traumatic injuries, should be more generally employed in congenital malformations of the hand. Most children with congenital malformations of the hand are mentally normal and can voluntarily control the placing of their hands. Mentally deficient children frequently do not have proper control of their limbs in space, and reconstruction of their malformed hands may not be helpful.

Many of the anomalies have a larger dynamic than static element in their nature, and it is often important to correct tendon anomalies early and provide better dynamic balance. Joint deformity and even bony deformity may be caused by soft tissue abnormalities. These must be corrected during the rapid growth period of the first 4 years of life. It is equally important to maintain frequent follow-up visits so that early changes of recurrence or of newly developing derangement can be detected and corrected.

Complicated tendon transfers should be delayed until at least 3 to 5 years of age because voluntary cooperation becomes more likely at this time; it is around this time that the children realize what is being done for them and volunteer the information that they have more hand function after a successful operation.

Early correction of skeletal alignment is vital. Extra bones, malpositioned bones, and fusions between adjacent bones must be dealt with to allow the child's growth to develop in a proper way. Osteotomies and fusions should, if possible, be left until a later date. Judgment in this area is probably the most difficult to learn, but my inclination has always been to try to align digital joint surfaces at 90 degrees to the long line of the ray early in the child's life. The longer this correction is delayed, the greater the difficulty in subsequent correction because the joint deformity usually develops in several planes.

There is no such thing as a standard malformation, and one must always anticipate unexpected anomalies to influence the plan of care. Because of this there can be no "cookbook" or "how-to-do-it" book on the surgical care of these anomalies. Nils Bohr has said that "a specialist is one who knows many of the worst mistakes that can be made and how to avoid them." The pages that follow summarize my specialized experience in this field and describe ways to avoid the many problems I have encountered.

CHAPTER 2

GENETICS AND INHERITANCE

The human arm bud appears and fully differentiates in a period of less than a month, between about the twenty-fifth and fiftieth days after fertilization. It would seem, therefore, that the great majority of congenital malformations must arise either at the time of conception, if genetic in origin, or during the second 25-day period, if caused by the developing limb bud's exposure to environmental teratogens. It is also possible that a combination of genetic and environmental factors can cause malformation. Many think that this combination is the cause of most common malformations.

Parents often hope that the smaller deformed limb will increase in size at a greater rate than the uninjured side until eventually both are the same size. This does not happen. It is probable that the limb malformations result from cell death or from abnormalities of fetal cellular growth and differentiation within the developing limb. These primary abnormalities in cellular growth persist into postnatal life, so the malformed limb continues to grow poorly and out of proportion with other, normal body structures. Thus the proportionate size difference will always remain and may even be accentuated if the normal limb hypertrophies from excessive use caused by the presence of the opposite, hypoplastic limb.

There is a continuing and fruitless debate as to whether the majority of congenital defects are attributable to genetic or environmental factors. Strong cases as to cause can be made in certain malformations by proponents of one or the other camp, but in about 60% of all cases no single cause can be assigned. Neel has estimated that of the remaining 40%, about 20% can be explained on the basis of simple dominant or recessive inheritance with full penetrance, 10% may be caused by environmental factors, and the other 10% can be attributed to major chromosomal abnormalities.

CLASSIFICATION OF DISORDERS

Genetic disorders are usually classified into three distinct groups:

Single gene disorders (mendelian)

Chromosome abnormalities

Multiple gene disorders (polygenic inheritance)

Mendelian disorders involve a single gene. Chromosome abnormalities involve absence or duplication of a much larger volume of nuclear material, often sufficient to allow their detection by routine methods. These defects may occur in the autosomes or in the sex chromosomes. Congenital abnormalities associated with autosomal defects are usually more severe than those associated with the X and Y sex chromosomes. A large number of genes are involved in chromosome abnormalities and many produce conditions that are incompatible with life. At least one third of all first trimester spontaneous abortions show chromosomal abnormalities. The polygenic inheritance group is also referred to as the multifactorial group because multiple genes are involved and environmental factors play a more significant part than they do in single gene disorders.

It is important to realize that any single defect may be produced in a variety of different ways by different causative factors. For instance, polydactyly can occur as an isolated malformation resulting from a single autosomal dominant mutant gene. It can also be seen as part of a larger group of malformations occurring in children with an extra 13 chromosome. It can also be seen in a number of patterns of unknown cause, which may be multifactorial or polygenic in origin.

The primary developmental lesion is always at the cellular level, since the genes are intracellular structures that control protein synthesis and are responsible for the primary deviation of the developmental process from the normal. The later, resultant lesion is the consequence of this disturbed early development and produces the congenital malformation. Malformations are most common toward the end of the limb and only in the mildest case is the abnormality confined to a single tissue. It is a safe generalization that defects of the soft parts roughly match the skeletal defects.

It was only in 1956 that Tjio and Levan showed that the diploid number of human chromosomes is 46. Since then our knowledge of heritable disorders has expanded enormously. Each of the 46 chromosomes is one of a homologous pair, one member of each pair coming from the mother and one from the father. Twenty-two of the pairs are identical in male and female and are known as autosomes. The sex chromosomes make up the twenty-third pair. In the female they are two similar X chromosomes, while in the male they are an X and a different Y chromosome.

Single gene disorders

The inheritance units, or genes, are segments of deoxyribonucleic acid (DNA) embedded in the protein, histone. There are estimated to be about 100,000 genes in the 46 chromosomes of a human cell. Each chromosome carries hundreds of genes, but an individual gene is so small it cannot be detected by current chromosome study techniques. Each one of us must carry some faulty genes, and it has been estimated that every person has from two to five serious mutant recessive genes. Fortunately the normal gene for a given trait transmitted at conception by one parent usually tends to overshadow the harmful effects of the faulty gene transmitted by the other parent.

Although individual genes can cause defects that appear in successive generations, it does not follow that the same gene in different individuals will produce the same defect. Wynne-Davies has stressed that the same gene in two different individuals may produce an absence of the scaphoid in one person and a profound disturbance, such as an absence of the radius and the thumb coupled with a congenital heart lesion, in another.

In several well defined categories such as polydactyly and polysyndactyly, the hereditary factors can be shown to follow mendelian laws in many instances. In other conditions, the regularity of the inheritance may be disturbed and the malformation may miss a generation, alter its degree or intensity, or demonstrate an atypical pattern. When multiple genes are involved in producing a defect, simple mendelian ratios, in which the risk is always the same, do not apply. Such disturbances of the expected pattern are explained away by geneticists with the aid of such terms as reduced penetrance, incomplete dominance, modifying genes, and variable expression. This unpredictable behavior makes it virtually impossible for the average physician or surgeon to issue competent genetic advice.

The calculation of genetic risks depends on having a proper diagnosis, the individual family history, and a profound knowledge of the literature combined with a lot of luck. The abnormality must be placed in one of four categories: environmental, single mutant genes, chromosomal aberrations, or multifactorial. The recurrence risk can then be derived from mendelian laws or by a guesstimate of the multifactorial factors.

The major genes causing malformations show the standard patterns of dominant, recessive, and X-linked inheritance. The terms dominance and recessiveness are relative in the sense that they are a function of the test used to detect the presence of the altered gene. Thus a gene is said to be dominant over the corresponding member in the pair if the altered gene is readily detectable when present in a single sampling.

In dominant inheritance the malformation appears in a heterozygote, that is, one in whom only one of a pair of genes is affected. All generations of either sex may be involved, and there is a 50% risk that each child will show the defect. However, there is an equal chance that a child will not receive the abnormal gene; thus he and his children would be free of the defect (Fig. 2-1). Examples of dominant inheritance in the hand are conditions such as syndactyly and polydactyly.

In recessive inheritance the abnormality is clinically apparent only in the homozygote, that is, one in whom both genes of the pair are abnormal. In this case both parents appear normal but by chance both carry the same harmful

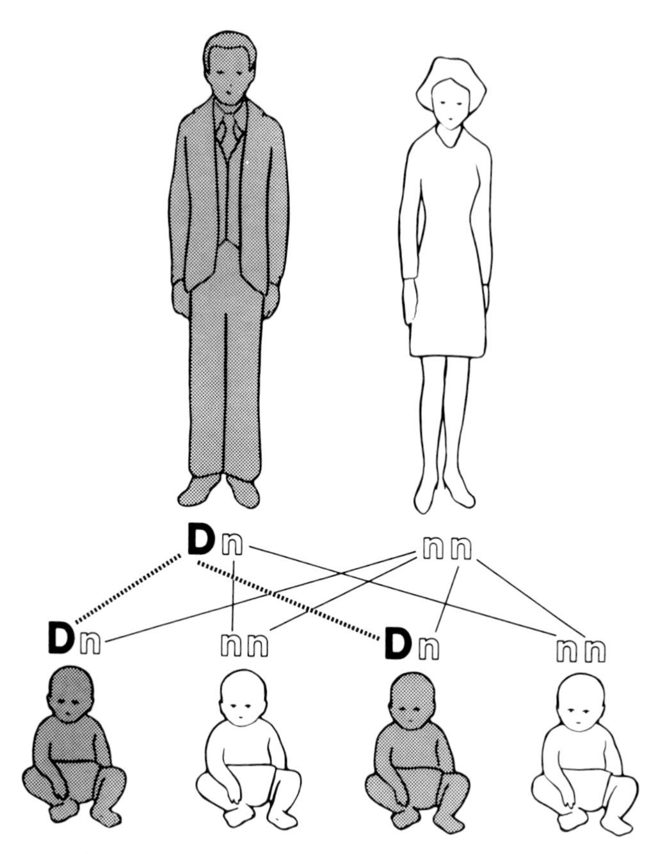

Fig. 2-1. *Dominant inheritance.* A single faulty gene D contributed by one parent dominates its partner in a pair of genes. There is a 50% chance that each child will inherit either the normal n or the abnormal D from the affected parent.

gene, although neither may know it. Each of their children has a 25% chance of showing the trait, each has a 25% chance of not inheriting the gene from either parent, and each has a 50% chance of receiving a single gene and thus becoming a carrier (Fig. 2-2). Any children of the malformed child, who carries both harmful genes, will be normal unless the child marries a similar carrier.

Consanguinity increases the chances that rare recessive traits may be present in both parents. The rarer the disease, the greater must be the frequency of parental consanguinity. Unfortunately, recessive abnormalities tend to be more severe than those caused by dominant inheritance. Malformations related to recessive inheritance are Carpenter and Laurence-Moon-Bardet-Biedl syndromes.

In X-linked inheritance the affected gene is in the X chromosome, and although the inheritance may be dominant, it is more commonly recessive. An X-linked condition can never pass from father to son since the male has only one X chromosome, which is inevitably passed to his daughters. When the mother carries a faulty gene on one of her X chromosomes, each son has a 50% chance

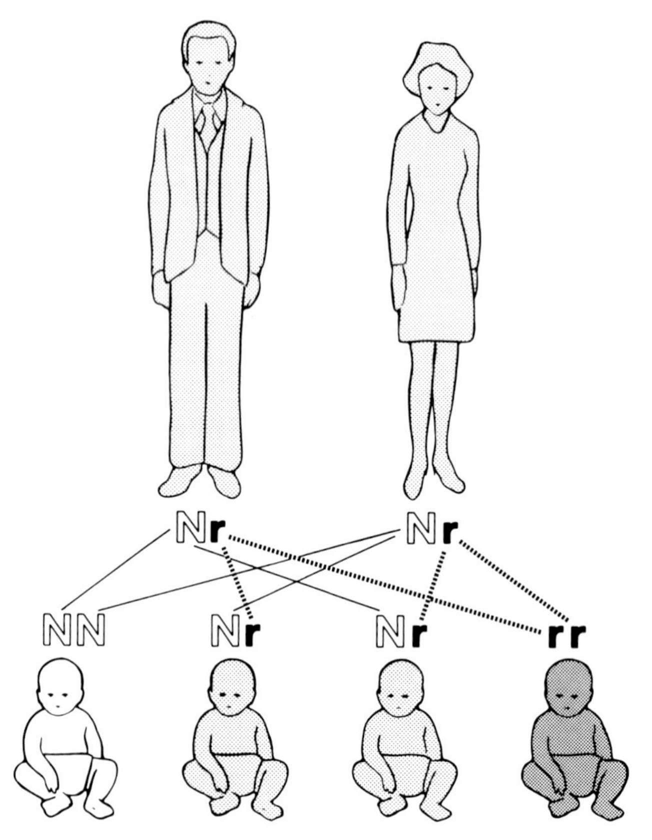

Fig. 2-2. *Recessive inheritance.* Usually both parents are unaffected because each carries N, a normal gene that "overshadows" its recessive faulty counterpart r. Each child has a 50% chance of being a carrier like its parents; each child also has a 25% chance of being abnormal by receiving two r genes or of being normal by receiving a pair of Ns.

of inheriting the gene and showing the disorder and each daughter has an equal chance of being a carrier like her mother, usually unaffected by the disorder but capable of transmitting it to the next generation of males (Fig. 2-3).

Chromosomal disorders

The most common sex chromosome abnormalities are the XO, the XXY, and the XYY syndromes; the most common autosomal disorders are trisomy 21, 18, and 13. An example of skeletal changes found in sex chromosome abnormalities is the short fourth metacarpal in Turner syndrome (45XO). In the XXXXY syndrome the extra three X chromosomes are associated with lack of pronation from elbow abnormalities and clinodactyly of the small finger. In trisomy 21, or Down syndrome, the hands are small and clinodactyly of the small finger is usually caused by an abnormal middle phalanx. More detailed discussion of these associations between chromosomal and skeletal abnormalities can be found in Poznanski's excellent book, *The Hand in Radiologic Diagnosis.*

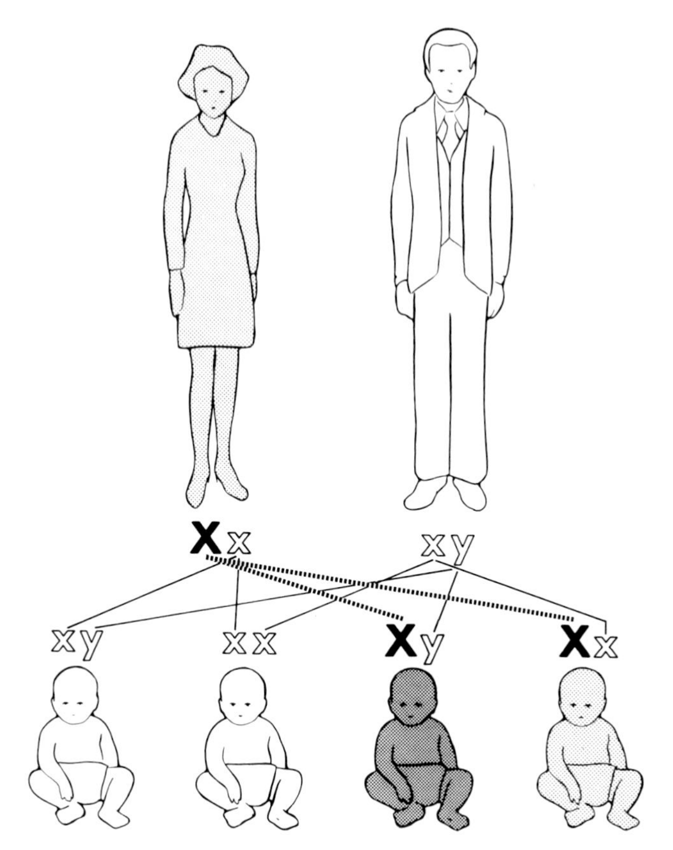

Fig. 2-3. *X-linked inheritance.* The affected gene is in the X chromosome, and although the inheritance may be dominant, it is more often recessive. When the mother carries a faulty gene on one of her X chromosomes, each son has a 50% chance of showing the disorder by inheriting the gene, and each daughter has a 50% chance of being a carrier, like the mother, and of transmitting the disorder to her sons.

Polygenic disorders

Polygenic inheritance combines disorders of multiple genes and varying influences by environmental factors. The inheritance pattern is ill defined, but it is known that the probability of recurrence is low. With one affected child in the family the chances of another having the same defect are 5% or less. Concentration of cases does occur in some families, and the more cases there are, the greater are the chances for an increasing incidence and an increasing severity in individual cases.

INHERITANCE

Reduced to their most elementary level, the risks that a child will inherit the mutant gene are:

1. Fifty percent for each child of a parent suffering from a dominant defect
2. Fifty percent for any child born to the child in 1

3. Fifty percent for any subsequent boys born to a couple who have had a child with an X-linked recessive defect
4. Twenty-five percent for any subsequent child born to a normal couple who have had a child with a recessive defect

Ruth Wynne-Davies has succinctly summarized the various possible causes when the firstborn child is deformed in a family in which the parents and all other members are normal. She lists six major possible explanations:

1. There may be no genetic or heritable basis, in which case the deformity is the result of some intrauterine environmental accident. Typical of this case are the characteristically unilateral deformities, the transverse congenital amputations, and the various hand malformations associated with ring constrictions. These seem to be truly sporadic, and one would not expect other cases in the family.
2. The cause may be a new mutation of a gene responsible for a dominant trait. One would be suspicious that this was so if the age of the parents was much above average since the number of mutations increases with age. Subsequent sibs would not be affected, but one in two of the children of this affected individual would carry the gene.
3. The cause could also be the first appearance of a homozygote in a disorder of recessive inheritance. However, none of the isolated hand deformities appears to be a recessive trait and so the disorder would have to be part of a syndrome, which may be easily identified.
4. If the hand malformation is associated with other congenital anomalies, it may result from a chromosome anomaly and an examination of the chromosomes would be worthwhile. These cases are almost always sporadic, and one would not expect a second child to be affected.
5. When the hand malformation clearly seems to be part of some syndrome that is unknown to the physician, it may be worth getting the help of a genetic counseling center since they possess a larger amount of up-to-date information than is easily available to the average surgeon.
6. One should not be too trusting about a sporadic case. The case may only seem to be sporadic in that the apparent father is not necessarily the real one and there may be a few more cases in the neighborhood.

COUNSELING

The families of children born with grossly or even mildly abnormal upper limbs are understandably distraught, and many are desperate for help. Mothers of malformed infants like to be told what is wrong with their infants in simple language; on the whole they prefer to be told the truth about their babies' condition. The initial counseling will make a lasting impression.

At this early time it is important, if possible, to at least outline what future medical or surgical steps can be taken to improve the child's ability to use his malformed limb.

Many mothers will not ask the question that bothers them the most: Will any normal children they already have and this abnormal baby inevitably produce abnormal children of their own in due course?

For genetic counseling to be effective and valuable, the following five requirements described by McKusick must be met: (1) the diagnosis should be clearly established, (2) genetic principles must be fully understood, (3) heterogeneity must be clearly defined, (4) a careful family pedigree and examination of even so-called normal members of the family must be carried out, and (5) knowledge of the condition must be adequate.

For a surgeon or family practitioner to attempt this is unwise; the better part of wisdom is to seek help when it is needed. Genetic counselors exist in many major cities, but currently there are no established standards for recognition of ability—some have medical degrees, some have degrees in genetics, a few have both. Two national organizations offer help in locating suitable counseling.* Clinical geneticists can be of great help and comfort to parents and in many instances can correctly discuss and predict the mathematical probabilities for deformity in future children.

Most parents have great difficulty in grasping the meaning of mathematical probability and certainly have difficulty in applying this to a decision. For instance, if the risks are described in fractions or percentages they can be easily misinterpreted. Parents who have produced a child with a malformation caused by recessive inheritance are told that there is a one in four, or 25%, risk that future children will be affected. Many parents think that this means that the next three children will not be affected; they find it hard to accept that the risk of genetic disease is the same for all future children.

Most counselors wisely avoid directly advising the parents on what they should or should not do. However, the parents' decision on having more children will be based on the counselors' advice, and it is often extremely difficult to reduce the complexities of genetics to a level at which it is certain the parents understand the information supplied and can relate their various options to this information.

---

*National Genetics Foundation, 250 West 57th Street, New York City 10019, and National Foundation-March of Dimes, 1275 Mamaroneck Avenue, White Plains, New York 10605.

CHAPTER 3

GROWTH, SIZE, AND FUNCTION

DEVELOPMENT

It is probable that our hands evolved from a tree-living primate. Embryologists claim to be able to trace the evolution of the human hand as far back as a primitive multirayed pectoral fin similar to that of the modern fish.

In the human embryo, limb differentiation occurs in a definite sequential order. The upper limb is first recognizable as a small bud of tissue on the lateral body wall about the twenty-sixth day after fertilization. At this time the embryo measures about 4 mm in length (Fig. 3-1). The budding limb grows rapidly in a proximal-to-distal direction with the arm and forearm appearing before the paddle-like hand plate. Scalloping of the edges of the hand plate soon occurs, and by the thirty-first day the hand segments are recognizable. The digital rays and fissures are formed by about the thirty-sixth postfertilization day (Fig. 3-2). Around this period the nerves of the limb bud segments grow into the bud mesenchyme for distribution to the dermis and the premuscle mass (Fig. 3-3). The muscles, like most tissues of the hand, are thought to develop in situ rather than to arrive by migration. Within another 10 to 12 days the digits are well differentiated and the thumb diverges from the rest of the hand.

During this time the muscles—with the exception of the intrinsic muscles—

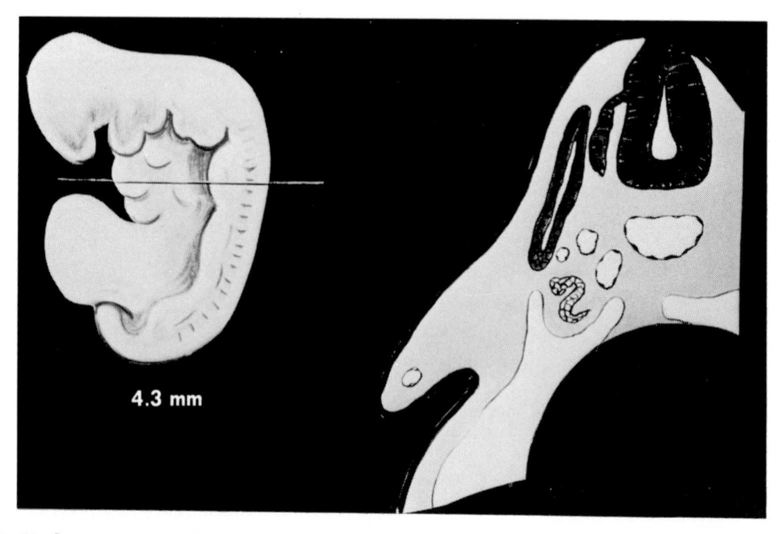

Fig. 3-1. *Embryo at 3 weeks.* The upper limb first appears as a small bud of tissue around the twenty-sixth day after fertilization.

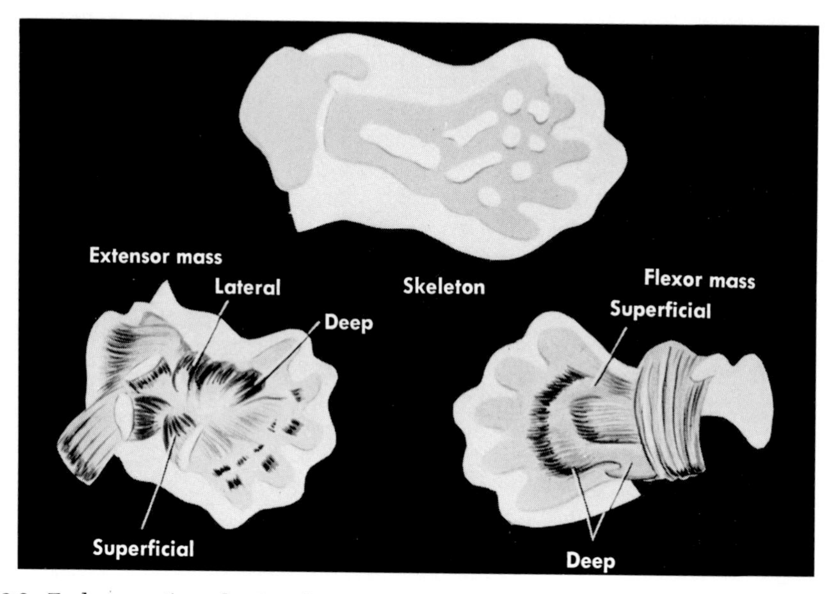

Fig. 3-2. *Embryo at 5 weeks.* Hand segments are now recognizable, and the digital rays and fissures are formed by about the thirty-sixth postfertilization day.

are defined, and the skeletal elements down to and including the proximal phalanges consist of cartilage. By the forty-eighth day all the skeleton of the hand is cartilaginous except the terminal phalanges, which still consist of condensed mesenchymal tissues (Fig. 3-4). After about the fiftieth day no further differentiation occurs. Subsequent changes are related to size increase and to the relative positions and proportions of the parts.

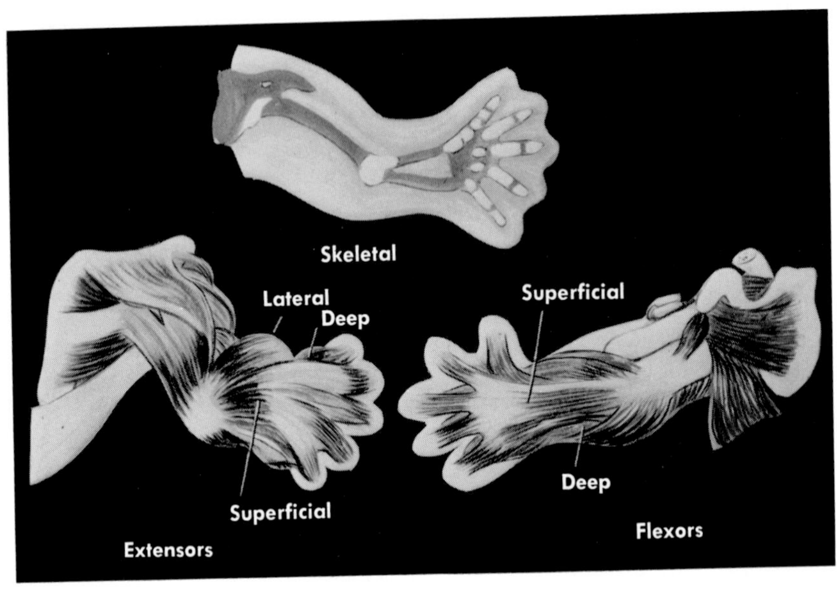

Fig. 3-3. *Embryo at 6 weeks.* The premuscle masses are forming, and the distal end of the limb bud is differentiating into digits.

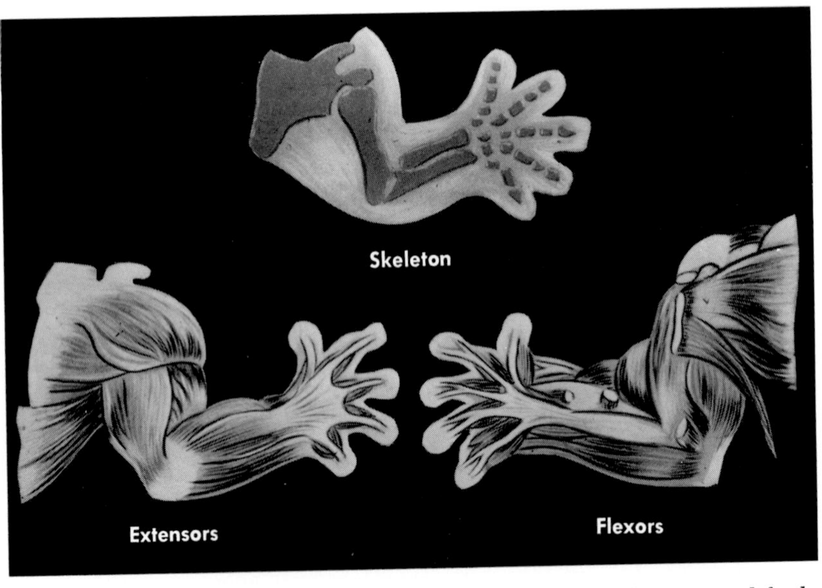

Fig. 3-4. *Embryo at 7 weeks.* The skeleton of the hand is now cartilaginous, and further differentiation has occurred in the muscles.

Joints form in the upper limb by a process of miniclefts developing in the interzones between future bones. About 48 days after fertilization, cavitation occurs in the wrist and the digital joints and, somewhat later, in the other joints. A joint cavity begins by coalescence of the minizones at the periphery of the interzone, but it cannot proceed to proper joint formation without embryonic muscular activity. There seems little doubt that movement is required throughout embryonic life—initially for the cavitation and final modeling of the joints and later for the maintenance of the joint cavities, even after they have reached a fairly late stage of development.

By definition, fetal life begins at the eighth week after conception. The joint flexure lines or skin creases of the digits are already present at the beginning of this period, which is before spontaneous digital motion has commenced. The dermal ridges do not differentiate until the third and fourth months of fetal life, and differentiation of the dorsal skin lags behind that of the palmar skin. The sweat glands begin to form in the fifth month, and their ducts reach the surface during the seventh month. Growth of a definitive nail does not begin until the fifth month, and even at full term, the tips of the nails just reach the end of their digits.

At the time of birth all major systems in the hand show their fully developed pattern except the nervous system. Usually myelinization is not complete until the child is approximately 2 years old. Thus subtle neurological testing, such as two-point discrimination and stereognosis, is valueless in very young children.

SIZE

During a child's growth, the head doubles in size and the legs become five times as long, the trunk three times as long, and the arms four times as long. No such generality can be made about the increase in hand size, but several studies, by both clinicians and anthropologists, have evolved tables that equate hand size to age. Most of these studies measure the hand in certain standardized directions (Fig. 3-5). Those commonly used are A, the length of the long finger; A + B, the length of the hand; C, the breadth of the hand; and D, the length of the thumb. Feingold has published tables showing that hand length nearly doubles in the first 2 years of life and then nearly doubles again in the remainder of the growth period (Fig. 3-6, *A*). In addition, he has shown that the average long finger length is about 42% of the total hand size (Fig. 3-6, *B*).

At the University of Iowa, our work in this area has included a study of over 1,700 school children from kindergarten through sixth grade.* For these age groups we have established norms of hand sizes and strengths for different age

*Burmeister, L. F., Flatt, A. E., and Weiss, M. W.: Size and strength development of the hand in elementary school children, Iowa City, Iowa, 1974, the Iowa State Services for Crippled Children, The University of Iowa.

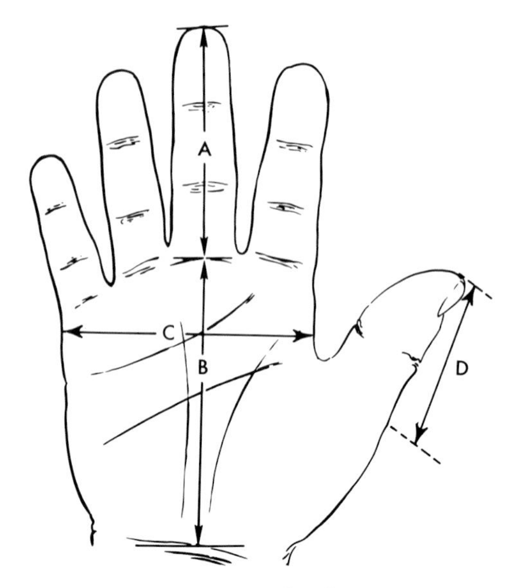

Fig. 3-5. *Hand size measurements.* The standardized measurements used are: *A*, The length of the long finger; *A* + *B*, the length of the hand; *C*, the breadth of the hand; and *D*, the length of the thumb.

groups in yearly increments. We also established separate norms for sex, hand dominance, ethnic groups, and geographical regions.

Other useful scales have been established to determine whether or not hand size is abnormal. Joseph and Meadow have employed Sinclair's method of establishing an adult metacarpal index to create an index for children up to 2 years of age. They have shown that this index is useful in diagnosing arachnodactyly and Marfan syndrome. The metacarpal index is determined by summing the lengths of the four finger metacarpals and dividing this figure by the sum of the breadths as measured at the midpoint of each metacarpal.

Radiologists have a particular interest in being able to establish normal bone lengths. Poznanski, Garn, and Holt have published a detailed study of thumb anomalies in congenital hand syndromes. They have established the normative values of the length ratios of the bones of the thumb to each other and also to the second metacarpal. The concept of using ratios is valuable because it is not dependent on the size of the individual or on the scale of the x-ray films. They have published an elaborate table of the mean and standard deviation ratios of the 12 possible bony combinations in both males and females. Tables such as these are useful in accurately establishing abnormalities of bone length but do not have any day-to-day use in ordinary clinical practice.

Of more general clinical interest is the work of the anthropologist, K. Hajnis, who has published a detailed study of the hands of 1,707 Czechoslovak school children between birth and 18 years of age. From this work he was able to establish two periods of time in which the growth of the hand is at its slowest

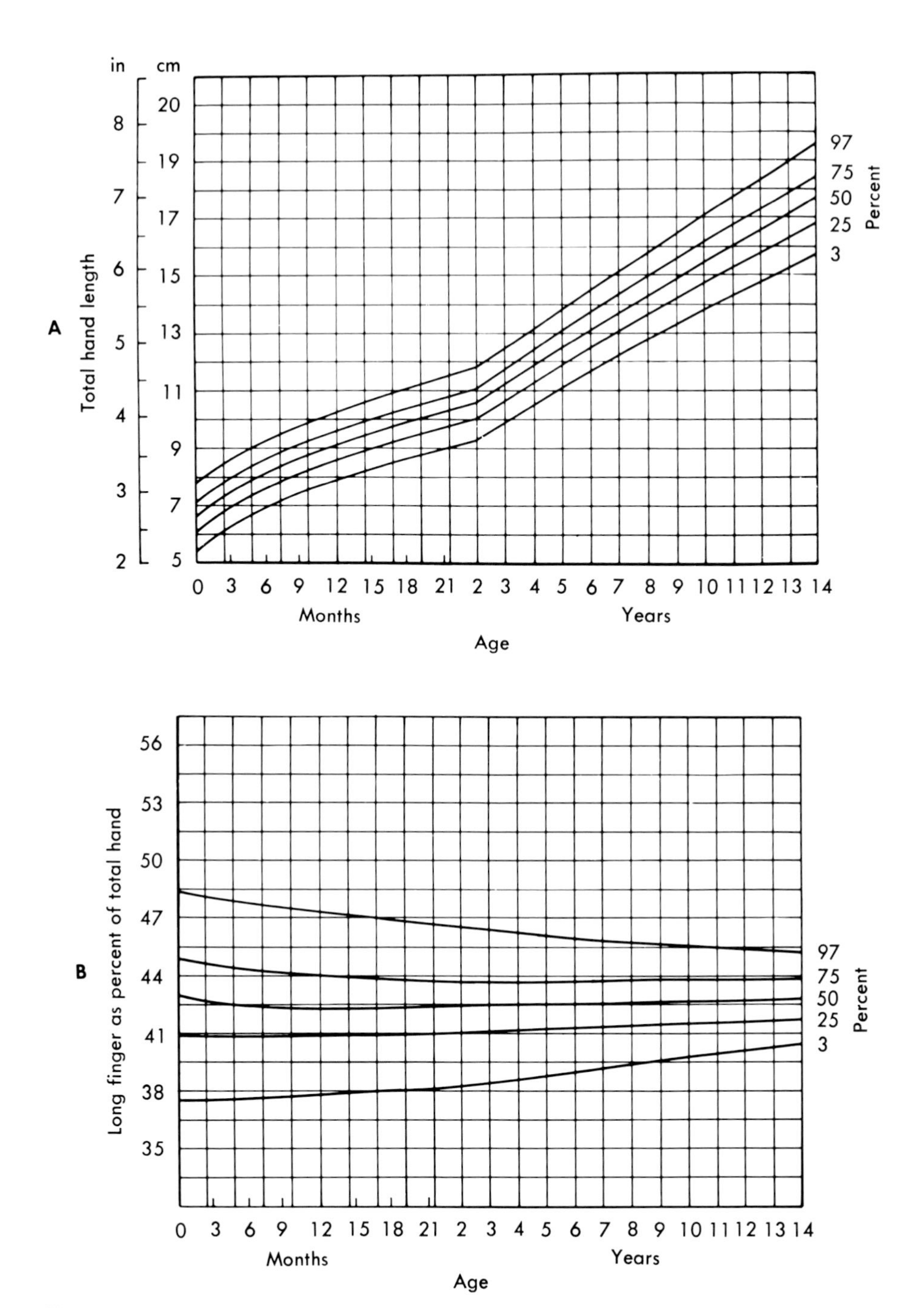

Fig. 3-6. A, Hand length. Length of the hand nearly doubles in the first 2 years of life and once again in the remainder of the growth period. **B,** Long finger length. The average length of the long finger is a little less than half of the total hand length. (From Feingold, M.: Clinical evaluation of the child with skeletal dysplasia, Orthop. Clin. North Am. **7**:291-301, 1976.)

(Table 3-1). He expressed the hope that these episodes of slower growth might be useful in surgical planning, but I have found that the functional demands of the hand do not coincide with these time periods. He also showed that there is no great difference bewteen boys and girls until puberty. The data confirm the clinical impression that the length and width of the hand reach their definitive size earlier in girls than in boys.

In the University of Iowa study of American school children we developed norms that can be used as standards in assessing the degree to which a child with abnormal hands deviates from the size and strength measurements expected of those in the same age, sex, and laterality group.

The size and strength measurements were statistically analyzed to determine the significant effects of age, sex, and hand dominance. All strength measurements were significantly affected by age and sex. However, only the left-hand strength measurements were affected by hand dominance; there was no difference in the strengths of the right hands of right-handed or left-handed children. This must be caused by the fact that the left-handed child has to develop his nondominant hand strength in order to compete in this right-handed world. When nondominant hands were compared, there was no significant difference between the right-handed and left-handed girls. The left-handed boys, however, obviously had more strength with their nondominant hands than did the right-handed boys.

The measurements of hand size show a significant relationship to age. Dominance of the hand apparently has little effect on hand size, with only the breadth of the left hand showing a significant difference between right-handed children and left-handed children. The effect of sex on hand size is apparently sporadic. There is strong evidence for a difference in breadth of boys' and girls' hands. Palmar length for both the right and left hands is also significantly affected by sex, although to varying degrees. Sex does not seem to have significant effect on digital length.

We have compared the size data of this study with those of the 1,700 Czechoslovak school children in Hajnis' study and have shown there is virtually no significant difference between the two groups with regard to hand size (Fig. 3-7).

Table 3-1. Period when the least growth of the hand and thumb takes place

	Age in years		
Boys			
Hand	2¼ to 2¾	8 to 10	17 and older
Thumb	1¼ to 2¼	8 to 10	15½ and older
Girls			
Hand	2¼ to 2¾	9 to 10	16 and older
Thumb	3 to 4	8 to 10	15 and older

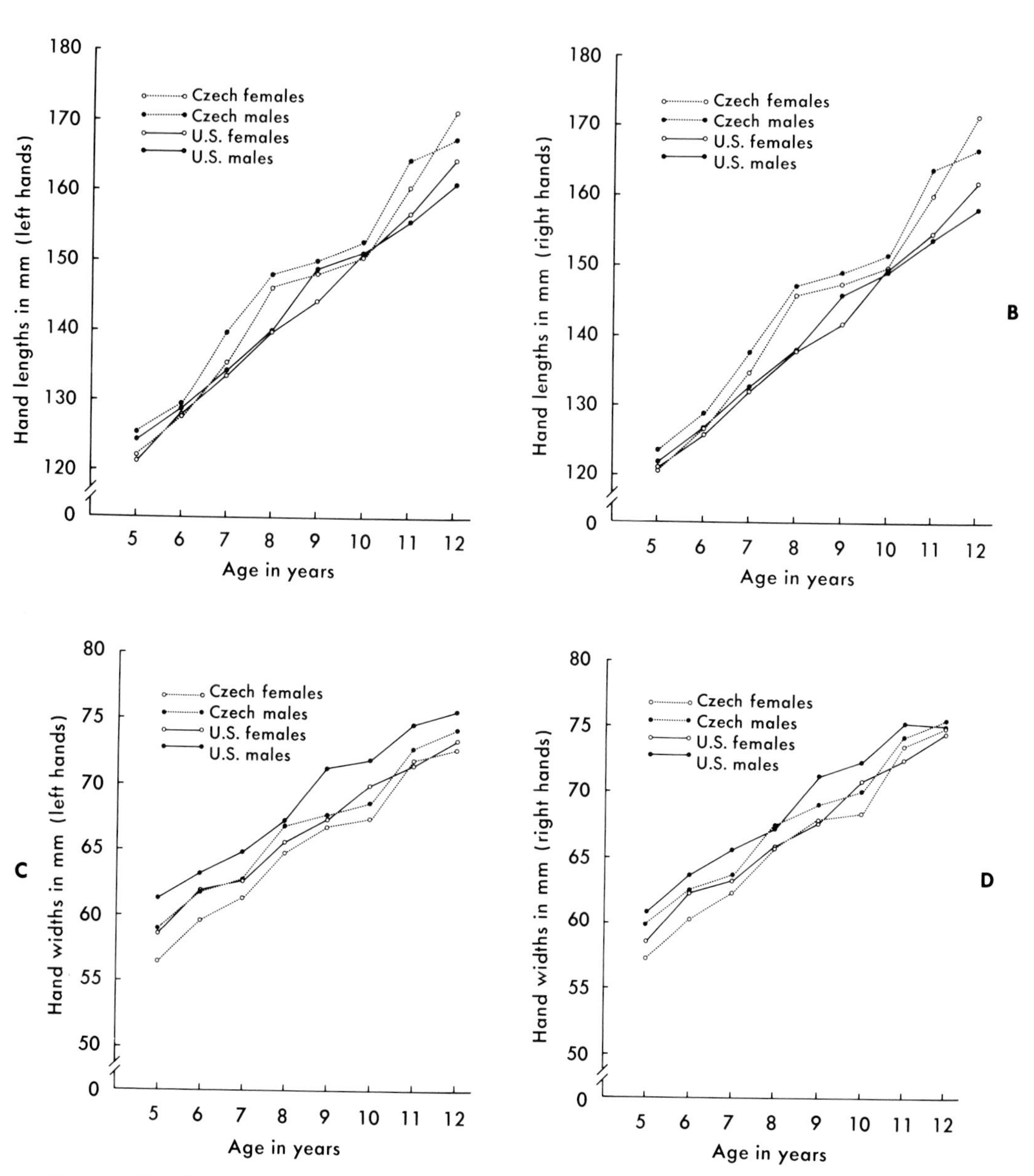

Fig. 3-7. *Hand size comparisons—U.S.A. vs. Czech children.* A comparison of our data with that published by Hajnis shows little difference in the growth curves of hand length (**A** and **B**) and width (**C** and **D**) for U.S.A. and Czech school children between the ages of 5 and 12 years.

The correlation of the size measurements with the strength measurements is sufficiently strong to indicate that the prediction of strength from size, age, sex, and hand dominance could be useful to the surgeon to indicate the degree of improvement that could be expected from surgery on the growing hand.

FUNCTION

A knowledge of the dynamic anatomy of the hand is fundamental in planning reconstructive procedures for the congenitally malformed hand. The functional patterns of prehension and use in both child and adult are based on the mobility of the skeleton, which is arranged in a series of integrated arches, the concavities of which face towards the palm (Fig. 3-8).

The depth of the concavities or cup is varied by the controlled mobility of the fingers and the two borders of the hand. The thumb contributes the greater part of the border mobility, because it can separate widely from the palm and swing around in front of, and oppose, any of the fingers. Mobility on the ulnar border is supplied by movement of the fourth and fifth metacarpals at their carpometacarpal joints. For control of the movement of the two borders, the metacarpals are slung from a centrally placed rigid pillar consisting of the metacarpals of the index and long fingers. The thumb is connected to this rigid pillar by the intrinsic muscles, and the ulnar fingers are connected by the transverse intermetacarpal ligaments.

Movement of the borders of the hand passes through the mobile transverse arch at the level of the metacarpal heads. It is the mobility of this arch that allows the palm to adapt to objects of various sizes. A more proximal transverse arch is present in the carpus and is permanent in shape. The mobile longitudinal arches are made up of all the digital rays. The metacarpal forms one side of the arch with the apex or keystone at the level of the metacarpophalangeal joint, and the phalanges form the other side of the arch. The longitudinal arches are more mobile than the transverse arches and can individually alter their shape in response to the demands of grasp.

The hand in use

Proper grasp is impossible without adequate sensory information from the skin of the palm. The direction of mobility toward the palm of the hand is associated with a great concentration of sensory nerve endings in this area. Over a fourth of all the pacinian corpuscles in the body are present in the pulp and skin of the hands.

The hand is used in two fundamentally different ways. The less common and completely unspecialized use is as a fixed end on a mobile arm. The hand acts as the passive transmitter of force produced by the arm muscles in such positions as the flattened hand or the clenched fist. The more common, skilled use of the hand is as a mobile organ at the end of a mobile limb. The types of movement

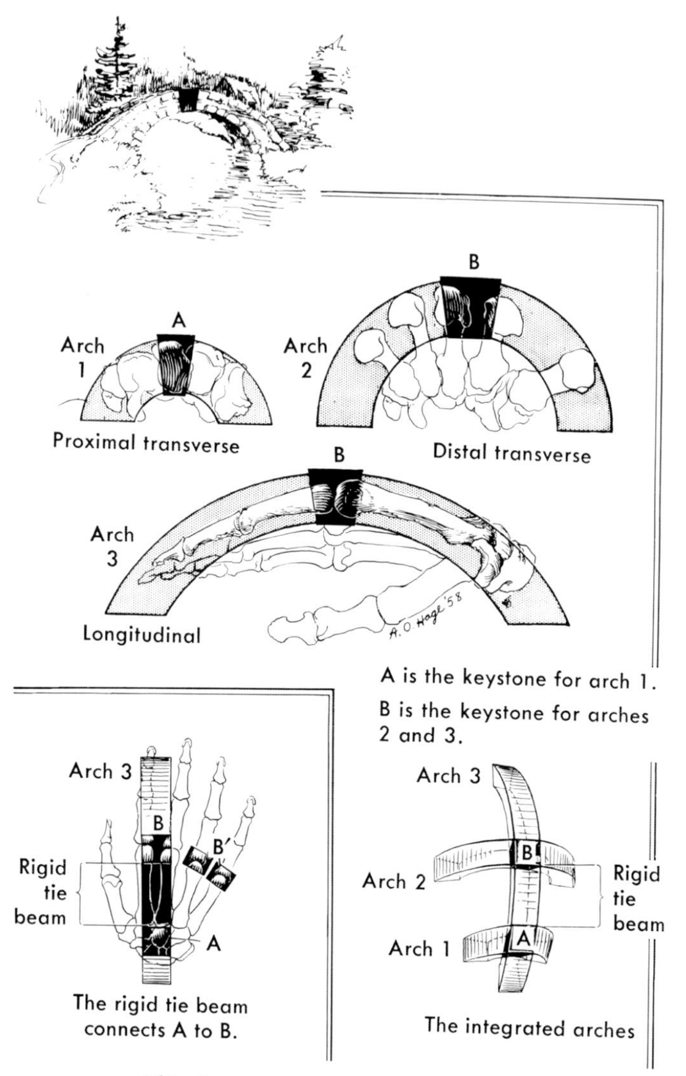

Fig. 3-8. *Arches of the hand.* The bones of the hand are arranged in three arches: one longitudinal and two transverse. The rigid proximal transverse arch passes through the distal part of the carpus. The keystone of this arch lies in the capitate. The mobile distal transverse arch passes through the metacarpal heads. The two transverse arches are held together by the rigid central pillar of the hand acting as a tie beam. The four fingers make up the complex of the longitudinal arch. The longitudinal and distal transverse arches have a common keystone, the metacarpophalangeal joint. (From Flatt, A. E.: The care of minor hand injuries, ed. 3, St. Louis, 1972, The C. V. Mosby Co.)

used by the hand have been grouped by Stetson and McDill into three major types:

1. Slow to rapid movements with control of direction, intensity, and rate
2. Ballistic or rapid repetitive motions
3. Fixations, including cocontractions yielding prehension

The slow-rapid movements consist of such actions as writing, sewing, and tying knots. Ballistic movements are usually repetitive rapid motions such as are used in typing or piano playing. Muscle power begins the movement and supplies momentum to the limb while the digits remain fixed in the required position. Fixations and prehensile movements are based on the so-called functional position. This is the position from which most prehensile actions of the normal hand develop, and it is this basic position that is modified when specialized activities are undertaken.

Anatomical patterns of prehension. In any prehensile activity the hand is considered to have two parts—the thumb and the rest of the hand. Most prehensile actions consist of an integration of the activities of the two parts, but occasionally pure finger prehension is used in a hook grasp. Thumb opposition is essential in all pinching or grasping movements. As the thumb opposes, it passes through a compound movement occurring at all three joints. The principal movements of abduction and rotation take place at both the carpometacarpal and metacarpophalangeal joints. This fundamental thumb position of abduction and rotation is constant in all grasping actions and will remain unchanged despite the size of the object held in the hand (Fig. 3-9).

Fingers in prehension. Because of the large number of bones and joints within the fingers, it would seem impossible to establish any form of coherent classification of finger movements. In fact, however, these finger movements fit into a relatively small number of patterns because of the overriding need for stability in the digital joints whenever prehension occurs.

The studies of Napier have established the basic patterns of use of the hand. He divides hand movements into two main groups. The first group is nonprehensile movement "in which no grasping or seizing is involved but by which objects can be manipulated by pushing or lifting motions of the hand as a whole or of the digits individually." The second is prehensile movement "in which an object is seized and held partly or wholly within the compass of the hand." In his full discussion of prehensile movements, Napier differentiates between power and precision grips and discusses factors influencing the posture of the hand including the influence of intended activity.

Landsmeer expanded on Napier's work by pointing out the importance of the dynamic phase of the grip. He then continued by further clarifying power grip and precision handling, a term he coined in preference to Napier's precision grip. Precision handling implies an active state in which manipulation is taking place. For manipulation to occur, an object must be in contact with the working surface of the digits. If the digit is in contact with any additional area proximal to the working surface, it cannot be manipulated. Power grip

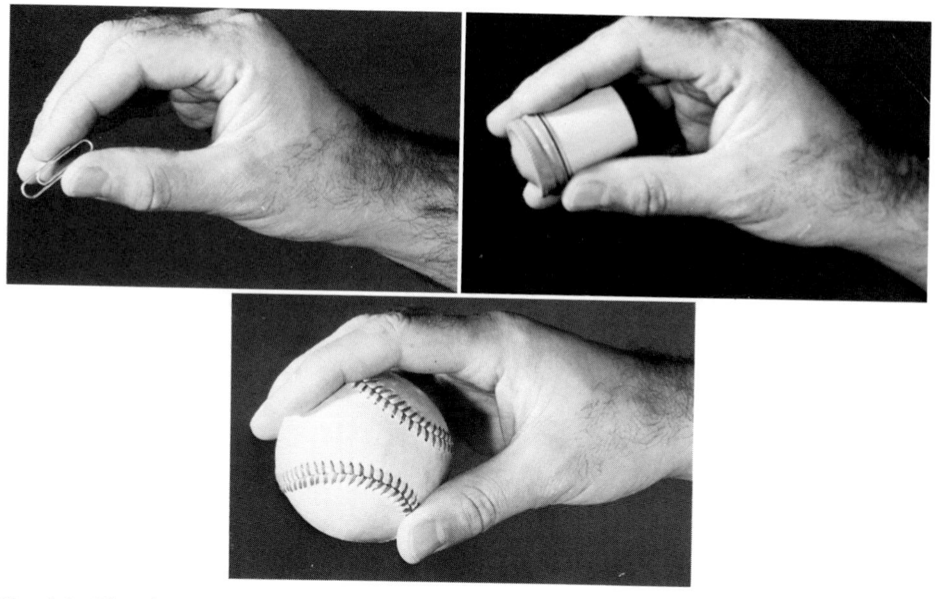

Fig. 3-9. *The thumb in prehension.* The fundamental position of abduction and rotation of the thumb relative to the palm is constant in all grasping actions. The basic position remains unchanged despite the variations in the size of objects held. (From Flatt, A. E.: Kinesiology of the hand. In the American Academy of Orthopaedic Surgeons: Instructional course lectures, vol. 18, St. Louis, 1961, The C. V. Mosby Co.)

is best considered a prehensile activity with no manipulation taking place. Thus the four key terms, nonprehensile, prehensile, power grip, and precision handling, encompass all basic hand functions.

Nonprehensile movements of the hand

Although nonprehensile motion of the hand contributes significantly to purposeful activity, I do not discuss it in detail because it is the active prehensile activities that need to be restored to the anomalous hand.

Prehensile movements of the hand

Hook grip. The most primitive type of nonmanipulative prehension is hook grip. Hook grip uses only the flexed fingers with the thumb being essentially nonfunctional. The hook grip may actively involve only flexion of the interphalangeal joints (Fig. 3-10), or it may also involve flexion of the metacarpophalangeal joints (Fig. 3-11). The latter joints become involved as the weight of the object being held increases.

This is a grip that requires relatively little muscular effort for maintenance. It is employed when precision requirements are minimal or when power needs to be exerted continuously for long periods, as when carrying a bucket. The heel of the palm may be incorporated into the hook grip to apply counter-

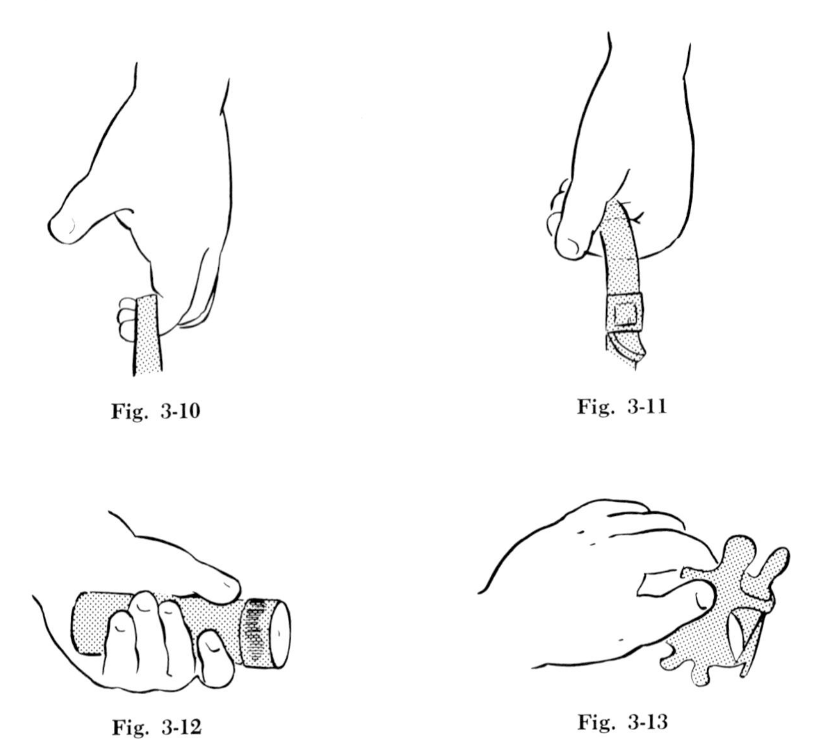

Fig. 3-10

Fig. 3-11

Fig. 3-12

Fig. 3-13

Fig. 3-10. *Hook grip.* This is the most primitive form of nonmanipulative prehension. It can involve only the interphalangeal joints.

Fig. 3-11. *Hook grip.* When the object held by a hook grip is heavy, the metacarpophalangeal joints are flexed.

Fig. 3-12. *Power grip.* The object is gripped between partly flexed fingers and the palm. Counterpressure is supplied by adduction of the thumb.

Fig. 3-13. *Lateral pinch.* This is a transitional form of grasp between power grip and precision handling. The object is held by the palmar surface of the thumb against the radial side of the index finger.

pressure to the flexed fingers, especially when the object being gripped is flat, such as a book carried at one's side.

Characteristics of hook grip are: the thumb relaxed or adducted to the radiopalmar surface of the index finger, the fingers adducted and flexed at the interphalangeal joints only or at both the interphalangeal and metacarpophalangeal joints, and the hand demonstrating a relatively flat distal transverse metacarpal arch.

Power grip. A more sophisticated type of nonmanipulative prehension involving the thumb is power grip. The object is gripped between partly flexed fingers and the palm. Counterpressure is applied by the thumb, which is wrapped over the dorsum of the middle phalanges of the fingers, thus reinforcing the power of grip. When larger objects are gripped, the thumb may not reach the dorsum of the fingers, but it remains in the same basic configuration (Fig. 3-12). The object being gripped is in contact with the entire flexor surface of the fingers, the thenar and hypothenar eminences, and the thumb. The tightly adducted and flexed fingers "hug" the object tightly to the palm. Placement of the object in the palm can vary from a position perpendicular to the long axis of the hand to an oblique position in the palm. The hand on the forearm, in power grip, is deviated ulnarward and lies in neutral between palmar flexion and dorsiflexion.

Landsmeer points out that the dynamics of gripping produce a particular grip, and the static concept indicates the final state of gripping. The dynamic and static phases can be clearly distinguished. The dynamic state comprises the following series of voluntary acts:

Opening the hand—This is a motion by which the interphalangeal and the metacarpophalangeal joints are moved simultaneously.

Choice of finger position—This is more or less simultaneous with the preceding phase. The fingers are adjusted metacarpophalangeally while the desired curving is carried out interphalangeally.

Approach—The movement of metacarpophalangeal joints is independent of the interphalangeal joints; as the fingers approach the object the metacarpophalangeal joints move while the interphalangeal joints change little.

Grip—As the fingers take hold of the object, they are powerfully drawn by the flexor digitorum profundus so that they remain in a position that ensures an efficient grip.

The choice of position and the final position in which static grasp is established are principally determined by the shape of the object.

An element of precision can be introduced to a power grip but not without sacrificing part of the power. When some precision is required, although the grip remains predominately one of power, the thumb adducts, aligning itself either along the long axis of the object being held or the long axis of the forearm. This position allows the thumb to make small adjustments controlling the direction in which the force is being applied. Napier makes it clear that the nature of the intended activity and not the physical form of the object is the factor determining the pattern of grip to be used, and the predominance of either precision or power requirements in any activity determines the posture of the hand.

Characteristics of power grip are: the thumb abducted and flexed around the object and over the dorsum of the fingers, the fingers adducted and flexed, and the hand essentially maintaining a flattened transverse metacarpal arch with the entire palmar working surface in contact with the object.

Lateral pinch. I regard lateral pinch as a transitional grasp because it represents a link or bridge between power grip and precision handling. It possesses some of the characteristics of both precision and nonmanipulative grasps but cannot be considered a pure form of either. In lateral pinch the object is held by the palmar surface of the distal phalanx of the thumb against the lateral side of the intermediate and distal interphalangeal joint of the index finger (Fig. 3-13).

The thumb functions as in precision handling, having the object in contact with its palmar working surface, but the object is not in contact with the palmar working surface of the index finger. Before manipulation, and therefore true precision activity, can take place the object has to be moved against the palmar surface of the index finger.

During the process of lateral pinch, a strong contraction of the first dorsal interosseus muscle causes abduction and radial rotation of the index finger as power becomes necessary. The finger also flexes at all joints, with the distal interphalangeal joint being in the least amount of flexion. When strong pressure is applied, the thumb does not abduct fully or rotate medially but its metacarpophalangeal joint extends and the carpometacarpal and interphalangeal joints flex strongly.

Characteristics of lateral pinch are: the thumb abducted and medially rotated at the carpometacarpal joint; the thumb extended at the metacarpophalangeal joint and flexed at the interphalangeal joint; the index flexed at the metacarpophalangeal, proximal interphalangeal, and distal interphalangeal joints; and the first dorsal interosseus contracted strongly.

Precision handling. The key factor in precision handling is that the object held can be manipulated by the fingers while the hand, in relation to the forearm, is held halfway between radial and ulnar deviation and in marked dorsiflexion. Another characteristic of this activity is the form of the distal transverse metacarpal arch, which is rounded in contrast to the relatively flat arch of power grip. This curvature of the arch is produced by adduction, flexion, and supination of the small and, to a lesser extent, the ring fingers.

Landsmeer's development of the concept of precision handling is based on the observation that precision grip does not involve any forceful grasping of the object. As soon as the "taking hold" is established, the pattern of the power grip is present regardless of how much power is exerted. The object is held between the tips of the fingers by an interdigital or contradigital antagonism that exerts a very small degree of power. Thus because a very small degree of power is exerted, the quality indicated by the word "grip" is absent from the movement pattern. Nor is there any possibility of a gradual transition to an increasingly more powerful grasp when the object is held in the fingertips. It is possible to hold the object immobile between the fingertips, but this is not typical for this form of grip, the chief purpose of which is to operate the object with precision by means of the fingers. This is the precision grip's cardinal difference from the power grip, in which the static phase is characterized by the fact that the gripping can retain a rigid relation to the wrist, elbow, or shoulder. In the precision grip there is no question of a static phase, since the fingers themselves manipulate the object, and there is no point in distinguishing dynamic and static phases in this movement pattern.

I believe it necessary to define the types of pinch and to point out their major characteristics to ensure a relatively standard method of observing and recording hand function. Pinch is most commonly thought of as involving the thumb and one finger—usually the index. However, Taylor has pointed out that a three-point pinch between thumb, index, and long fingers is one of the most frequently used pinches of the normal hand.

Characteristics of precision handling are: the digital metacarpophalangeal and interphalangeal joints move independently; with profundus activity, the terminal phalanx moves in coordination with the middle phalanx; with superficialis activity, the distal interphalangeal joint is beyond muscular control and may hyperextend; and the interphalangeal joint of the thumb has no standard posture.

The two recognizable types of precision handling are palmar and tip pinch.

Palmar pinch. Palmar pinch is the meeting of the distal phalanx pad of the abducted and medially rotated thumb with the pad of one or more fingers. It occurs primarily with the index and thumb, with the two digits forming a tapered oval from the radial view (Fig. 3-14). The carpometacarpal joint of the thumb is in full abduction and medial rotation so that it is perpendicular to the palm. The thumb metacarpophalangeal joint is slightly flexed, and the interphalangeal joint is extended. All joints of the index finger are flexed, and it lies in ulnar deviation. The metacarpophalangeal joint is in the greatest degree of flexion, and the distal interphalangeal joint is only slightly flexed. When pressure is applied, the thumb and index finger remain in the same position, with the exception of an increase in the degree of rotation of the index finger. Also, with the application of increasing force, one of the following two situations commonly occurs: (1) the distal interphalangeal joint of the index finger hyperextends, while the proximal interphalangeal joint flexes further, and the interphalangeal joint of the thumb flexes, or (2) the interphalangeal joint of the thumb hyperextends while the distal interphalangeal joint of the index finger flexes. Marked rotation of the phalanges occurs in this type of pinch.

Three-point palmar pinch is the standard two-digit palmar pinch that in-

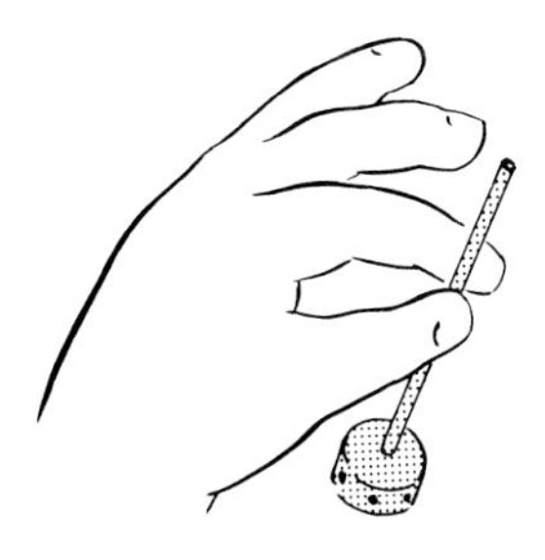

Fig. 3-14

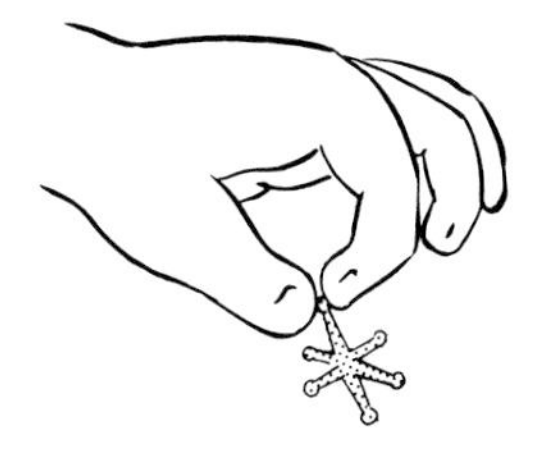

Fig. 3-15

Fig. 3-14. *Palmar pinch.* In palmar pinch the distal palmar surface of the abducted and medially rotated thumb meets with the distal pad of the index or long finger or both.

Fig. 3-15. *Tip pinch.* In this form of prehension the tip of the thumb meets the tip of any one finger—usually the index.

cludes the long finger in addition to the index. It is the primary manipulative pinch used in the normal hand. The long finger takes on the same posture as the index finger and reinforces the pinch.

Tip pinch. Tip pinch is the meeting of the tip of the thumb with the tip of any one finger, primarily the index. The thumb and index finger form a perfect circle from a radial view (Fig. 3-15). In a true tip pinch, all joints of the index finger are flexed and the proximal phalanx goes into ulnar deviation. This deviation becomes more pronounced as force is applied to the pinch. The thumb is abducted and extended at the carpometacarpal joint as the position of tip pinch is being assumed, but as pressure is applied, it flexes and abducts perpendicular to the palm with simultaneous extension at the metacarpophalangeal joint.

Although tip pinch is a manipulative type of pinch, it is frequently used to pick up very fine objects such as pins, needles, and threads, which after being seized are transferred to a palmar pinch for manipulation.

Cylindrical and spherical grips. Both Napier and Landsmeer comment on cylindrical and spherical grasp. I have eliminated these terms from this discussion because it is my contention that these terms are descriptive of the object being held rather than of the grip being used. Both of these activities can be either true grips or precision pinches, depending on the size and shape of the object and where contact is taking place between the object and the holding hand (Fig. 3-16, *A* and *B*).

Manipulative and nonmanipulative prehension

As a summary of the foregoing discussion, I would stress that an object being held in a nonmanipulative grip must be shifted to a manipulative position prior to precision handling. Objects already held in a manipulative position may be immediately maneuvered by the hand with no additional shifting.

The features necessary for a prehensile grip to be manipulative are: (1) abduction and medial rotation of the thumb, (2) curved distal transverse arch,

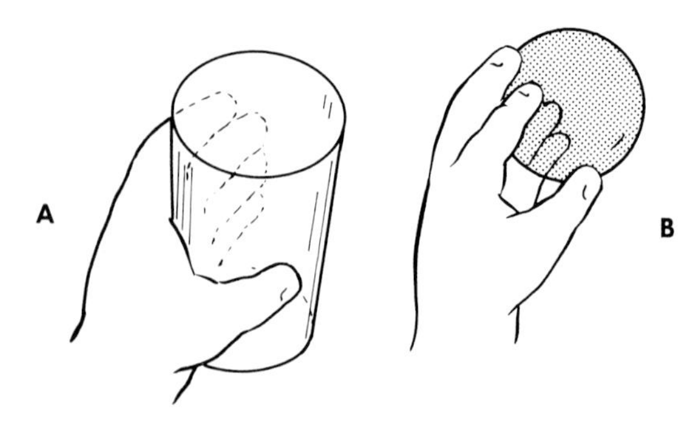

Fig. 3-16. *Cylindrical and spherical grips.* These grips are more descriptive of the shape of the object being held than the type of grip. The action is prehension, but the posture of the fingers is dictated by the object held.

and (3) distal location of contact—the object being held must be in contact with the palmar surface of the distal phalanges of the digits.

The features that limit manipulation of an object maintained within the hand during a prehensile grip are: (1) adduction or relaxation of the thumb, (2) a flat distal transverse arch, and (3) proximal location of contact—contact between the object being held and the palm or the proximal palmar surfaces of the digits or both limits the possibility for manipulation.

EVALUATION OF FUNCTION

I have provided this detailed analysis of normal hand function because I have found that it is the essential background knowledge in evaluating function of the congenitally anomalous hand. Equally important is a knowledge of the frequency of use of the various forms of grasp. The two most common prehensile movements are the picking up of an object and the holding of it for use. Keller analyzed the type of grip used in these two forms of prehension and found that in both actions the palmar, or long grip, pattern is the one most commonly used. This predominance of palmar prehension was significant in both actions. The actual figures recorded are given in Table 3-2.

The overriding importance of palmar prehension must be appreciated in all reconstructive plans for the congenitally malformed hand. Tip pinch is the most precise of all forms of prehension but requires the greatest degree of control and is used the least often. The transitional grip of lateral pinch is relatively useful compared with tip pinch but is a poor second when compared with palmar pinch. I therefore always give priority to providing a good palmar-type pinch whenever possible.

With this knowledge of the normal mechanisms of hand use, it becomes easier to understand the substitutionary prehension patterns used by children with malformed hands. These children frequently substitute one adaptation pattern for another and often accomplish the same task as that of a normal hand amazingly well by using a different pattern.

There is no virtue in drilling a child into following the normal patterns of prehension when the hand is abnormal. Equally there is no value in attempting to surgically reconstruct a normal hand from a malformed one when reinforce-

Table 3-2. Comparative frequency of the three types of grip used in prehensile movements

Prehensile movement	*Palmar*	*Lateral*	*Tip*
Picking up	50%	33%	17%
Holding	88%	10%	2%

Reported in Taylor, C., and Schwarz, R.: The anatomy and mechanics of the human hand, Artif. Limbs 2:22-35, 1955; based on data frpm Keller, A. D., Taylor, C. L., and Zahm, V.: Studies to determine the functional requirements for hand and arm prosthesis, 1947, Department of Engineering, University of California at Los Angeles.

ment of a substitutionary pattern would yield a more significant increase in function.

Our study unit has tried for a number of years to develop an evaluation procedure that will analyze the child's function in terms of normal functional ability and of functional adaptative and substitutionary patterns of function. The resultant test is administered by a therapist, who is given unlimited time in which to evaluate the child. We have published, in the *American Journal of Occupational Therapy*, the full details of the method.

First, measurements of hand size are recorded for comparison with the tables of normal measurements. Palmar creases are used as the anatomical landmarks. Palmar width is the distance from the radial side of the midpalmar crease to the ulnar side of the distal crease. Palmar depth is the distance between the proximal end of the thenar crease to the center of the proximal flexion crease of the long finger (Fig. 3-5). The length of the digit is measured from the proximal flexion crease to the tip. An example of the relevance of this information is that a frequent finding among congenitally abnormal hands is a finger and thumb length inadequate to span an object, as would be necessary in a cylindrical grip.

Next is the evaluation of the types of pinch and grip available to the anomalous hand. Variations from the normal should be carefully described, using terminology that accurately describes any unique qualities. Additional information may be necessary to indicate the mechanics of abnormal function. The amount of active (and passive, if the two differ) range of motion in each joint is recorded as one aspect of analysis. This determines if lack of joint mobility or anatomical lack of musculature is the cause of any abnormal grasp and pinch patterns.

The most commonly employed, preferred patterns of usage must be observed. Objects of various shapes and sizes requiring different manipulative processes are presented during the test. A hand may be capable of grasping an object using the normal pattern of prehension but may regularly adopt a different pattern. It should be noted which digits are used predominantly and the manner in which they are used. For example, a patient may be able to pick up an object with the thumb and index finger upon request. However, if observed at free play, he or she might do it with the long or ring finger and thumb. Further evaluation might reveal that the index finger is unusually narrow and that the small pad provides insufficient stability, or measurement of pinch strength may point out weakness in flexion power of the index finger.

Pinch strength is a valuable measurement in analyzing hand usage patterns. A standard pinch gauge is used to measure palmar pinch between thumb and index, thumb and long, thumb and ring, and thumb and small finger together with the strength of lateral pinch of the thumb against the radial side of the distal interphalangeal joint of the index finger.

The five key points of the evaluation procedure are:

Analysis of pinch and grip patterns

CONGENITAL HAND STUDY

FUNCTIONAL STUDY SECTION

EVALUATION OF BASIC HAND FUNCTION

Carrie . S.

Evaluation Date: 1-2-77
Evaluation: Pre-op ______ . Post-op # ______
Dominant Hand R
Age 3 YR 4 months

A. PINCH
1. Tip: thumb to index NA, long NA, ring NA, small NA
2. Palmar: thumb to index 1000 gms, long 1000, ring 1000, small 800
3. Lateral: thumb to index 1000, long 900, ring 800, small 600
4. Thumb touch base of small finger? Yes
5. Most predominantly used pinch: palmar
Comments Pt unable to understand tip prehension

B. POWER GRIP, thumb abducted:
1. 1 pos., 16 lbs.; 2. 2 pos., 10 lbs.; 3. 3 pos., 6 lbs.
Comments:

C. PRECISION GRIP, thumb abducted:
1. Cylindrical
a. 1 pos., 7 lbs.; b. 1 pos., 5 lbs.; c. ______ pos., ______ lbs.
d. largest diameter 2 1/2 inches; e. smallest diameter 2 inches
Comments:

2. Spherical:
a. largest diameter 14 lbs.; b. smallest diameter 10 lbs.
c. largest diameter 2 7/8 inches; d. smallest diameter 1 7/8 inches
Comments:

D. HOOK GRIP
1. Proximal (M.P.) Extended ; 2. Distal (I.P.) flexed
Comments: Thumb flexed into palm

E. Describe the resting position of the hand: wrist dorsiflexed, MP's extended, PIP's slightly flexed, DIP's slightly flexed.

F. SENSATION
1. Pick-up test: (sec.) sighted, L ______, R ______; blindfolded, L ______, R ______
2. Object Identification: a. intact ✓, b. impaired ______, c. absent ______
3. Position sense, comment: WNL
4. Light touch, comment: WNL
5. Pin prick: (see diagrams below)
6. Hand size

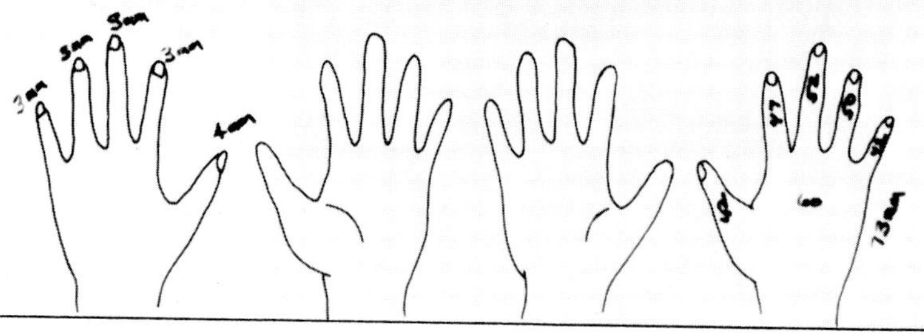

3mm 3mm 3mm 3mm 4mm

13mm

Fig. 3-17. *Evaluation of basic hand function.* An example of the data sheet on which is recorded all the basic measurement data.

Measurements of active range of motion and passive range of motion if the two differ

Observations of preferred patterns of usage

Measurements of pinch strength

Measurements of hand size

Several forms are used to record the findings. The one reproduced here is the Evaluation of Basic Hand Function on which are recorded all the basic measurement data (Fig. 3-17). We recognize that this test is far from perfect; it is still being modified in detail and we would welcome suggestions for its further improvement.

CHAPTER 4

CLASSIFICATION AND INCIDENCE

CLASSIFICATION

No comprehensive classification of congenital malformations of the hand exists, and probably one should never be allowed. Classifications are only conveniences, and it is not possible to precisely classify all hand malformations. The value of a classification is that it enables comparisons to be drawn and incidence rates to be established. The problem has always been to develop a classification midway between one so general that it is valueless and one so detailed that its use becomes impossible. An additional problem is that any classification will be slanted towards the particular interests of the anatomist, geneticist, surgeon, or radiologist who devised it. Kelikian lists 21 tables of classification based on such factors as (1) discrepancies in number and size, (2) simplicity, (3) embryopathy, (4) endogenous and exogenous deformities, (5) skeletal deficiencies, and (6) comprehensive and anatomical peculiarities.

In 1837 Isidore Geoffrey Saint-Hilaire published a classification of congenital anomalies introducing such terms as phocomèle, hemimèle and ectromèle. Over the years confusion grew up around the correct usage of these words and as Kelikian stated, a conglomeration of names that neither describe patterns nor conjure clear pictures was introduced. The confusion was compounded by the use and abuse of both Latin and Greek roots. Many of these confusing terms are still used and, as an aid in their "decoding," I have included a glossary at the end of this book.

The pioneer work of Frantz and O'Rahilly published in 1961 introduced a concise comprehensive system that has formed the basis for the current classification. The American Society for Surgery of the Hand, the International Federation of Hand Societies, and a multinational working group sponsored by the International Society for Prosthetics and Orthotics have now agreed on a new descriptive terminology that makes it possible to classify these malformations simply and precisely.

A full and well illustrated discussion of this classification has been published by Swanson in the first issue of the *American Journal of Hand Surgery* and is partially reprinted here.* He points out that Greek and Latin terms such as hemimelia or ectromelia have been eliminated, eponyms are not employed, and a simple descriptive terminology is used. This classification allows the recording of common clinical entities with minimal confusion, and yet it permits the full categorization of complex cases. The method is based on the ability to group cases according to the parts that have been primarily affected by certain embryological failures. The malformations are categorized in seven groups:

I. Failure of formation of parts (arrest of development)
II. Failure of differentiation (separation) of parts
III. Duplication
IV. Overgrowth (gigantism)
V. Undergrowth (hypoplasia)
VI. Congenital constriction band syndrome
VII. Generalized skeletal abnormalities

This grouping takes note of the similarities in certain large groups of deformities. Subclassification of these categories recognizes the basic differences and expresses the degree of severity of the type of deformity. As an example, the so-called radial clubhand is a failure of radial longitudinal formation. In addition to the fact that the radius is either partially or totally absent, the condition can also be associated with hypoplasia or aplasia of the thumb, carpal bones, or musculature on the preaxial or radial side of the limb. Each group can be subclassified according to its degree of severity as desired for completeness.

Failure of formation of parts (arrest of development)

The grouping of failure of formation of parts is divided into two types—transverse and longitudinal. The transverse defects include all so-called congenital amputation conditions. The second major group, longitudinal deficiencies, includes all deficiencies that are not in the transverse category.

Transverse deficiencies. Transverse deficiencies appear as amputation type stumps and are classified by naming the bony level at which the remaining limb terminates, it being understood that all functionally significant elements

---

*Modified text and figures from Swanson, A. B.: A classification for congenital limb malformations, J. Hand Surg. **1**:8-22, 1976.

distal to the level named also are absent (Fig. 4-1). The deficiencies in this group represent congenital amputations ranging from aphalangia to amelia. The stump usually is well padded with soft tissue and may show rudimentary digits or a dimpling at the end of the stump. It represents an arrest of formation in the limb anlage (Fig. 4-2).

One of the most common types of transverse deficiency is one at the level of the upper third of the forearm or a short below-elbow defect.

Longitudinal deficiencies. All skeletal limb deficiencies in the failure of formation category, other than the transverse type, are placed arbitrarily in the longitudinal group. In identifying longitudinal deficiencies, all completely or partially absent bones are named. Any bones not named as being absent are understood to be present (Fig. 4-3). The deficiencies in this group reflect the separation of the preaxial and postaxial divisions in the limb and include longitudinal failure of formation of entire segments (phocomelias) or of either the radial, central, or ulnar components of the limb.

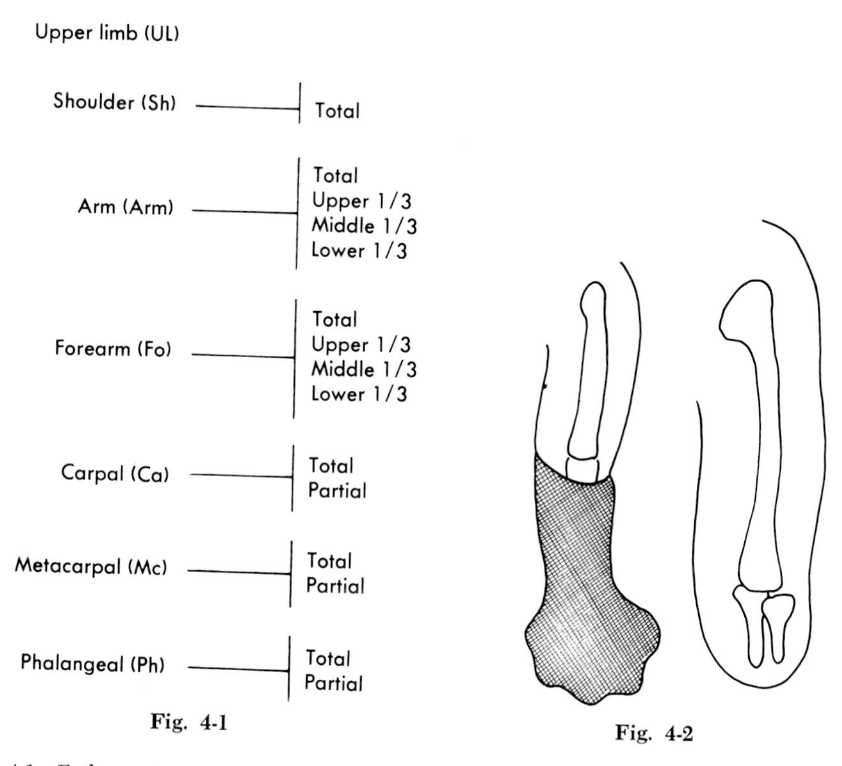

Fig. 4-1

Fig. 4-2

Fig. 4-1. *Failure of formation of parts, transverse (congenital amputations).* The digits are named by expressing the anatomical level at which the limb terminates. Note an oddity that apparently occurs: total absence of the shoulder and all distal elements would be a transverse defect. However, if only a portion of the shoulder were absent, the deficiency would be of the longitudinal type. The level at which the remaining stump terminates is named using the suggested abbreviations.

Fig. 4-2. *Failure of formation of parts, transverse.* The crosshatched area shows the area of embryological insult.

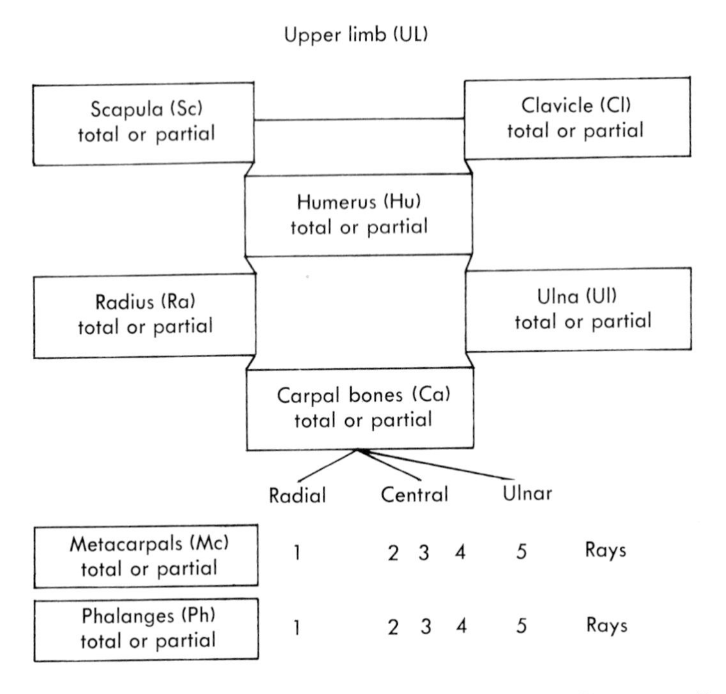

Fig. 4-3. *Failure of formation of parts, longitudinal.* All bones totally or partially absent are named using the suggested abbreviations. Any bone not named is presumed present.

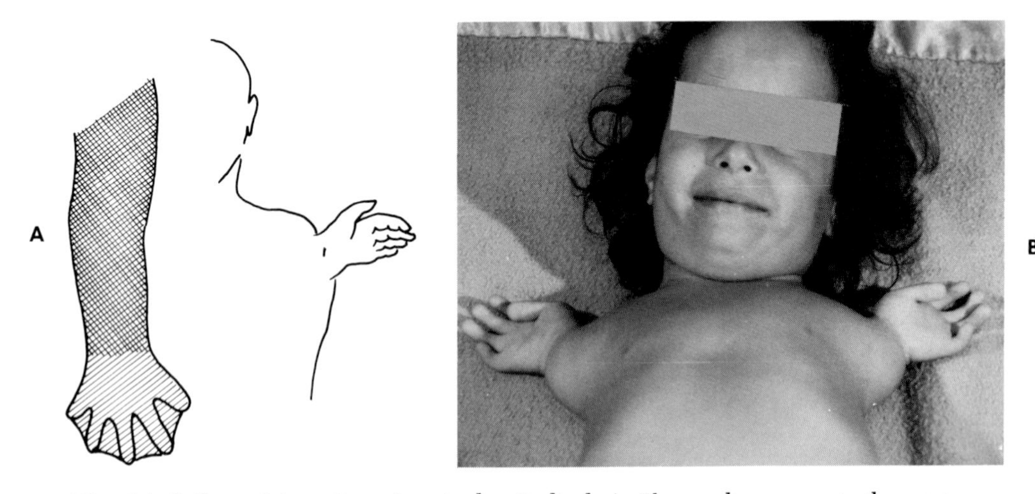

Fig. 4-4. *Failure of formation of parts, longitudinal.* **A,** Phocomelia represents the most profound expression of longitudinal failure of formation of parts. The crosshatched area shows the maximal area of embryological insult. However, some defects of the terminal limb are also always present. **B,** An infant with complete phocomelia. Note the associated hand anomalies.

The most profound expression of the longitudinal reduction of the limb is phocomelia, which demonstrates the proximodistal failure of development. Phocomelias are classified as: (1) complete phocomelia—the hand is attached directly to the trunk, demonstrating the failure of development of the proximal segments of the limb (Fig. 4-4); (2) proximal phocomelia—the proximal segment (arm) is missing and the hand is attached to the forearm, which is attached to the trunk; and (3) distal phocomelia—the forearm is absent and the hand is attached directly to the arm.

The absence of parts or deficiencies on the radial (preaxial) side of the

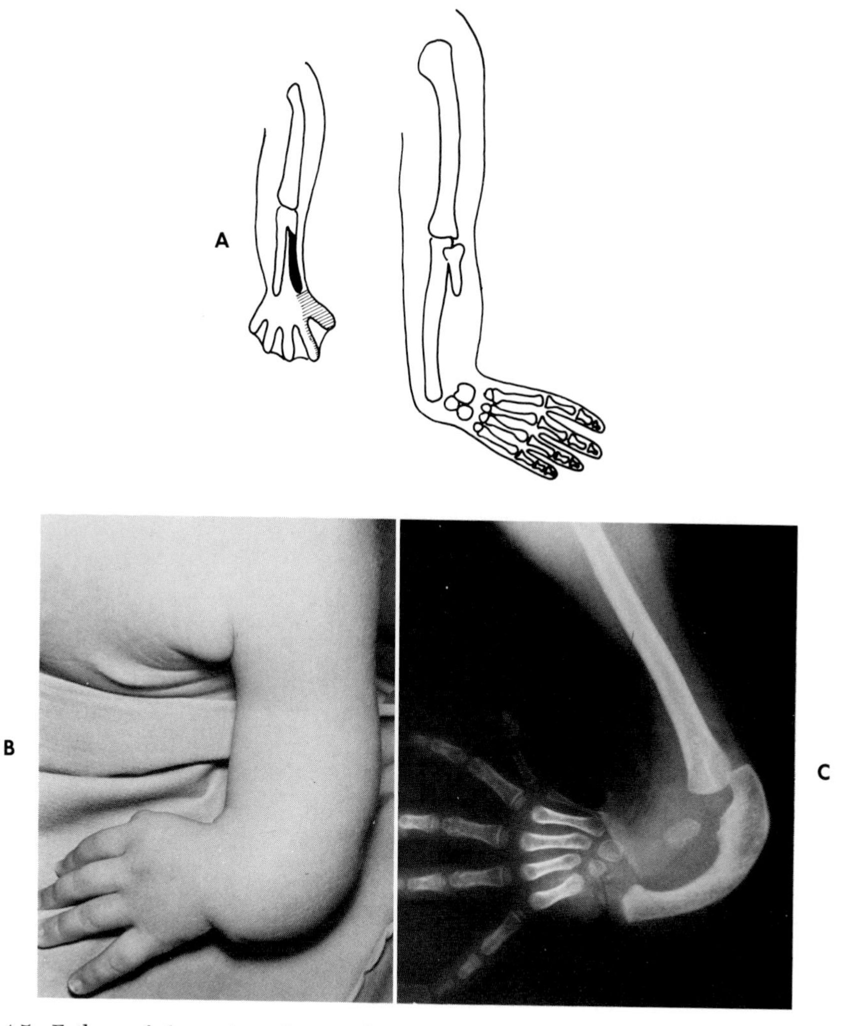

Fig. 4-5. *Failure of formation of parts, longitudinal-radial.* Radial ray deficiencies (or the so-called radial clubhand or radial hemimelia). **A,** The areas of embryological insult. **B** and **C,** Clinical record of a radial ray deficiency.

limb may vary from deficient thenar muscles to a short, floating thumb and from deficient carpals, metacarpals, and radius to the classical so-called clubhand, which are preferably described as radial ray deficiencies or radial deficiencies (Fig. 4-5). These deficits may involve the areas of the radius and thumb, radius only, or thumb only.

In central ray or central (carpal or digital or both) deficiencies, the second, third, and fourth of the so-called central rays of the hand, which are embryologically differentiated at a separate time from the radial and ulnar rays, may be involved (Fig. 4-6). The long finger or metacarpal may be missing as in the so-called lobster claw hand. In severe central deficiencies, the central digits may be reduced to nubbins. In a severe case the defect may result in complete suppression similar to a transverse arrest resulting in aphalangia.

In deficiencies of the ulnar side of the limb, the small or ring fingers may be deficient with or without associated deficiencies in the ulna and carpal bones (Fig. 4-7). These deficits may involve the areas of the ulna and fingers, ulna only, or fingers only and are described as ulnar ray deficiencies or ulnar deficiencies.

Failure of differentiation (separation) of parts

Failure of differentiation or separation of parts is that category in which the basic units have developed but the final form is not completed. The

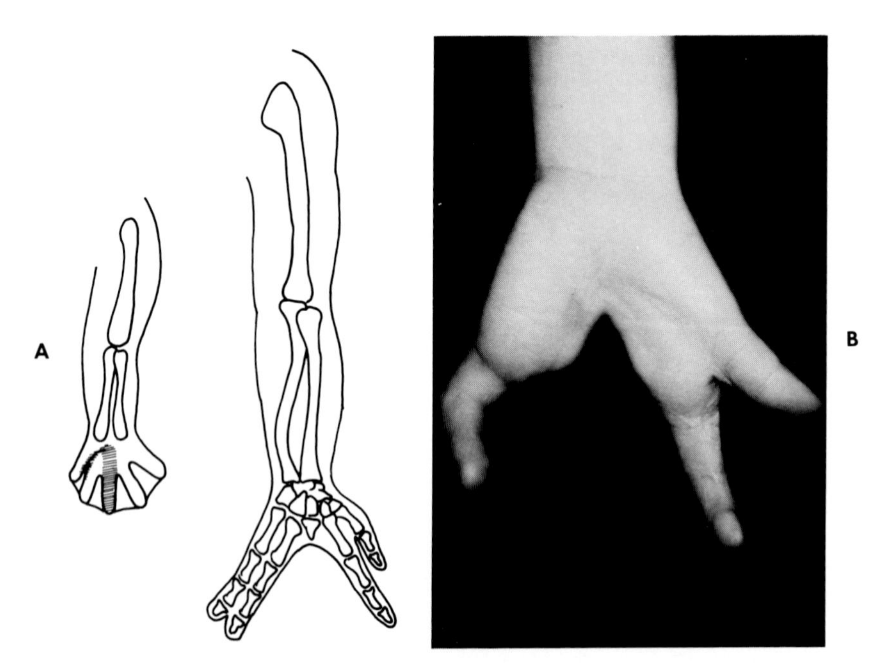

Fig. 4-6. *Failure of formation of parts, longitudinal-central.* **A,** The areas of embryological insult. Note the absence of the middle ray and associated failure of separation of parts between the small and ring fingers resulting in a cleft hand with syndactyly of the small and ring fingers. **B,** A central deficiency or cleft hand.

homogenous anlage divides into separated tissues of skeletal, dermomyofascial, or neurovascular elements found in normal limbs but fails to completely differentiate or separate. For example, these defects may include the following:

A. Shoulder
    1. Undescended shoulder
    2. Absence of pectoral muscle or muscles
B. Arm
    1. Elbow synostosis
C. Forearm
    1. Soft tissues
    2. Synostosis
        a. Proximal radius or ulna (Fig. 4-8, *A*)
        b. With or without radial head dislocation
D. Hand
    1. Carpal
        a. Deformities
        b. Synostosis
    2. Metacarpal
        a. Deformities
        b. Synostosis
    3. Digital
        a. Deformities
        b. Symphalangia

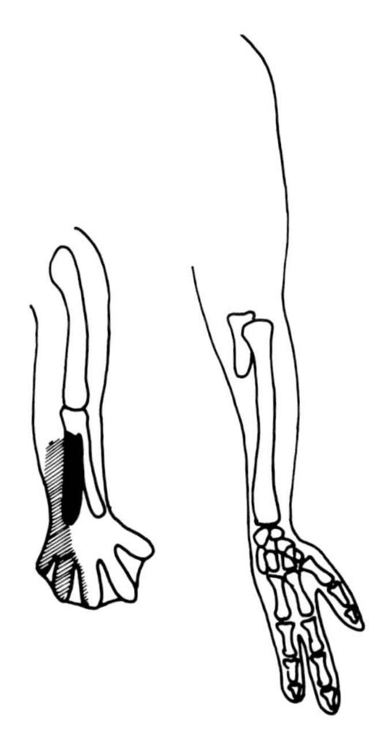

Fig. 4-7. *Failure of formation of parts, longitudinal-ulnar.* Ulnar ray deficiencies or ulnar deficiency (or so-called ulnar clubhand or ulnar hemimelia).

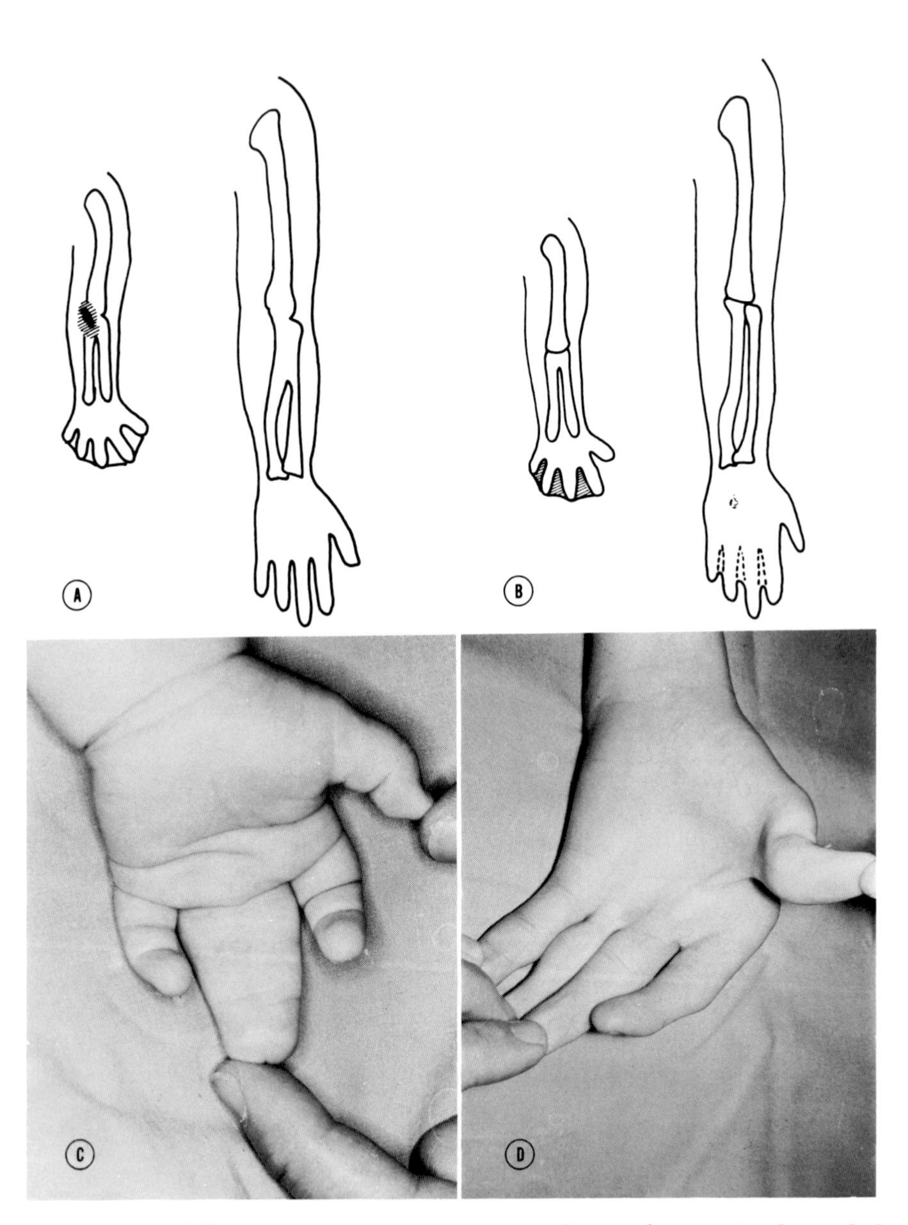

Fig. 4-8. *Failure of differentiation of parts.* **A,** Synostosis, elbow and proximal radius and ulna. **B** and **C,** Syndactyly. **D,** Example of failure of differentiation of skin, muscle, ligaments, and capsular structures of the hand resulting in so-called arthrogrypoid hand.

c. Syndactyly (Fig. 4-8, *B* and *C*)
  (1) Simple
  (2) Complicated
    (a) Soft tissue
      (*1*) Nails (synonychia)
      (*2*) Other
    (b) Skeletal
      (*1*) Fusions
        (*a*) Phalanges (side-to-side)
        (*b*) Acrosyndactyly
        (*c*) Apert syndrome
      (*2*) Disarray
      (*3*) Brachysyndactyly (short, webbed fingers)

d. Contracture resulting from failure of differentiation of muscle, ligaments, and capsular structures
  (1) Soft tissue
    (a) Thumb-web space
    (b) Arthrogryposis (Fig. 4-8, *D*)
    (c) Camptodactyly (flexion contracture of small finger)
    (d) Trigger digit
  (2) Skeletal
    (a) Clinodactyly (lateral deviation or displacement caused by asymmetrical abnormalities of the digits)

Duplication

Duplication of parts probably results from a particular insult to the limb bud and ectodermal cap at very early stages of their development so that splitting of the original embryonic part occurs (Fig. 4-9). These defects may range from polydactyly to twinning or mirror hands and are classified as follows:

A. Skeletal
  1. Humerus
  2. Radius
  3. Ulna
  4. Carpals
  5. Metacarpals
  6. Phalanges

B. Skin and nails

C. All tissues
  1. Whole limb
  2. Partial limb
  3. Polydactyly
    a. Radial
    b. Central
    c. Ulnar
  4. Mirror hand

Overgrowth (gigantism)

The whole limb may be affected by overgrowth. Some cases appear to be caused by skeletal overgrowth with normal-appearing soft parts. Others show excess fat; lymphangioma or angioma may be present. In digital gigantism the abnormality usually is greatest at the periphery. Deformities may increase be-

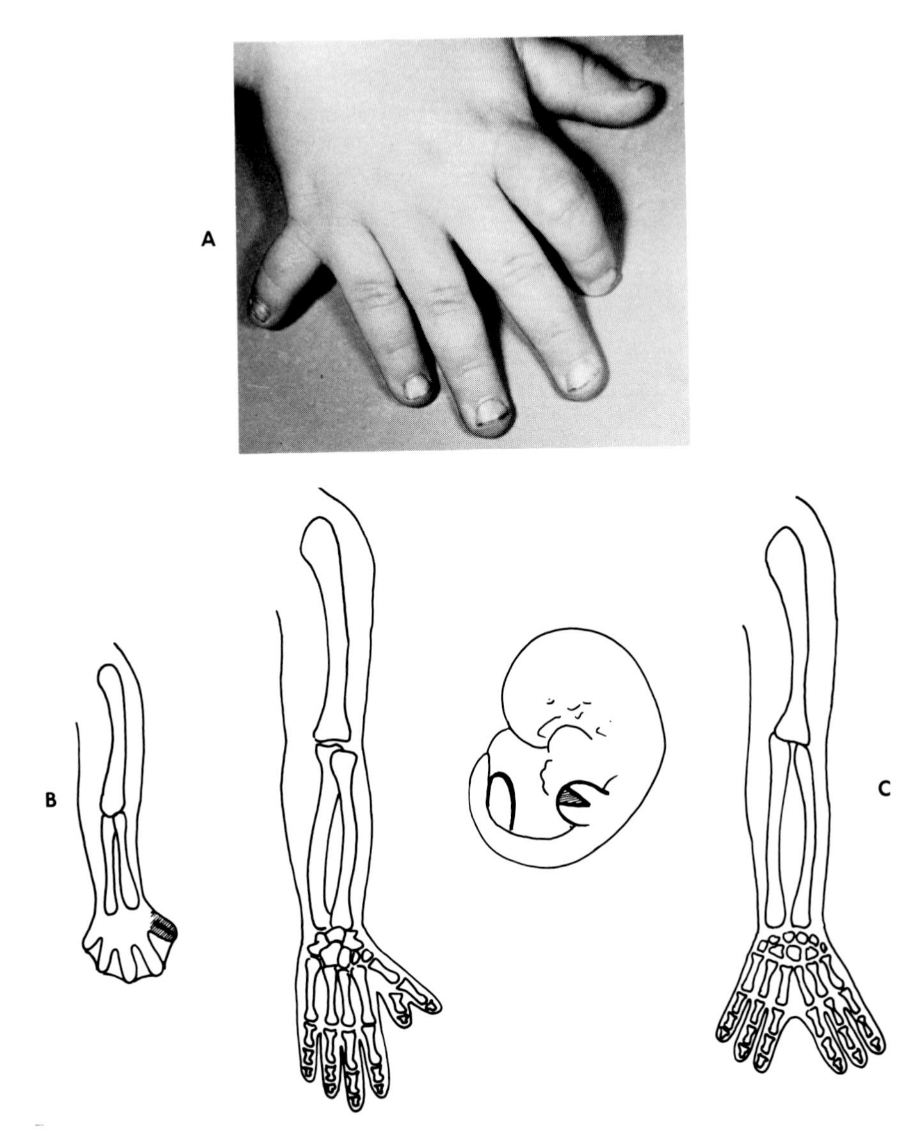

Fig. 4-9. *Duplication.* Duplications occur as a result of splitting of the original embryonic parts. **A** and **B**, phalanges; **C**, so-called mirror hand.

cause of asymmetrical growth of the part (Fig. 4-10). Gigantism may involve the following: (1) arm, (2) forearm, (3) hand, and (4) digit.

Undergrowth (hypoplasia)

Undergrowth, or hypoplasia, denotes defective or incomplete development of the parts. This may occur in the entire extremity or its divisions (Fig. 4-11). Hypoplasia may be used as a supplementary description for remaining skeletal elements in cases of failure of formation of parts (Category I). However, this

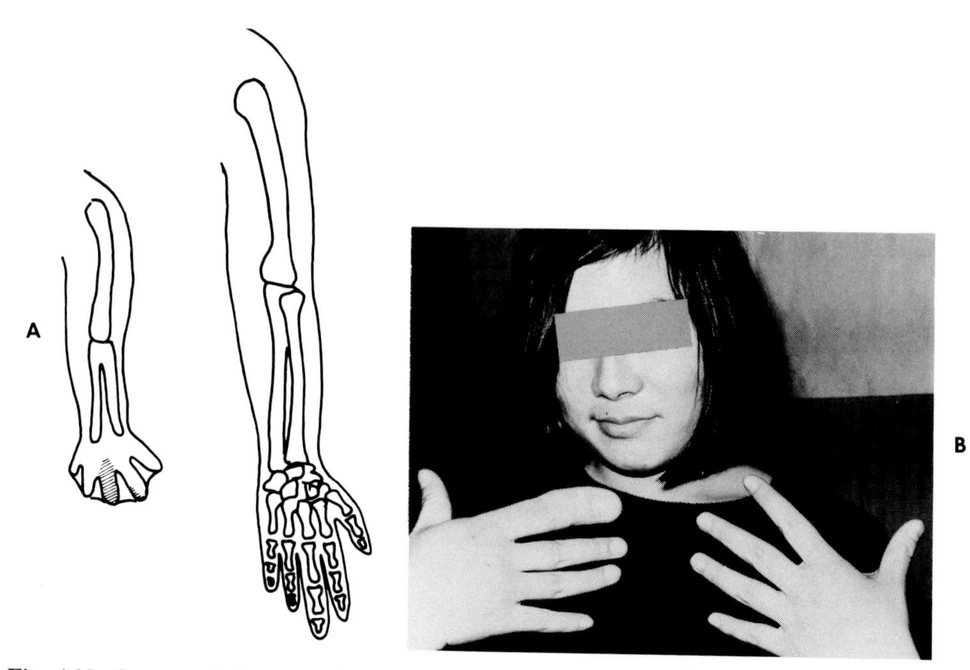

Fig. 4-10. *Overgrowth (gigantism).* **A,** Involvement of the second and third digits. **B,** Example of involvement of the index finger.

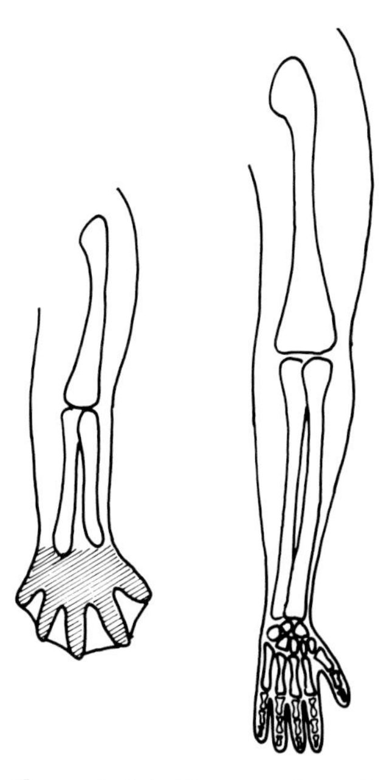

Fig. 4-11. *Undergrowth (hypoplasia).* This can be a local or more generalized involvement. Diagram illustrates the involvement of the entire hand.

category is represented separately in the classification because of the prevalence of hypoplasia. It may involve any of the following systems:

A. Skin and nail
B. Musculotendinous
C. Neurovascular
D. Extremity
  1. Arm
  2. Forearm
  3. Hand
    a. Whole hand
    b. Metacarpals (brachymetacarpia)
    c. Phalanges (brachyphalangia)

Congenital constriction ring syndrome

There is no general agreement as to the cause of these rings, but it is generally accepted that they are best categorized as a separate entity (Fig. 4-12).

Generalized skeletal abnormalities

Defects in the hand may be manifestations of generalized skeletal defects, such as dyschondroplasia, achondroplasia, Marfan syndrome (with arachnodactyly), and diastrophic dwarfism (Fig. 4-13).

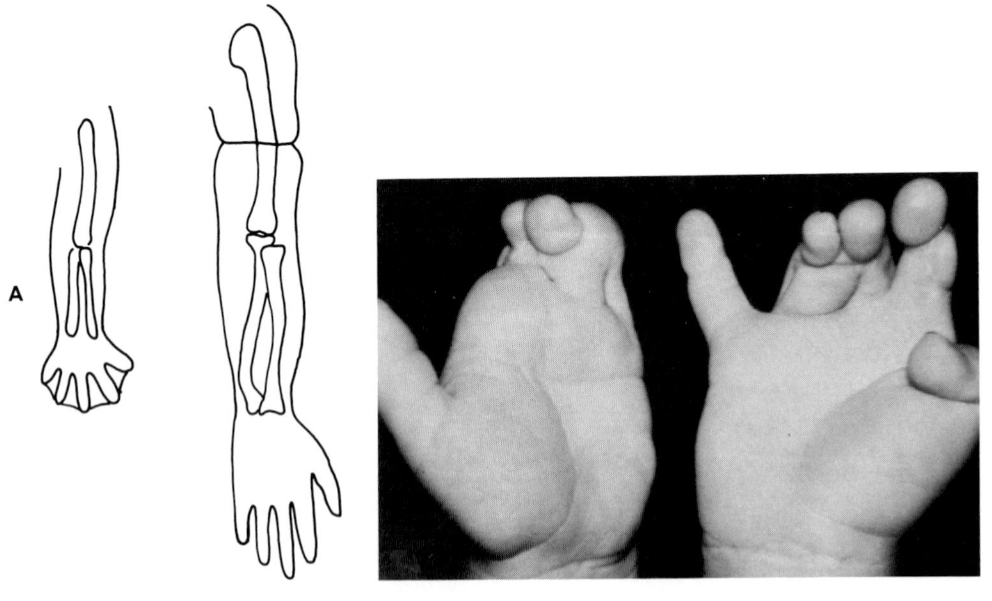

Fig. 4-12. *Constriction ring syndrome.* **A,** At the middle of the arm. **B,** At the hand, resulting in pseudosyndactyly. Most frequently the web spaces are intact, and the fused mass of digital substance is the result of intrauterine healing of the necrotic areas involving the digits. This should not be confused with true syndactyly resulting from failure of differentiation of parts.

In his paper Swanson gives very detailed instructions in the use of this classification system. He discusses such factors as abbreviations, supplemental conditions, order of classification, and the recording of surgical corrections.

Occasionally the insult to the limb may cause an apparent mixed involvement; for instance, only rudimentary radial and ulnar digits may be present on a hand that otherwise represents a transverse arrest. When there is such variation from the classical patterns, judgement should be used, and the patient should be placed into the category that represents the predominant deficiency.

A common problem in using this and all classifications is that when a child is young x-ray examinations cannot show which bones are yet to ossify. Thus early diagnosis must be subject to later revision.

Another problem is the impossibility of identifying some abnormally shaped bones or bony remnants early; again, judgement must be used but reasonable doubts should be entered on the record card. There should be little problem in international use of this classification system. Some years ago I carried out a comprehensive clinical test of an earlier edition and showed that it was properly conceived and could be used accurately by a variety of physicians to whom English was a foreign tongue.

It would be of great value if some uniform classification such as this could be adopted throughout the world to monitor congenital malformations of the hand. Its adoption would be of fundamental help in research regarding causative factors, preventative measures, and methods and results of treatment. It would also permit comparisons of incidence between different areas and different countries.

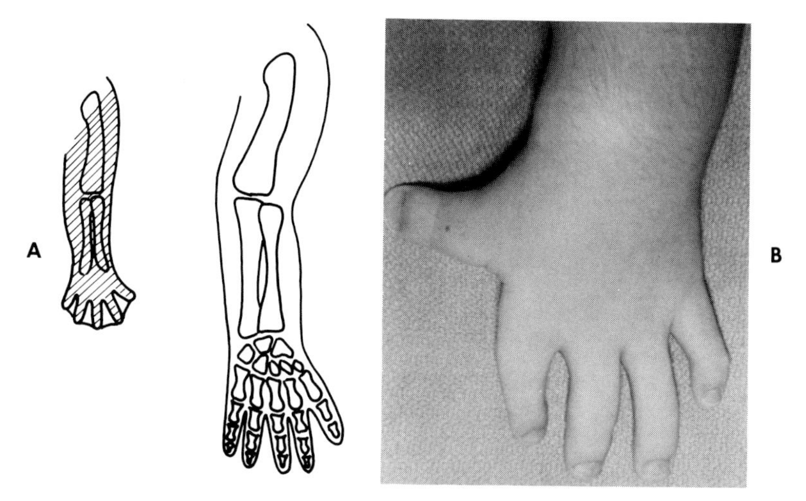

Fig. 4-13. *Generalized skeletal defects.* **A,** Achondroplastic extremity. **B,** Typical hand of diastrophic dwarf.

INCIDENCE

It is impossible to establish accurate figures for the incidence of congenital malformations because of variations in the definition of malformation and varying degrees of accuracy in reporting the data. It is generally said that between 1% and 2% of all live infants are recognized to be malformed at birth or during the neonatal period. Despite the fact that the hand is readily visible, the true incidence of hand malformations is unknown since many of the changes are minor and unless specifically sought can easily be missed. Attempts have been made to establish incidence figures by examining birth certificate data, but these are notoriously inaccurate in the details they record.

I have made no attempt to establish a percentage incidence for the general population but I am interested in the relative frequency of the various anomalies as related to each other. There are now 1,476 patient records in the files of the University of Iowa's Congenital Hand Project, and all of the hand malformations recorded have been classified. These data show that in a cross-section of a midwest population the most common malformation is syndactyly. The relative frequency of the more important diagnoses in our series is shown

Table 4-1. Primary diagnoses in Iowa series

Diagnosis	*Total*	*Percent*
Acrosyndactyly	66	3.1
Amputation—total	156	7.2
Amputation, arm or forearm	61	2.8
Amputation through wrist	24	1.1
Amputation through hand or digits	71	3.3
Brachydactyly	113	5.3
Camptodactyly	140	6.5
Central defects	60	2.8
Cleft hand	49	2.3
Clinodactyly	104	4.8
Clubhand, radial	81	3.8
Clubhand, ulnar	22	1.0
Constriction rings	48	2.2
Macrodactyly	19	0.9
Madelung deformity	37	1.7
Musculotendinous defects	42	1.9
Polydactyly—total	320	14.8
Polydactyly, radial	143	6.6
Polydactyly, central	42	1.9
Polydactyly, ulnar	135	6.4
Symphalangia	12	0.6
Syndactyly	413	19.1
Syndromes (Apert, etc.)	94	4.4
Synostosis, radioulnar	29	1.3
Thumb, absent	74	3.4
Thumb, adducted	12	0.6
Thumb, triphalangeal	13	0.6
Trigger digit	47	2.2

NOTE: 2,159 affected hands of 1,476 patients are recorded. Over 82% of the hands are included in this table. The remainder could have been included under "other."

in Table 4-1, which includes over 82% of the abnormal hands. In this table are the primary diagnoses of the affected hands of 1,476 patients. Since nearly half of the patients had bilateral deformities, there are 2,159 hands included in this count. The commonest diagnoses in descending order of frequency are syndactyly, polydactyly, camptodactyly, clinodactyly, and radial clubhand. This order of diagnosis is different from that derived by Yamaguchi and his colleagues in Yokohama who studied 256 congenitally malformed hands seen in the years 1961 to 1972. Their most common diagnosis was polydactyly followed by syndactyly. The various incidence rates in their patients are shown in Table 4-2.

Study of our figures shows that the relative incidence of the various diagnoses are somewhat different from those usually published. I feel that the Iowa figures are more probably nearer the truth because our unique situation of a largely stable population has allowed the collection of a large number of patients who have been carefully studied.

No inferences can be drawn from these figures in relation to absolute incidence in the population at large. All that can be said is that given a malformed population of over 1,475, the percentage incidence of any particular deformity can be determined.

Table 4-2. Diagnoses in Yokohama patients*

Type of anomalies	*Number of cases*	*Percent*
Syndactyly	23	9
Polydactyly	65	25
Brachydactyly	19	7
Brachysyndactyly	10	4
Symphalangia	1	—
Annular grooves	3	1
Ectrodactyly		
Cleft hand	12	4.6
Ectrosyndactyly	17	6.6
Amputation	16	6
Microdactyly	5	2
Floating thumb	5	2
Hypoplasia of the thumb	3	1
Five-finger	2	0.8
Monodactyly	1	—
Floating little finger	1	—
Defect of fifth metacarpus	1	—
Macrodactyly	3	1
Arthrogryposis	29	11
Clinodactyly	3	1
Clubhand	14	5
Phocomelia	2	0.8
Others	21	8
Total	256	

*Department of Orthopaedic Surgery, Yokohama University School of Medicine, Yokohama, Japan.

Data from Proceedings of the Sixteenth Annual Meeting, The Japanese Society for Surgery of the Hand, Fukuoka, 1973.

SECTION TWO

THE THUMB

CHAPTER 5

THE INADEQUATE THUMB

SMALL THUMBS

A small, poorly functioning thumb is a relatively common abnormality, and in many instances its presence is a signal that other congenital abnormalities may be present. The tip of the normal thumb usually reaches nearly to the middle or proximal interphalangeal joint of the finger. Thumbs that are significantly shorter than this can be caused by hypoplasia of any or all of the three constituent bones.

When the metacarpal is short and slender, it may be part of a syndrome such as Fanconi, Holt-Oram, or Juberg-Hayward. It can also be associated with abnormalities of the heart, spine, or esophagus.

When the metacarpal is short and broad it can be associated with the Cornelia de Lange syndrome, the hand-foot-uterus syndrome, or diastrophic dwarfism. I have also seen it in myositis ossificans progressiva, although classically it is the great toe that is short. A hypoplastic metacarpal is often associated with a more proximal growth disturbance of the preaxial structures such as the thenar muscles, the radial styloid, the scaphoid, and the more radial extrinsic forearm muscles. In more advanced degrees the scaphoid and even the trapezium may be absent.

Shortening of the proximal phalanx of the thumb can be found in brachydactyly of all digits when the middle phalanges of the fingers are also affected. The distal phalanx can be short in a variety of brachydactyly types, either as an isolated finding or in combination with shortening of more proximal bones. The distal phalanx can also be broad as well as short, and several syndromes, such as the Rubinstein-Taybi, Apert, Carpenter, and hand-foot-uterus, are associated with a short and broad terminal phalanx. The thumb is usually radially deviated, hence the nickname, hitchhiker's thumb. The isolated stub thumb has acquired a variety of labels including potter's thumb and murderer's thumb.

In contrast to the stub thumb, the distal phalanx can be slender and hypoplastic. This deformity is associated with a number of congenital disorders, particularly the preaxial hypoplasia syndromes such as Fanconi and Holt-Oram.

No matter what the cause of the thumb shortening, the practical effect is to significantly interfere with grasp and prehension. If the carpometacarpal joint is stable, apparent but effective lengthening of the thumb can be obtained by recessing the web proximally. Although this can be achieved by using a two-flap Z-plasty, a more effective deepening and a more rounded edge to the web can be obtained by using the four-flap Z-plasty.

FLOATING THUMB

When the thumb is slender and unstable at its base it is usually called a "pouce flottant" or floating thumb. The classic floating thumb is characterized by a slender thumb with a nail and two phalanges but absence of part of or the entire metacarpal. There is no carpometacarpal joint since the trapezium, and often the scaphoid, is missing. More often than not, control of the thumb is impossible because of the lack of extrinsic tendons (Fig. 5-1). These scrawny little thumbs are usually placed too far distally and too far radially to provide reasonable opposition even if a stable base could be provided. The majority, in fact, are useless and should be removed. Parents seem to develop a marked attachment to these miniature thumbs, and many different operations have been devised to stabilize the thumb and provide some function. Although these operations are ingenious, the average result is unimpressive, and I strongly believe that early amputation followed by pollicization of the index finger should be the treatment of choice.

This is particularly true in bilateral cases since a child must have at least one thumb for prehension. I therefore do one side and wait for the decision regarding the second hand. In most cases the parents or the child asks that the second thumb be made. In unilateral cases I do not feel justified in pressing the issue, but I arrange for the parents to meet with a child who has had a pollicization and his or her parents and encourage the parents to exchange views. Usually the decision is for an operation but not always.

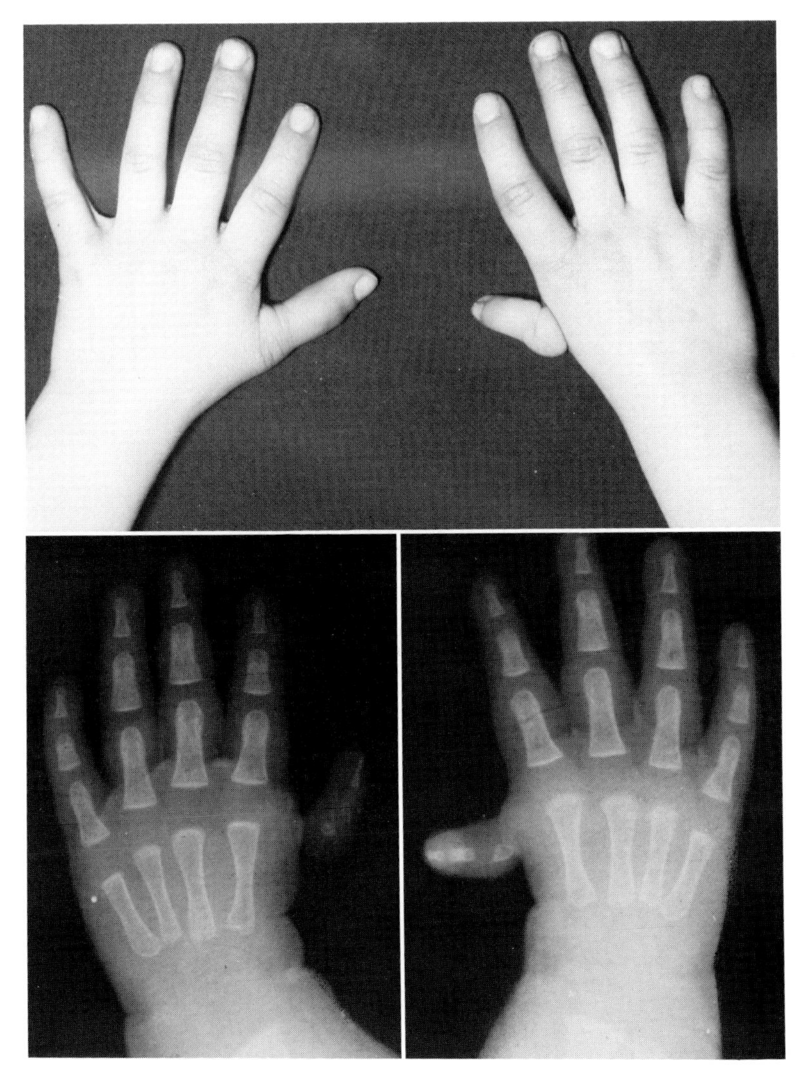

Fig. 5-1. *Floating thumb.* Three examples of floating thumbs or "pouce flottant." There is a slender thumb with a nail and two phalanges, but the first metacarpal is missing.

Parents are naturally reluctant to consider "amputation" of the index finger even though through it a thumb will be provided. Unfortunately, even the best results of any alternative operations do not compare with the results of pollicization.

Stabilization by using the fourth metacarpal and its epiphysis as a graft or by whole joint transplant from a toe or the building up of length by stacking several toe phalanges on each other have all been tried with varying success. Occasionally suitable graft material can be obtained from the same or the other hand if the congenital deformities present justify the amputation. Probably every hand surgeon has listened to the pleas of parents at least once and tried one of the procedures. I have done so, I regretted it, and I now feel justified in standing fast in my recommendation of pollicization for the truly floppy, useless floating thumb.

Occasionally the thumb is not completely floppy, and small degrees of jerky movements can be seen as if tendons were working. I have explored some of these thumbs and have never found adequate tendons. Even if these thumbs were to be anchored by a bone graft passing from the underdeveloped thumb metacarpal to the second metacarpal, the result will be a rigid post sited too far distally and radially for effective function. The use of local flaps or skin from a distance to allow better siting of the base of the thumb increases the operative problems without a significant increase in thumb function. Pollicization is still the best advice.

FLEXED THUMB

Persistent flexion of the thumb in an infant or child can be caused by abnormalities on either the flexor or extensor surfaces. Trigger thumb and clasped thumb are the most common causes, while varying absences of the extrinsic extensor tendons are rare causes for the deformity.

Trigger thumb

Congenital trigger thumb is not common, but it is significantly more common than trigger finger in children. Trigger thumb can occur on one or both hands and is thought to be hereditary in some families (Fig. 5-2). It has been described as a symptom in trisomy 13 syndrome.

The child does not complain of the fixed flexion of the interphalangeal joint and is usually brought to consultation because the parents thought "it looked funny."

It should be appreciated that the normal rest position for a child's thumb is with flexion of the interphalangeal joint. In trigger thumb the joint is usually held in 20 to 50 degrees of flexion and attempted passive extension may produce pain.

When one attempts to passively straighten the thumb, the metacarpophalangeal joint tends to go into hyperextension, and as the pressure is increased the interphalangeal joint finally yields with perhaps a palpable click at the metacarpophalangeal joint and a cry of protest from the child. Occasionally the condition is fixed, and extension of the thumb is impossible. In an adult with trigger thumb one would expect to notice the click; in infants and children it is unusual to find it.

The block to voluntary extension is produced by a constricting fibrous or fibrocartilaginous band in the sheath of the flexor pollicis longus at the level of the metacarpophalangeal joint. There is also a nodular thickening in the tendon so that the incongruity of size between thickened tendon and narrowed sheath produces a mechanical block to tendon movement.

The basic problem in infantile trigger thumb is not establishing the diagnosis but deciding whether or not one should operate. If the interphalangeal joint can be passively extended without causing significant distress, then splinting in exten-

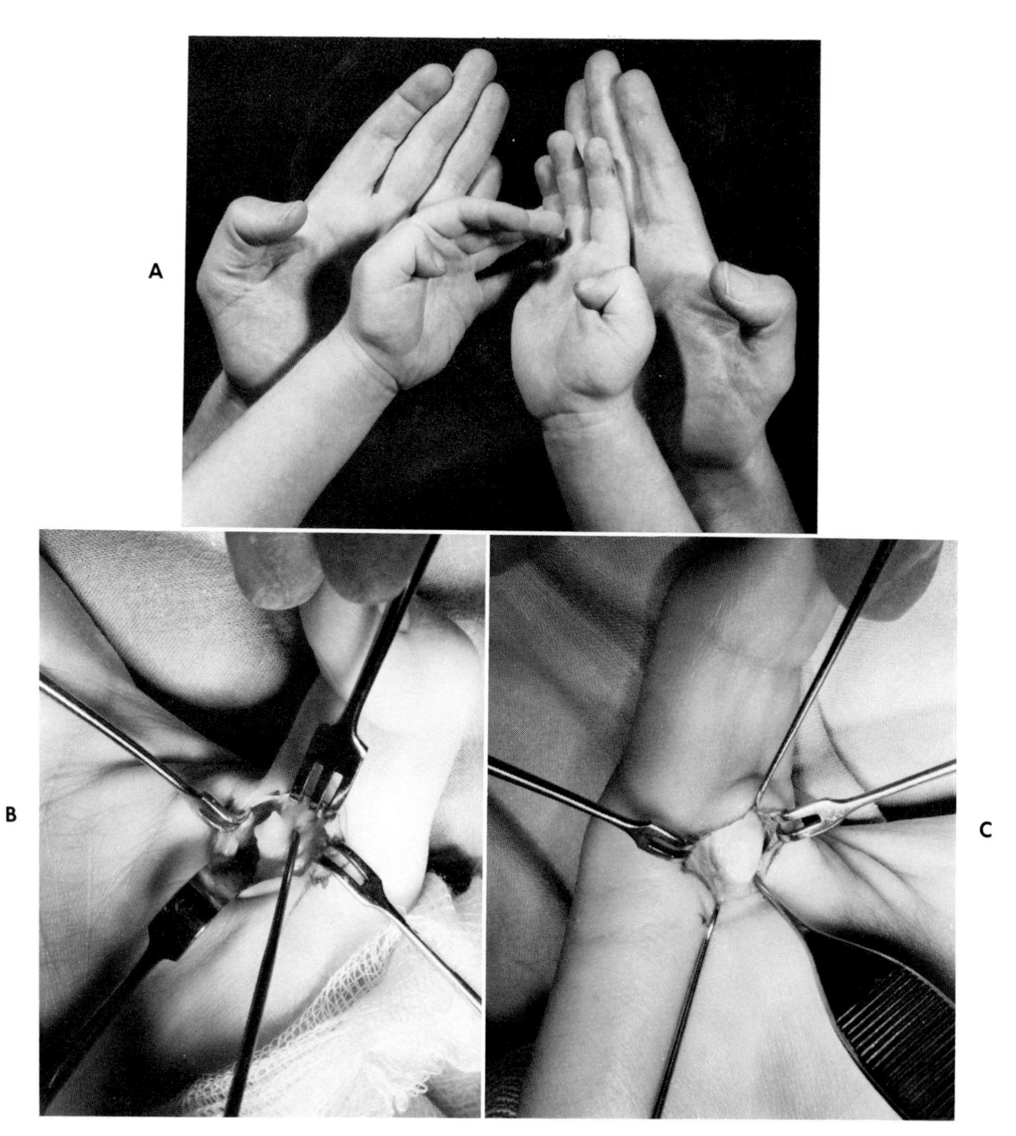

Fig. 5-2. *Trigger thumb.* Congenital trigger thumb is occasionally inherited and is more common than trigger finger in children. **A,** A mother and her son with trigger thumbs. As a child the mother was told surgery would be too dangerous. **B** and **C,** The mother's right thumb and her son's left thumb during surgical release. Note the large irregular nodule in the mother's thumb.

sion for about 3 weeks can be tried. Splinting is not uniformly successful, and more often than not one is once again faced with the decision of whether to operate or not. Durham and Meggitt have published a thorough review of over 130 trigger thumbs in children. Their figures showed that trigger thumbs that are present at birth can be safely left unoperated for the first year of life since at least 30% will show spontaneous recovery. They also showed that if the diagnosis is not made until between 6 and 30 months of life, one can still wait a further 6 months because about 12% of these patients can be expected to recover. Operation is certainly recommended if the child is over 3 years of age when first seen, but they did not see any cases of permanent contracture of the interphalangeal joint provided the operation was done before the child reached the age of 4 years.

Surgical release is done under general anesthesia through a skin incision made transversely in the flexion crease at the metacarpophalangeal joint (Fig. 5-3). A longitudinal incision over the line of the tendon produces a bad, contracted, and often painful scar. The two vital digital nerves lie relatively subcutaneously on either side of the tendon, and the skin incision must be shallow enough to incise skin only. After longitudinal skin retraction, the thickened tendon sheath is exposed and should be incised over about 1 cm length; this is usually sufficient to provide full tendon movement. I usually excise a small portion of the thickened tendon sheath and probe both proximally and distally within the intact tendon sheath to free up any adhesions which may be present. I do not, and believe one should not, excise any of the tendon nodule. I think that the primary lesion is the sheath constriction and that nodule formation is secondary. The nodule, if left unexcised, will therefore slowly be absorbed, whereas surgical excision of the nodule must create raw surfaces that can become adherent to surrounding tissues.

Clasped thumb

Unlike trigger thumb the characteristic deformity of congenital clasped thumb is extreme flexion of the metacarpophalangeal joint and adduction into the palm. It is an unusual condition that does not have a single cause. The posture is the result of flexion-extension imbalance from several different causes. Weckesser, Reed, and Heiple aptly subtitled their useful paper on these problems "A syndrome, not a specific entity." The following classification they devised takes into account the varied causes: Group I, deficient extension only; Group II,

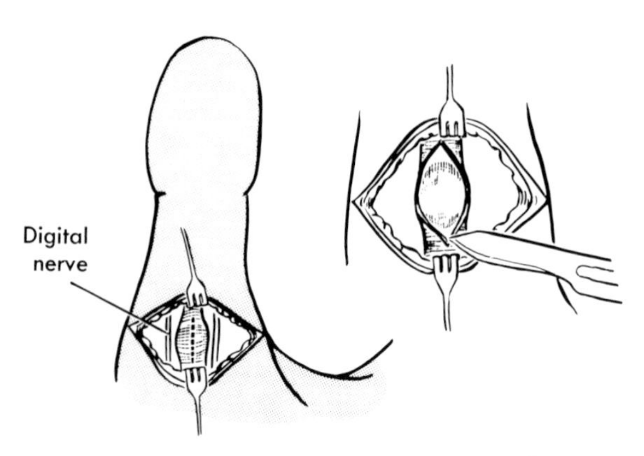

Fig. 5-3. *Operation for trigger thumb.* The skin incision is made transversely in the flexion crease at the metacarpophalangeal joint. The tendon sheath is opened by a longitudinal incision.

flexion contractures combined with deficient extension; Group III, hypoplasia of the thumb including tendon and muscle deficiencies; and Group IV, the few cases that do not fit into the first three groups.

Group I cases are three times as common as Group II cases, while Groups III and IV are five times less frequent than Group II cases. The condition is almost twice as common in boys as in girls and is nearly always bilateral. A familial incidence is reported in about one third of the cases and the posture has been described in the whistling face syndrome and in arthrogryposis. A proper physical examination will exclude both these possibilities as well as a spastic cerebral palsy. The fully developed condition is probably best called congenital clasped thumb, but it has also been named pollex varus, infant's persistent thumb-clutched hand, congenital flexion-adduction deformity of the thumb, congenital absence of the extensor pollicis longus, and congenital aplasia or hypoplasia of the thumb extensors.

The diagnosis can be hard to make in the first few months of life, but it is in this time period that the best results are obtained from treatment. During the early weeks of life, an infant frequently clutches the thumb but does release it intermittently for spontaneous motion. It is during these times that one must watch for active extension of the metacarpophalangeal joint. In the classical case, the thumb is held in the palm by the grasp of the overlying fingers, but when they are released, full passive extension is easily obtained even if no spontaneous motion has been observed. It is not until after 3 months of life that the infant discovers its thumbs and begins to use them in grasp. If after this time the thumb continues to lie flexed and adducted, the diagnosis of clasped thumb can be considered established.

The underlying cause is hypoplasia or absence of the extensor pollicis brevis muscle, but sometimes the extensor pollicis longus muscle is also involved. The majority of cases are in Group I and, if seen early in life, respond very well to immediate but prolonged immobilization in extension and abduction. White and Jensen used a simple elastic-webbing abduction splint for immobilization, but Weckesser and his colleagues, who have reported the largest series of these cases, use a plaster of Paris splint, which they change every 6 weeks to allow for growth of the hand. Immobilization needs to be continued for at least 3 months and may even be necessary for up to 6 months. The longer period of immobilization is justified if at the end of 3 months some active extension is possible, even if there is a droop toward flexion in the posture at the metacarpophalangeal joint.

The long-term results of treatment with corrective plaster casts have been shown to remain good if the response to primary treatment was good. No adverse affects on growth of the hand have been noted.

If at the end of the 3-month period there is absolutely no improvement in the posture and the deformity returns, then it is reasonable to assume that the extensor pollicis brevis is either absent or nonfunctional. No harm is done by continuing the immobilization for a further 3 months, and it is possible, but unlikely, that spontaneous extension will occur after this extra immobilization.

Extension can be restored by a tendon transfer to the base of the proximal phalanx.

Operative treatment is not always easy because it may be difficult to find a suitable tendon to transfer. The nearest and best tendon would be the extensor indicis proprius. The common extensor tendon to the index can also be used as a transfer but only after it has been demonstrated that there is an effective extensor indicis proprius muscle. I have also used the brachioradialis for this transfer procedure. I prefer to put the transfer in fairly taut, obtaining a tenodesis effect to resist the strong flexor pollicis longus muscle. Patients in whom there is also weakness or absence of the extensor pollicis longus usually show an equivalent weakness of the extensor indicis proprius since both arise from adjacent muscle masses in the forearm.

The flexed thumb resulting from absence of the extensor pollicis longus and weakness or absence of the extensor pollicis brevis is almost always associated with a characteristic droop of the index finger because of the additional absence of the extensor indicis proprius (Fig. 9-25, page 169).

I have found, in the majority of cases that I have explored during the course of tendon transfers, that the tendon in question is not completely absent but is represented by a very thin threadlike tendon. This tendon narrows proximally and eventually ends in fibrofatty tissue with no real muscle tissue attached to it. Occasionally by doubling up the vestigial tendon one can have enough stock for the attachment of the new proximal motor. By threading the tendon on a large-eyed needle the transfer can literally be sewn into the base of the phalanx using the vestigial tendon instead of a foreign body. When a motor, such as a prime wrist mover, is used it will have to be prolonged either by using a free graft such as palmaris longus or a toe extensor or by doubling up the vestigial tendon.

Conservative treatment has a very real value in treating the infant and very young child. Good to excellent results can be anticipated in about two thirds of the cases because it appears that the weak or hypoplastic muscles and tendons will become functional if they are properly protected in early life against the powerful flexor forces. Night splints alone do not give adequate protection. If patients are not seen until about age 2 years or older, the secondary contractures around the joint and in the overlying skin are additional factors that adversely affect the results of conservative treatment. Z-plasty release of the skin contracture or even incision and full-thickness skin grafting may be needed to allow full correction of the joint after the capsular tissues have been released. When this is done, I prefer to hold the metacarpophalangeal joint in extension for 3 to 4 weeks by transfixing it with a Kirschner wire. The older the child, the more likely that it will be necessary to relieve fascial and joint contractures. In adolescents and teenagers it will always be necessary.

Patients in Group II, who have significant flexion contracture of their fingers, are probably suffering from mild to moderate arthrogryposis. Splinting is a useful preoperative measure to obtain maximum correction, and joint fusions later in life may give some increase in function, but tendon transfer is not usually effective. Skin grafts or Z-plasties may also be necessary, particularly in the thumb web.

Group III patients usually have small hypoplastic thumbs with little or no muscle control of extension or opposition. Tendon transfers usually have to be prolonged by a graft to the base of the terminal phalanx of the thumb for extension, and an additional motor has to be found for an opposition transfer.

In the Group IV patients the problem is usually one of varying degrees of polydactyly with associated tendon weakness, rather than absence of tendons. Prolonged casting of the thumb in abduction and extension after removal of the duplication will often restore good extension to the remaining thumb.

WEB SPACE PROBLEMS

The thumb web may be too long, producing a partial syndactyly that results in a relative shortening of the effective length of the thumb. Apparent lengthening can be created by operations designed to set back the leading edge of the web.

Adduction of the thumb and narrowing of the web space is usually caused by unopposed action of the adductor pollicis resulting from congenital absence of the thenar musculature. This deformity is best treated by supplying some form of active abduction force and relieving the contracture of the adducted skin web. Total absence of the web space is seen in the so-called five-fingered hand, in which major surgery is needed to release the radial digit and create a proper web space.

Abduction of the thumb is usually caused by a congenitally aberrant flexor pollicis longus tendon pulling the thumb into abduction. Release and replacement of the tendon usually gives satisfactory results.

Deepening the web space

It is vital to provide full mobility for the thumb, and any degree of tethering by a too distal web should be removed. Several procedures using local flaps can provide good deepening and separation of incomplete syndactyly in the first web space.

Traditionally, two-flap Z-plasty has been used to deepen the thumb web. The two flaps do provide a widening of the span of the web but tend to sit back proximally in a narrow V-shaped apex. The normal rounded contour of the distal edge of the thumb web is provided by using a four-flap Z-plasty. When the web is far distal, adequate deepening can be provided only by using properly designed rectangular flaps based on the thumb and index finger. Usually there is not sufficient skin to provide adequate coverage for both digits and a supple web space; skin grafting will be necessary.

Operative procedures

Four-flap Z-plasty. The optimum gain in length with the least tension on the flaps is provided in two-flap Z-plasty by making the apex of the flap at an angle of 60 degrees. In the four-flap procedure the tip angle should be 45 degrees.

Two pairs of flaps will be transposed, and the resultant distal edge of the web will become curved and lax.

When planning the flaps, the basic principle in all Z-plasty work, equal length of all flap sides, must be followed. Therefore, the length of the distal edge of the thumb web will establish the length of side for all flaps. Having established this line, I draw two right-angled lines at each end and then bisect these 90 degree angles creating the four flaps. Confusion sometimes arises, after all four flaps have been raised, as to which changes place with which. It is sensible to identify each flap with a letter or numeral. I letter each successive flap A, B, C, and D and remember that the transposed pattern is represented by CADB (Fig. 5-4).

After the flaps have been raised, the digital nerves and vessels should be identified and freed. When the thumb is abducted as much as possible, the tight dorsal fascial bands and any tight muscles will be seen. These must be carefully released in a proximal direction until the flaps lie comfortably in their new sites; the flaps must not be sewn into place with their bases draped over tight underlying muscles or fascia.

When sewing the flaps together I use No. 6-0 ophthalmic catgut for infants and toddlers and No. 5-0 monofilament nylon in older, more cooperative children. The edges of the flaps tend to roll under, and I believe it is essential to use mattress sutures to obtain the proper eversion of the skin edges.

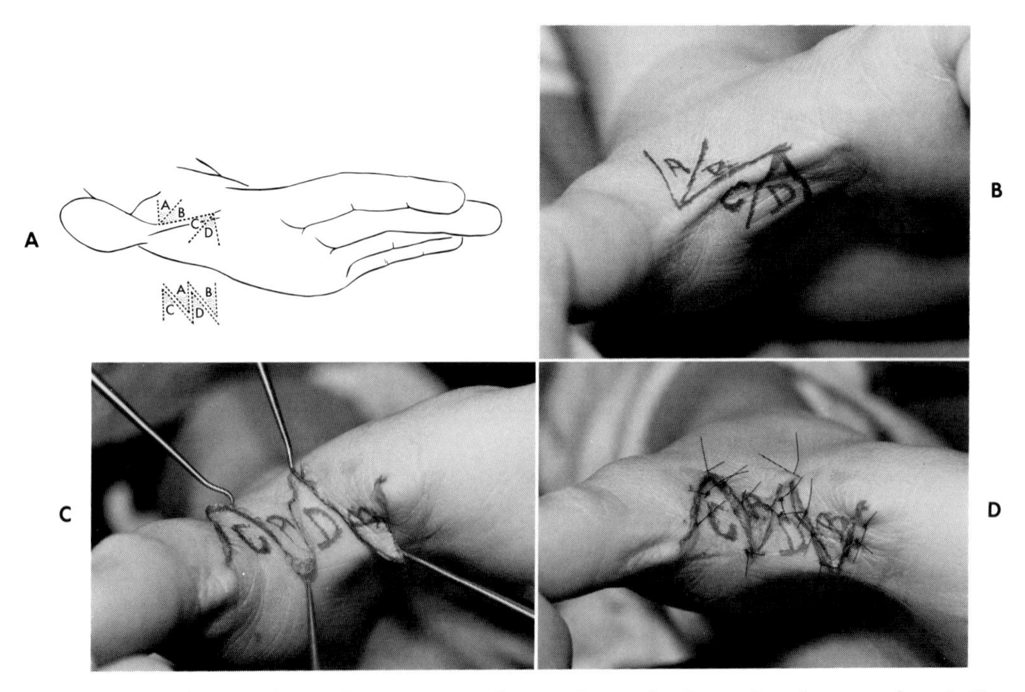

Fig. 5-4. *The four-flap Z-plasty.* **A,** Four flaps with equal sides and with tip angles of 45 degrees are raised. They are identified as *A, B, C, D,* and after transposition the letters read *CADB.* **B,** At surgery. **C,** After transposition. **D,** Wound closed.

Dressings and postoperative care are routine and I usually free the hand of all dressings at about 2 weeks postoperatively.

Dorsal-palmar flap combination. When the thumb web extends to the level of the interphalangeal joint, it is too far distal to benefit from a four-flap Z-plasty procedure. In these cases I usually use a modification of the standard syndactyly operation. A dorsal flap is outlined with its distal edge on the thumb web's distal border. It is usually a fairly broad flap that will set back and provide good width to the web. A palmar flap based on the thumb is designed so that its width and depth will cover the whole of the denuded ulnar side of the thumb. Inevitably the distal portion of this flap will encroach on the radial and palmar side of the index finger so that a full-thickness graft will be necessary to cover the defect on the radial side of the index finger. The size of the graft will vary, but it is wrong to compromise the size of the thumb-based flap because of concern for the size of the index finger graft (Fig. 5-5).

The distal edge of the dorsal flap is sewn into the proximal palmar edge of the thumb web that remains after the thumb-based flap is raised. Mattress sutures should be used throughout. Next the flap based on the thumb is brought around the ulnar side of the thumb and sutured to the dorsal thumb skin and the radial side of the transposed dorsal flap. Finally a pattern is cut of the index finger defect and a full-thickness graft cut to pattern from the elbow crease or groin. Postoperative dressing and care is routine and no special precautions are necessary. The thumb can usually be freed about 2 weeks postoperatively, but very young children and infants should have dressings applied for a further week. The former are likely to inadvertently split apart the healing edges, and the latter usually macerate the thumb flap by constant sucking.

There are many variations to the basic concept of using local flaps in the thumb web. Some prefer to cover the radial side of the index finger with a flap and the dorsoulnar defect on the thumb with a graft. This procedure works perfectly satisfactorily, but I prefer to provide a thumb with total full-thickness coverage in which all its skin has mobility on the deep tissues and an intact nerve supply.

The creation of a thumb-web space for a severely adducted thumb is a considerable undertaking and several different methods have been described. An

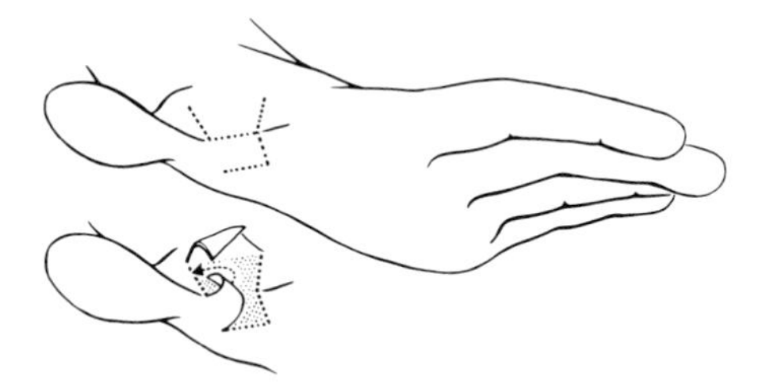

Fig. 5-5. *Dorsal-palmar flap combination.* The dorsal flap is drawn with its distal edge on the thumb web distal border. Beneath it a palmar flap is based on the thumb so that it will wrap around the lengthened thumb. The resultant raw area on the index is covered by a skin graft.

important clinical distinction to be made is whether the adducted thumb is truly a thumb or whether the child in fact has what is called a five-fingered hand with the nails of all five digits lying in the same plane. The technique for creating a thumb in the five-fingered hand is described on pages 72 to 75.

ADDUCTED THUMB

If the metacarpal shaft is adducted, it is usually because the absence of the thenar intrinsic muscles allows an unopposed action of the adductor pollicis. Strauch and Spinner have pointed out that in these patients the flexor pollicis longus is also usually absent. When examining the thumbs of these children, the physician must note the state of the first web space in addition to whether or not the thumb metacarpal can be passively abducted. If there is hypoplasia or absence of the median innervated thenar muscles, the thenar eminence will be flattened and thumb function, such as opposition, tip and lateral pinch, and most grasping activities, will be seriously impaired.

Most of these children will also show complete absence of the flexor pollicis longus tendon. They will, however, be able to flex the metacarpophalangeal joint to some degree by using the ulnar-supplied deep head of the flexor pollicis brevis or the attachment of the adductor pollicis into the extensor mechanism. This lack of powerful flexion of the metacarpophalangeal and interphalangeal joints also affects the appearance of the thumb because the normal flexion crease at the interphalangeal joint will be absent or very reduced in size, giving the thumb a peculiar tapering appearance (Fig. 5-6). Most of these thumbs are slightly shorter than normal, but they are usually of adequate functional length. It is thought that this abnormality is probably an autosomal dominant trait; therefore, there is a 50% chance that each child of an affected parent will inherit the deformity.

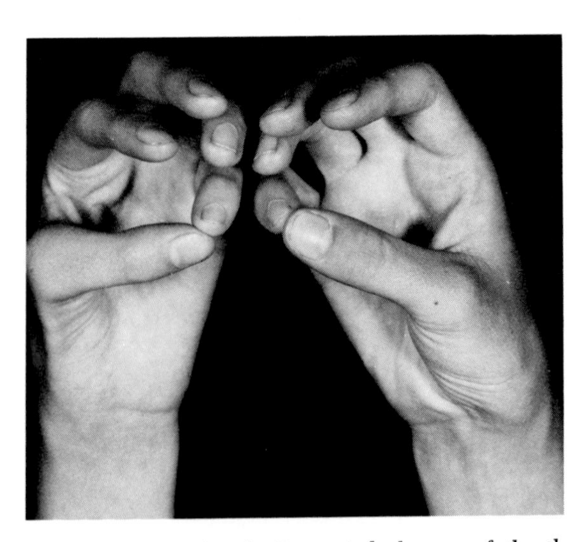

Fig. 5-6. *The adducted anomalous thumb.* Congenital absence of the thenar muscles in the left hand leads to adduction. Absence of the flexor pollicis longus accounts for the smooth tapered appearance of the thumb.

Restoration of opposition

Opponens transfer. One of two procedures is usually selected to restore opposition. The choice lies between an opposition tendon transfer and a transfer of the hypothenar muscles to replace the missing thenar group. A tendon transfer is the normal operation and the ring finger superficialis tendon is the tendon of choice. It should be passed through a pulley constructed from a slip of the flexor carpi ulnaris, passed subcutaneously across the palm and thenar surfaces, and inserted by two slips, one to the normal site of insertion of the abductor pollicis brevis and the other to the extensor expansion. Strauch and Spinner recommended that the second slip not go to the extensor expansion but be passed across the dorsum of the metacarpal and used to reinforce the collateral ulnar ligament (Fig. 5-7). By this means they complete the reconstruction in one

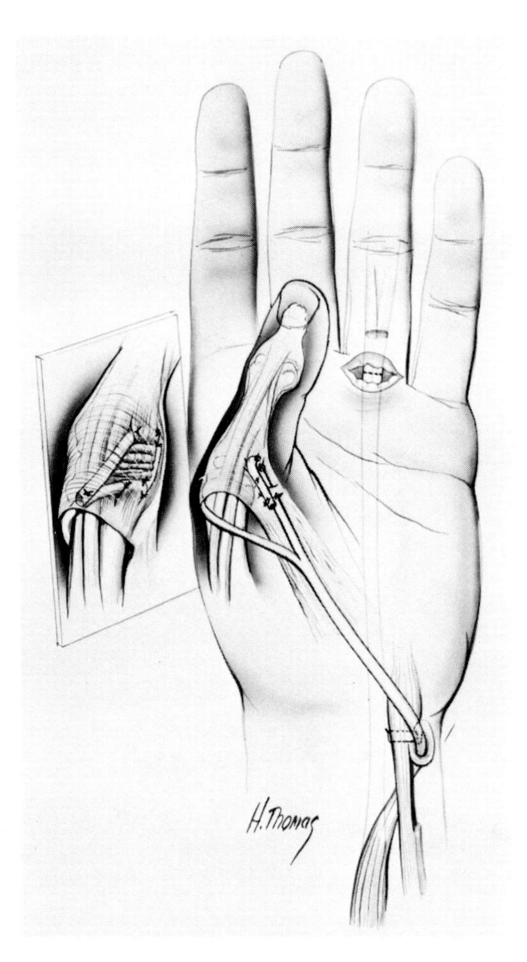

Fig. 5-7. *Restoration of opposition.* The transferred ring finger superficialis tendon is attached by two tails; one sutured to the normal insertion of the short abductor and the second passed dorsally over the thumb and used to reinforce and reef the ulnar collateral ligament. (From Strauch, B., and Spinner, M.: Congenital anomaly of the thumb: absent intrinsics and flexor pollicis longus, J. Bone Joint Surg. [Am.] **58:**115-118, 1976.)

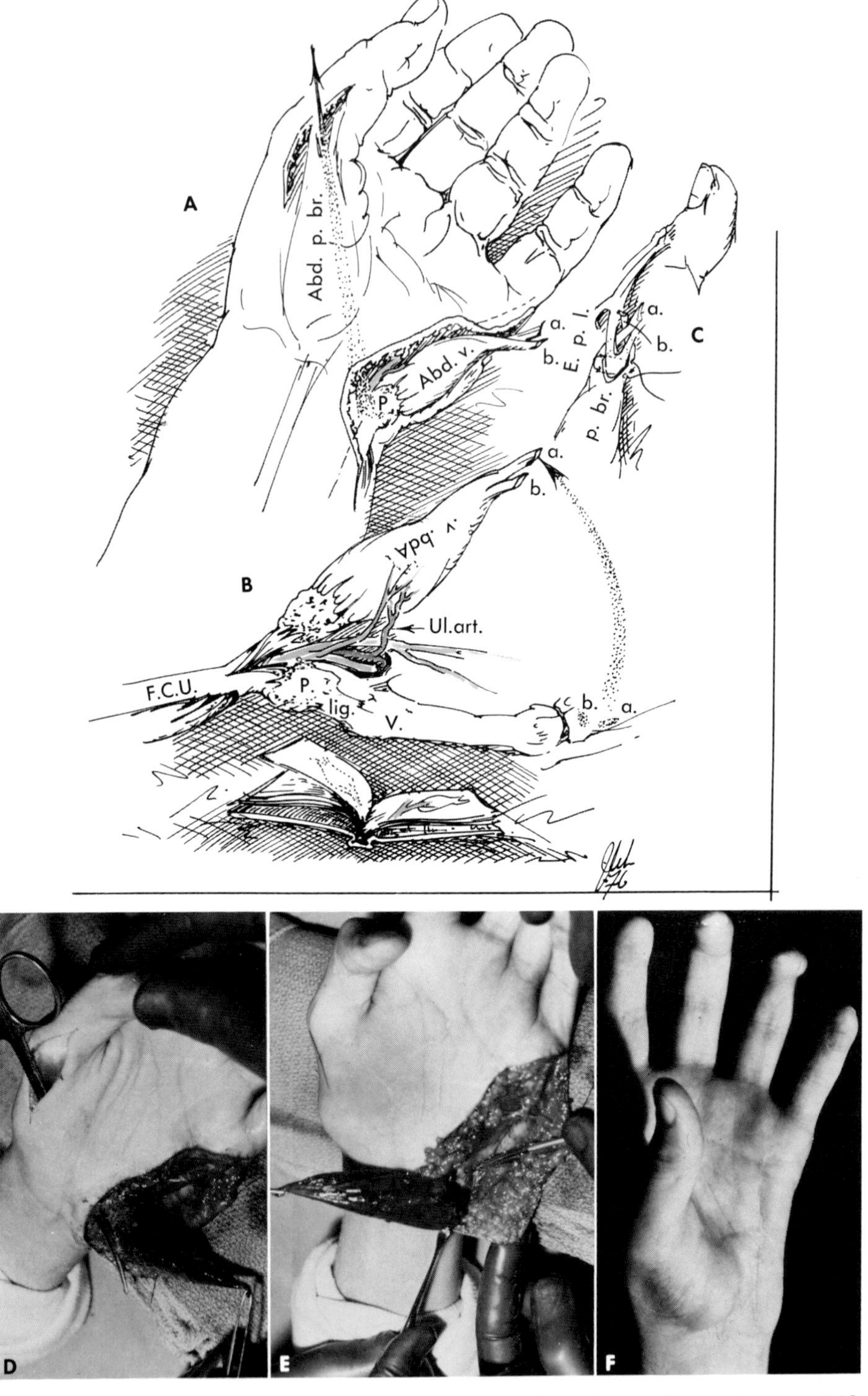

Fig. 5-8. *Hypothenar muscle transfer.* In this operation the abductor digiti minimi is folded over 170 degrees and passed subcutaneously to the thenar area to act as an abductor pollicis brevis. **A** and **D,** A capacious thenar pocket is made to receive the transferred muscle. **B** and **E,** Mobilization and isolation of the abductor digiti minimi on its neurovascular pedicle. **C,** The two slips of insertion *a* and *b* are sutured to the thumb metacarpophalangeal joint aponeurosis radial to the tendon of the extensor pollicis longus. **F,** The transfer in action providing thumb abduction and rotation. (Drawing and photographs courtesy of Dr. J. William Littler.)

operation. They report excellent results with six of the seven thumbs operated upon being able to oppose to the small finger with proper nail to nail orientation. The seventh thumb reached the ring finger. I have not tried their method, preferring to correct the adduction deformity and laxity of the joint in one procedure and supply opposition at a second procedure.

Hypothenar muscle transfer. The standard opponens transfer using the ring finger superficialis gives good results in terms of function but does nothing to restore the bulk and cosmetic appearance of the thenar area. Littler has recently reintroduced the Huber procedure in which the whole hypothenar muscle group with its neurovascular bundle is moved over as a substitute for the absent thenar muscle (Fig. 5-8). The muscle origins are left attached to fibers of the flexor carpi ulnaris to provide a firm proximal anchor. The advantage this procedure has over the standard opponens transfer is that an intrinsic muscle with a contraction amplitude similar to the thenar muscles is used. The tension of the transfer is automatically correctly adjusted and the muscle bulk produces a cosmetic appearance far better than that provided by extrinsic tendon opponens plasties.

Mere transfer of the hypothenar muscles is not in itself sufficient. It is also essential to totally release the adduction contracture. If this is not done, failure must occur since the transferred muscles are not strong enough to stretch out the adducted web space. This is a formidable procedure for the beginner, and neurovascular damage can be caused by inept dissection. In the hands of an experienced surgeon the operation gives a pleasing result.

Skin cover

Various methods can be used to supply needed skin cover, but in small children, and for that matter in children of any age, the use of local skin flaps is far more satisfactory than the use of flaps or tubed skin from a distance. The basic plan is to raise a proximally based flap, either from the thumb or index finger, that will rotate into the web space and leave raw a less important area, which will then need to be skin grafted.

Whatever method is used, careful dissection and excision of dorsal fascia over the web space are essential. One or two Kirschner wires, passed between the first and second metacarpals, will immobilize the thumb and maintain the abduction that has been obtained. This immobilization of the thumb is also helpful when suturing in the skin flaps or grafts at the end of the procedure.

Strauch has described a flap that uses practically the whole dorsal skin of the thumb. The distal end of this flap should extend several millimeters beyond the dorsal crease of the interphalangeal joint (Fig. 5-9). The ulnar incision should be carried to about the midpoint of the web and then turned proximally along the middorsal line of the web; it should extend at least half way to the carpometacarpal joint. On the palmar surface the incision should extend from the junction of the web with the side of the index finger proximally to the point of

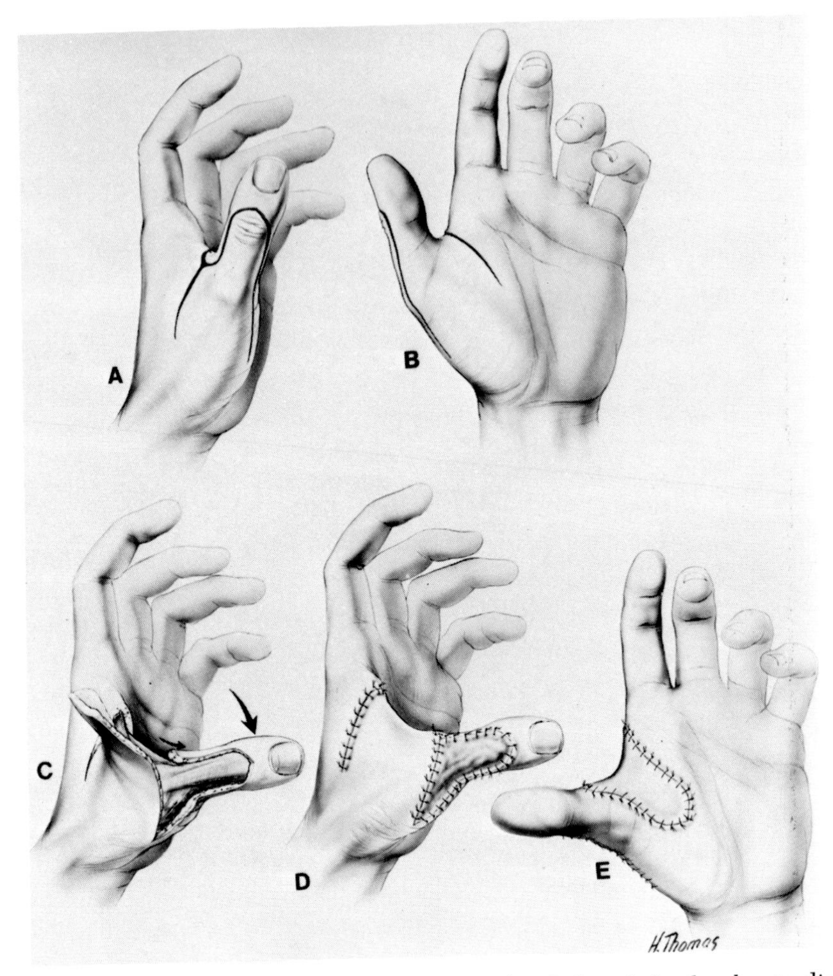

Fig. 5-9. *Web deepening—thumb flap.* **A,** The entire dorsal skin of the thumb extending 3 to 4 mm beyond the IP joint crease is moved. **B,** The palmar incision extends from the web to the junction of the first and second metacarpals. **C,** Abduction of the thumb allows insertion of the flap. **D** and **E,** The flap sutured in place and a split-thickness skin graft covering the donor site on the dorsum of the thumb. (From Strauch, B.: Dorsal thumb flap for release of adduction contracture of the first web space, Bull. Hosp. Joint Dis. **36**:34-39, 1975.)

junction of the first and second metacarpals. After the thumb is pinned in abduction, the flap will cover the opened up web space and leave a raw area on the dorsum of the thumb, which will need coverage from a thick split-thickness skin graft.

Strauch and Spinner have illustrated two ways in which a flap based on the radial side of the index finger metacarpophalangeal joint can provide web coverage. The first is the sliding flap proposed by Brand, in which a round-ended flap is moved proximally leaving a raw area on the radial side of the base of the proximal phalanx and the metacarpophalangeal joint of the index finger (Fig. 5-10). In order to get adequate release of the skin flap and of the adducted web

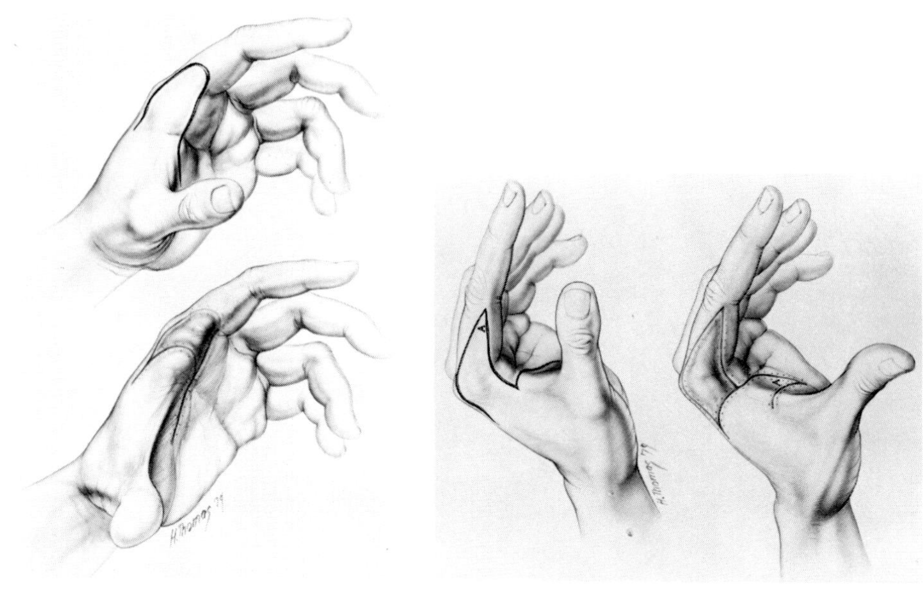

Fig. 5-10 **Fig. 5-11**

Fig. 5-10. *Web deepening—sliding flap.* A sliding flap, as proposed by Brand, is raised on the radial side of the index finger. The dorsal and palmar incisions must extend to the level of the junction of the thumb and index metacarpals in order to properly release tissues at the apex of the web space. A split-thickness skin graft is used to cover the defect on the side of the index finger. (From Strauch, B., and Spinner, M.: Congenital anomaly of the thumb: absent intrinsics and flexor pollicis longus, J. Bone Joint Surg. [Am.] **58:**115-118, 1976.)

Fig. 5-11. *Web deepening—transposition flap.* A relatively sharp-pointed flap is raised on the radial side of the index finger and transposed into the web space, which has been widened by a straight palmar incision. A split-thickness skin graft covers the index finger donor site. (From Strauch, B., and Spinner, M.: Congenital anomaly of the thumb: absent intrinsics and flexor pollicis longus, J. Bone Joint Surg. [Am.] **58:**115-118, 1976.)

space, the dorsal and palmar incisions must extend proximally to the level of the thumb carpometacarpal joint.

The second method is the use of a relatively sharp-pointed flap that is transposed from the radial side of the index finger into the web space, which has been widened by a straight palmar incision (Fig. 5-11).

FIVE-FINGERED HAND

Total adduction of the thumb with corresponding absence of the thumb web is seen in mitten hands and five-fingered hands. In patients with this condition the thumb nail lies in the same plane as the fingers and opposition is not possible (Fig. 5-12).

Some years ago I described the use of multiple dorsal rotation flaps, which provide adequate skin for both the dorsal and palmar surfaces of the first web space. The thumb is released and immediately rotated into its normal position at 90 degrees to the plane of the fingers (Fig. 5-13).

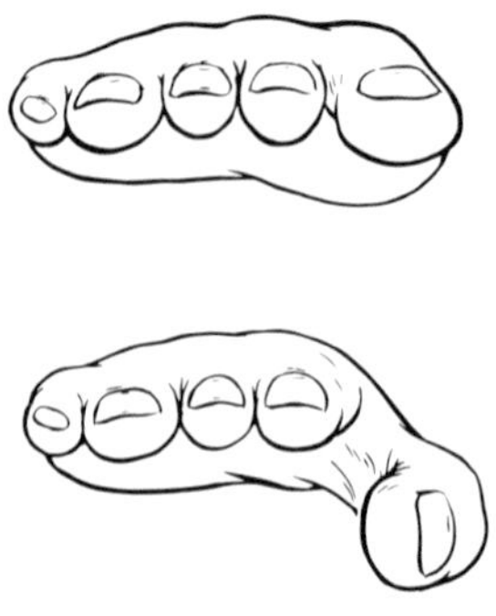

Fig. 5-12. *Five-fingered hand.* The position and plane of the thumb before and after surgical release has allowed abduction and rotation. (From Flatt, A. E., and Wood, V. E.: Multiple dorsal rotation flaps from the hand for thumb web contractures, Plast. Reconstr. Surg. **45:** 258-262, 1970, © 1970, The Williams & Wilkins Co., Baltimore.)

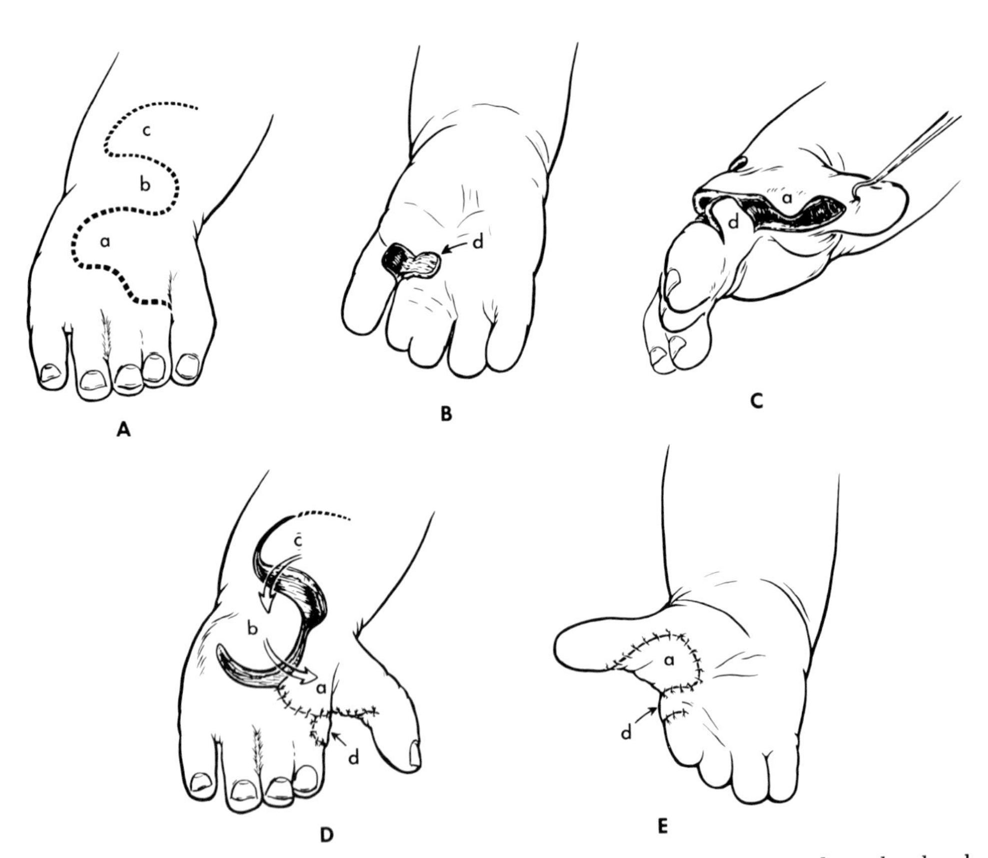

Fig. 5-13. *Multiple dorsal rotation flap technique.* The first flap, *a,* is based on the thumb and rotated into the opened web space. The second flap, *b,* is rotated into the defect left from *a;* the third flap, *c,* is rotated into the defect left from *b;* the defect from *c* is higher up on the forearm where there is more loose skin, so it can be closed in a curvilinear fashion. A small, square palmar flap, *d,* based on the index finger is used to cover the proximal radial portion of the index. (From Flatt, A. E., and Wood, V. E.: Multiple dorsal rotation flaps from the hand for thumb web contractures, Plast. Reconstr. Surg. **45:**258-262, 1970, © 1970, The Williams & Wilkins Co., Baltimore.)

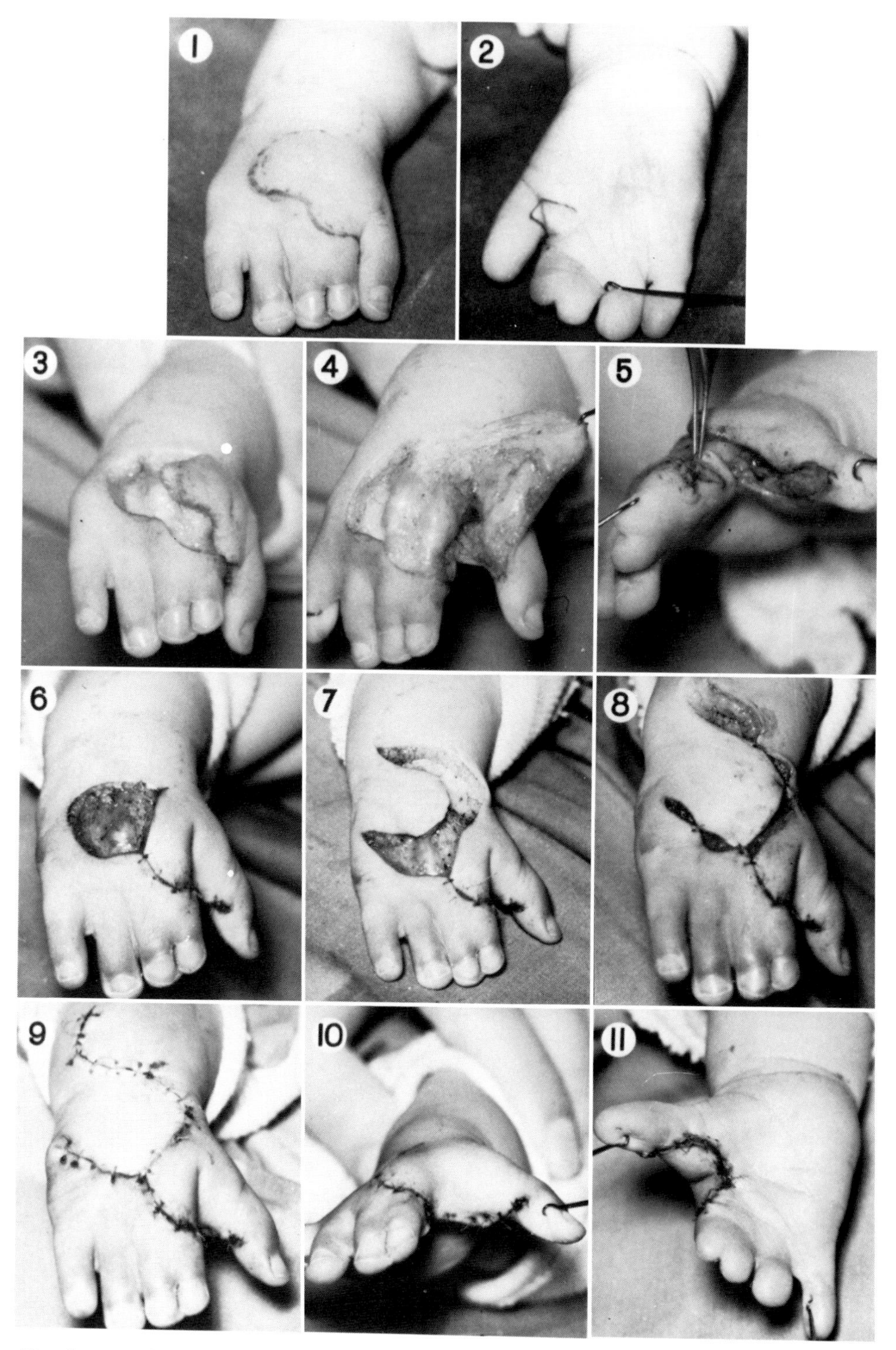

Fig. 5-14. *Multiple dorsal rotation flap.* The various stages of the operation are shown. Note how the plane of the thumb nail changes after development of the first dorsal flap shown in stages 1, 3, and 6. (From Flatt, A. E., and Wood, V. E.: Multiple dorsal rotation flaps from the hand for thumb web contractures, Plast. Reconstr. Surg. **45**:258-262, 1970, © 1970, The Williams & Wilkins Co., Baltimore.)

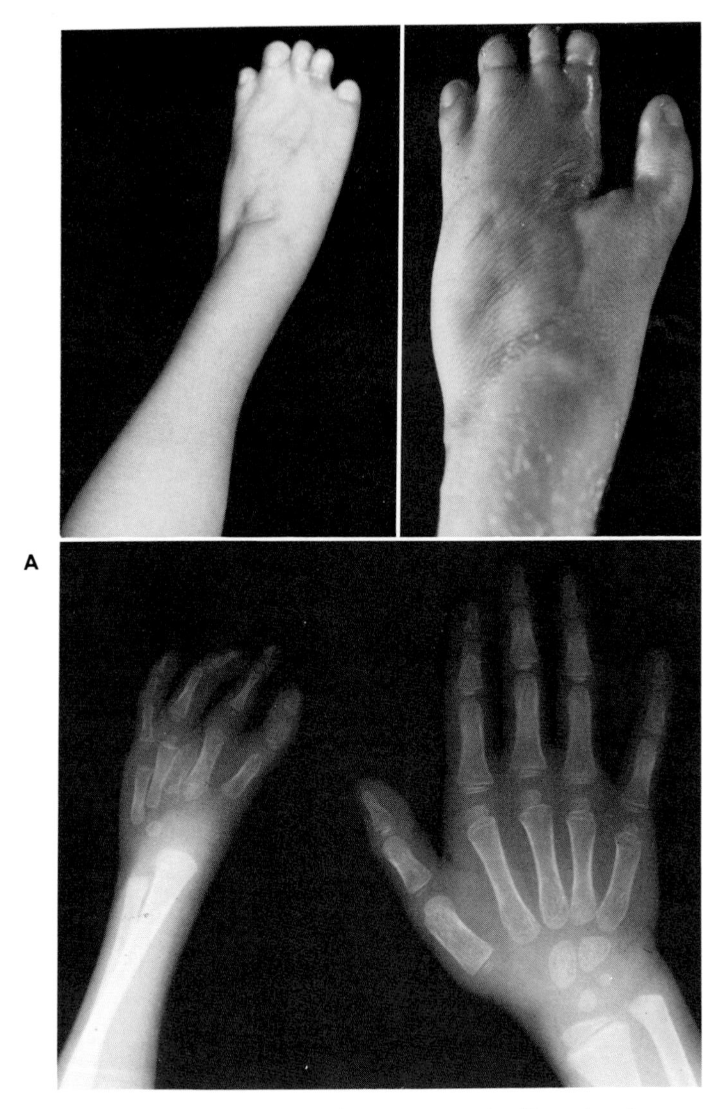

Fig. 5-15. *Multiple dorsal rotation flap results.* **A,** A typical adducted and unrotated thumb of the hypoplastic mitten hand. Correction is obtained by the multiple dorsal rotation flap technique. The degree of hypoplasia is shown by the x-ray film of the other, normal hand. **B, C,** and **D,** Other representative results; note that the dorsal scarring is sometimes unattractive. (**A,** From Flatt, A. E., and Wood, V. E.: Multiple dorsal rotation flaps from the hand for thumb web contractures, Plast. Reconstr. Surg. **45**:258-262, 1970, © 1970, The Williams & Wilkins Co., Baltimore.)

The first, or most distal, wide dorsal flap is based on the radial side of the dorsum of the thumb so that the venous drainage and radial nerve sensibility supply are intact. After this flap has been raised, a small square flap of palmar skin based on the index finger is raised; this flap will later cover the proximal and radial aspect of the index finger (Fig. 5-13, *B, d*). The thumb web space can now be cleared of all tight fascia, the thumb can be widely abducted, and the

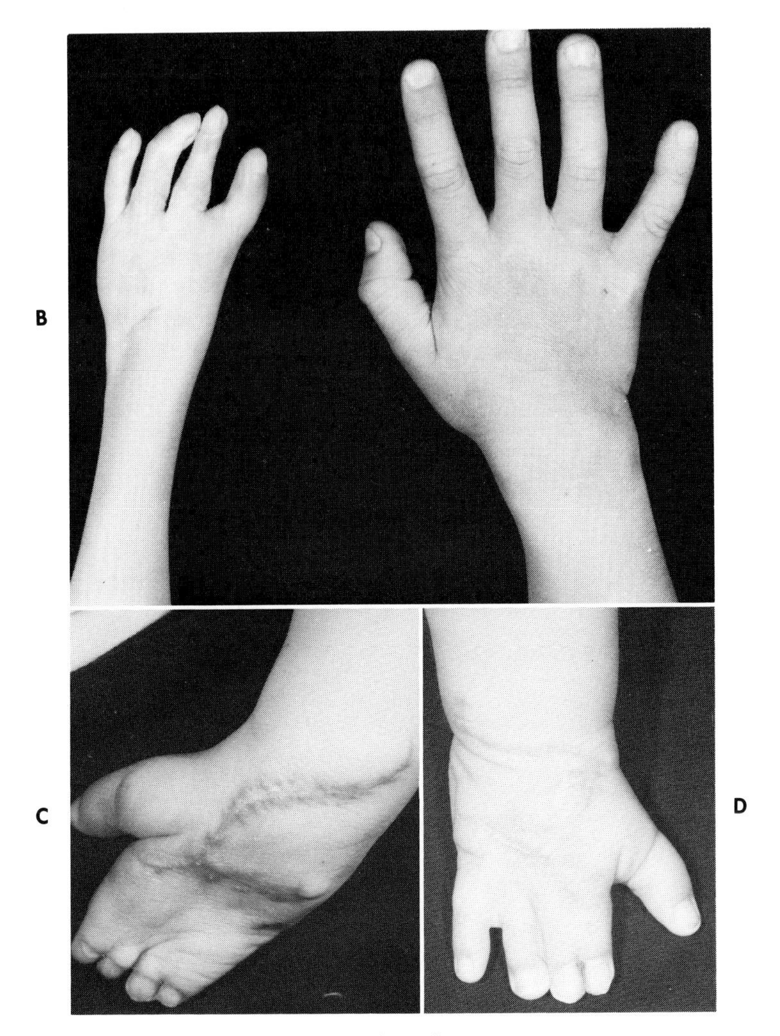

Fig. 5-15, cont'd. For legend see opposite page.

dorsal flap can be laid in place. Both the dorsal flap and the square index finger flap should be completely sutured in place before the second dorsal flap is raised. This second flap should be as wide as the others but based on the ulnar side of the wrist. It is then rotated distally into the defect on the dorsum of the hand (Fig. 5-13, *D, b*). A third and final dorsal flap is needed to fill the defect on the dorsum of the wrist created by the rotation of the second flap. This third flap is based radially on the most distal portion of the dorsal forearm.

Undermining of the surrounding tissues and careful lining up of the skin edges will allow total closure of the area, resulting in a final lazy-S scar running from the dorsal-radial side of the index metacarpophalangeal joint to the lower portion of the foream (Fig. 5-14).

I have used this technique for a number of years and have been pleased with the functional results seen on follow-up (Fig. 5-15). It is not an operation for the tyro; it demands respect for skin as a specialized tissue and experienced judgment in its use.

ABDUCTED THUMB

Marked abduction of a child's thumb is uncommon. The most likely cause of a radial abduction at the metacarpophalangeal joint is the condition described by Tupper as pollex abductus.

In children so affected there is usually a hypoplastic thumb but the thenar muscles, flexion creases, and flexor pollicis longus are present. In the cases described by Tupper, an aberrant flexor tendon could be felt to actively contract in a line passing proximally from the distal phalanx along the radial midlateral border of the proximal phalanx to the palmar-radial aspect of the wrist. The tendon crossed the metacarpophalangeal joint and lay directly over the anterior aspect of the thenar muscles. The tendon did not bowstring at the wrist during active contraction, but the abduction deformity did increase.

In this condition there is usually a lack of full extension of the interphalangeal joint, and surgical exploration has shown that the tendon inserts in a Y-shaped manner. There is an extra insertion slip that passes dorsally to join the insertion of the extensor pollicis longus (Fig. 5-16). This tendon lies in its normal plane

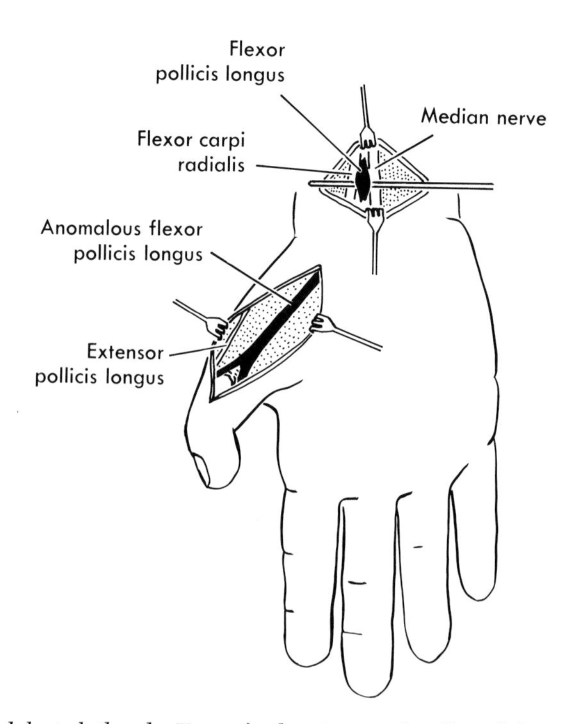

Fig. 5-16. *The abducted thumb.* Tupper's description of pollex abductus shows that there is an extra insertion slip of an anomalous flexor pollicis longus, which passes dorsally to join the extensor pollicis longus.

except when the thumb lies in full abduction, when it tends to palmarward subluxation.

Treatment for these abducted thumbs consists of relieving the primary deforming force and reinforcing the secondarily stretched ulnar collateral ligaments to overcome the instability of the metacarpophalangeal joint.

UNSTABLE JOINTS

Interphalangeal joint

Any of the three joints of a thumb can be deviated and unstable and cause functional disability. The interphalangeal joint is most commonly affected by a delta phalanx. The instability is made worse by simple excision of the abnormal phalanx, and ligament tightening operations are usually necessary.

Metacarpophalangeal joint

It is the metacarpophalangeal joint that is most frequently unstable. In both adducted and abducted thumbs, unusual forces work on the collateral ligaments; they yield, and instability results. Hypoplastic thumbs often have lax joint capsules.

The usual clinical picture is of a mildly hypoplastic thumb with significant abduction at the metacarpophalangeal joint, which can be passively fully corrected (Fig. 5-6). It is the ulnar collateral ligament that is lax in both adducted and abducted thumbs.

During surgery, the ligament is approached through an oblique dorsoulnar incision over the joint and the attenuated intrinsic hood is either incised or retracted to allow access to the ligament. Repair can be made either by ligamentous reconstruction or by muscle transfer. When the ligament is to be reconstructed, a graft should be taken from the palmaris longus or the plantaris and sewn into the origins of the lax capsular structures on both bones. This tendon graft should be crossed in an X fashion and pulled up tightly before sutures are placed. Holes can be drilled in the bones, but they must be placed sufficiently far from the articular surfaces to avoid the growth plates. In adults this operation has an undeserved reputation for failure; certainly it will fail if the X of the graft is loose, but a tautly pulled, reconstructed ligament gives good, long-lasting support both in children and in adults.

A more dynamic support can be provided by a transfer of the insertion of the adductor muscle. It must be dissected free of its insertion into the sesamoid on the flexor aspect of the joint and of its attachment to the extensor expansion. As much soft tissue as possible should be retained at the insertion so that it can be passed through a hole drilled into the ulnar side of the base of the proximal phalanx. The transfer should be pulled up tight rather than loose, and it will need protection by plaster immobilization for at least a month. After this, a night and resting splint will be needed for at least a further 3 months.

Both of these procedures will provide a stabilized metacarpophalangeal joint and a thumb much better able to resist the abduction forces of prehension.

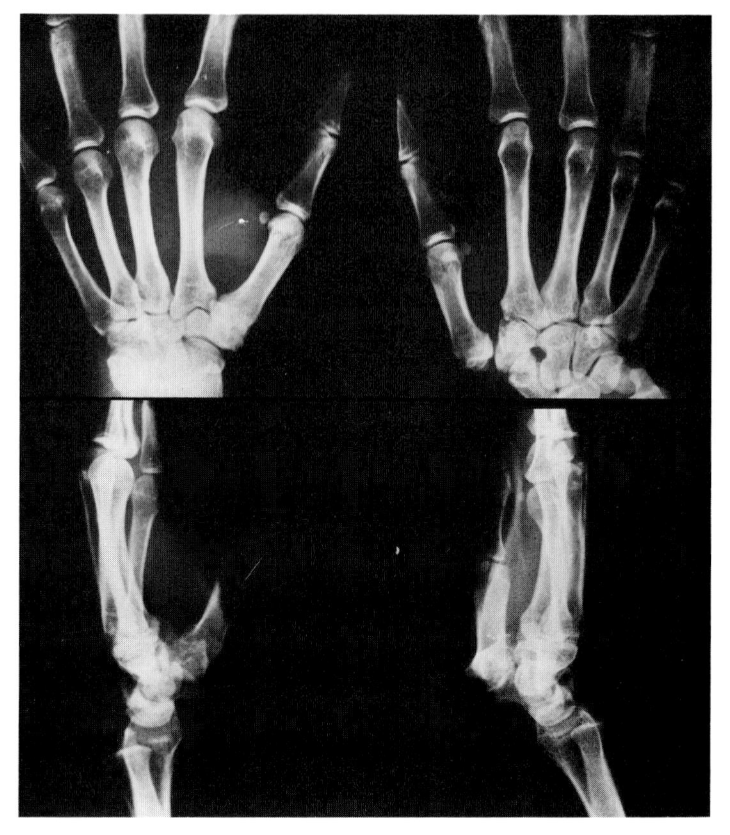

Fig. 5-17. *Unstable carpometacarpal joint.* Lack of development of the trapezium leads to gross instability because the normal saddle-shaped joint is absent. Fusion gives good stability. This patient was in her early 20s and was unable to write for any length of time. The hand on the left has had a carpometacarpal fusion; that on the right has not been operated on.

Carpometacarpal joint

Occasionally the carpometacarpal joint is unstable and thumb function is seriously embarrassed. The instability is caused by lack of development of the trapezium so that the normal saddle-shaped joint does not exist. The strong extrinsic extensor and flexor tendons are unopposed and the base of the thumb dislocates proximally.

The diagnosis is hard to establish in a young child because the trapezium and trapezoid bones do not ossify until the fourth to sixth years of life. In older children the diagnosis is easily established by the complaints of "weakness" when using the thumb, the proximal telescoping of the thumb on passive motion, and the characteristic absence of the trapezium on x-ray examination.

The treatment is fusion of the base of the thumb to the trapezoid when there is sufficient bone to allow good union without risking growth arrest (Fig. 5-17). Girls mature earlier than boys and in the former the operation can usually be done safely at about 12 years of age. Boys may well have to wait a further 2 years.

CHAPTER 6

THE ABSENT THUMB

Concepts of treatment
- Recession of the index

Pollicization
- Choice of digit
- Timing of surgery
- Development of the operation

Operative procedure
- Six major operative steps
- Skin incisions
- Neurovascular mobilization
- Mobilization of extrinsic and intrinsic muscles
- Positioning the new thumb
- Reattachment of muscles
- Closing the incisions
- Dressings

Postoperative care

Pollicization and the floating thumb

Pollicization in complicated deformities

The child with a grossly deficient or totally absent thumb is severely handicapped in social and educational development and in future ability to earn a living. Absence of the thumb can occur as an isolated anomaly but is usually associated with other congenital anomalies. Some of these can also severely handicap the potential development of the child.

An absent thumb may be associated with a normal, hypoplastic, or absent radius, but conversely an absent radius is almost always associated with an absent thumb. The exception to this is the thrombocytopenia-radial aplasia syndrome in which, in spite of the absence of the radius, the thumb is present.

Absent thumbs and, more commonly, hypoplastic thumbs may occur in many syndromes. In the ring D chromosome abnormality, absence of the thumb may be the characteristic finding. Other conditions associated with absence of the thumb are Fanconi (pancytopenia-dysmelia), Holt-Oram, trisomy 18 with multiple malformations, Rothmund-Thompson, Treacher Collins, and pseudothalidomide or S-C syndromes and true thalidomide embryopathy.

Children born with absence of one or both thumbs are usually brought for consultation shortly after birth, and nowadays the parents can be reassured that surgical reconstruction can provide significant functional restoration.

CONCEPTS OF TREATMENT

For the child born without thumbs, the dilemma in treatment is not whether to create a thumb but whether to create two thumbs. I believe it correct to make both thumbs for most individuals with bilateral absence. The problem for the child with one normal hand is whether or not the parents consider it reasonable to sacrifice a finger to make a thumb. I do, because a hand without a thumb has lost at least 40% of its usefulness, whereas loss of a finger deprives a hand of only 10% to 20% of its use. In bilateral cases the question is often raised—which thumb should be made first? The answer is simple—this is a right-handed world and even children who have a left-handed parent do well with a right thumb. In fact, left-handed parents are aware of their awkward situation and some ask for their child's right thumb to be made first. The implication of "first" is that subsequently they almost invariably ask for the second thumb to be made.

The absence of a thumb is compensated for by developing a reasonably accurate lateral pinch grip between the index and long fingers. Quite a strong grip may develop between the ulnar side of the index nail and the pad of the long finger, and this is often associated with a considerable degree of rotation of the index finger. At best, this is a very poor substitute for normal tip or pulp pinch; it could perhaps be tolerated if the other hand was entirely normal, but in bilateral loss this type of prehension is totally inadequate compared with the grip that can be provided by transposing a finger.

Recession of the index

When a child has developed a lateral pinch type of prehension and the parents cannot accept the concept of pollicization, the operation of recession and rotation of the index metacarpal offers some increase in function while maintaining the appearance of four fingers. The operation recesses the index to make it more resemble a thumb and provides a wider gap between index and long fingers. Two incisions are required. A short dorsal incision over the space between index and long finger metacarpal heads allows one to cut the juncture between the extensor tendons, the dorsal deep fascia, the intermetacarpal ligament, and the palmar fascia. A second curved dorsoradial incision near the base of the index metacarpal allows one to excise about 1.5 cm of metacarpal shaft (Fig. 6-1). The index can now be mobilized, recessed, rotated, and placed in radiopalmar abduction. It should be held in its new position of 20 degrees of radial abduction, 35 degrees of palmar abduction, and about 110 degrees of rotation by a Kirschner wire passed through the base of the metacarpal. A second wire placed more distally and passed into the second metacarpal is often helpful to maintain position. The hand should be protected in a plaster cast for 6 to 8 weeks, by which time the osteotomy site should be united.

This operation does in fact improve prehension and under the strict limitations imposed by the parents it is probably the best that can be done. There are also several other operative methods for restoring thumb length or substituting for

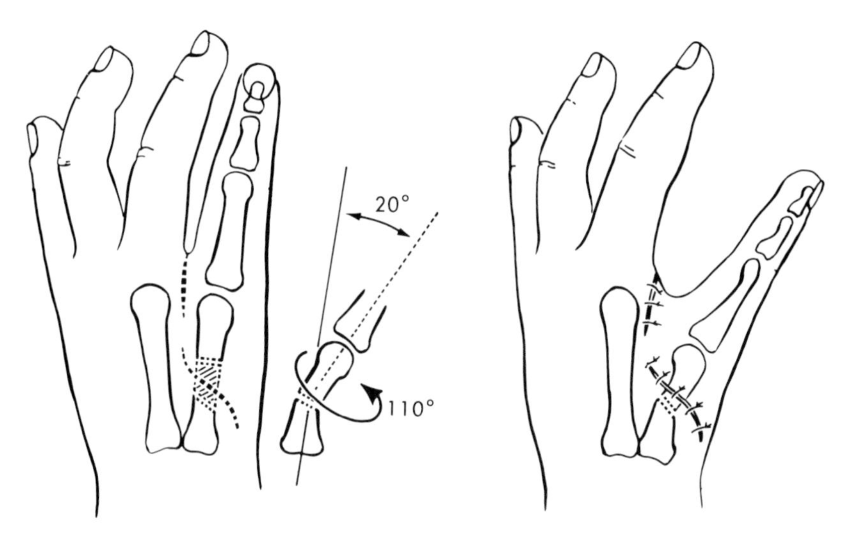

Fig. 6-1. *Recession of the index.* Two incisions are required. The more distal is used to cut the intermetacarpal ligament, and osteotomy of the index metacarpal is done through the curved more proximal incision. The distal portion of the index should be rotated 110 degrees and abducted 20 degrees dorsally.

the absent thumb. However, all of these are significantly inferior when compared with the result that can be obtained by pollicization.

POLLICIZATION

Pollicization is a term that is generally used specifically for thumb reconstruction by transfer of a finger and that excludes other methods. This concept of mass transplantation of bone, joint, tendon, and skin, together with neurovascular supply was described in detail by Littler in 1953. Since this pioneer work by Littler, refinements have been made in the basic technique until the operation as practiced today contains contributions by a significant number of hand surgeons. The history of the surgical provision of a thumb is fascinating and has been well reviewed by Iselin in 1955 and more recently in 1973 by Ahstrom. The caution of earlier years has given way to a procedure that can now produce in the very young child a thumb that works and looks like a normal thumb. Several of the later refinements have been contributed by Riordan and by Buck-Gramcko. The latter has had the unique opportunity of making over 200 thumbs for children born deformed as a result of the thalidomide tragedy.

Choice of digit

There is honest dissent as to which finger should be transposed to make the thumb. It is surgically possible to move any one of the four fingers and have it survive. I am out of sympathy with those who prefer the small or ring finger because I believe it wrong to disturb the normal ulnar border of the hand. I have taught for years—and honestly believe—that the finger I would be most loath to lose would be my small finger and the one I could spare with the least loss of dexterity or power would be my index finger. The ring and small fingers are vital to all power grip activities, but whenever one injures the index finger, all precision activities are automatically moved to the long finger with little or no loss of skill. To move the long finger risks depriving it of essential dorsal venous drainage, while the index can be readily jogged across to the thumb position while retaining all of its vital neurovascular components. I am therefore a staunch supporter of the index finger as the prime choice in pollicization.

Occasionally in radial aplasia the fingers show varying degrees of stiffness, and some surgeons feel that this should be a bar to pollicization. Total lack of motion in the interphalangeal joints does not provide a good thumb, but usually the metacarpophalangeal joint has adequate or good motion. Since this will become the carpometacarpal joint and therefore provide good motion at the key basal joint of the new thumb, relative stiffness at the interphalangeal joints can be tolerated. If the stiffness persists past the growth period, then secondary fusions of the joint in more flexion may increase the function of the new thumb.

In a child with a five-fingered hand the only satisfactory way of providing opposition is to pollicize the most radial finger and thereby move the first ray out of the palmar plane of the other digits.

Timing of surgery

There is also honest dissent as to the best time to perform the operation. The most important factor is the skill and experience of the surgeon. Although the logical steps of the operation have now been worked out, the procedure is far from easy, and in general most surgeons work down to earlier ages as they become more proficient in the operative details. A case can be made for both early or late timing of the operation. Probably the fundamental factor in favor of early surgery is that it will help the child avoid acquiring undesirable motor patterns because of an uncorrected deformity and will allow him or her to learn new and proper prehension patterns at an early age. Edgerton, Snyder, and Webb have discussed the psychological problems and point out that the ease with which a child views his or her own hands makes the deformities an even greater source of psychological insult than facial deformities. Thus an early operation reduces the time during which a child may establish an unfavorable body image from the visual impact of the deformed hands. A child first becomes aware of his or her thumbs a little after 3 months of life, and prehensile movements of the thumb and forefinger are properly established before the end of the first year of life. The fine coordination of hand movements is normally not developed until about 3 years of age, but the earlier the maximum potential function can be provided, the more likely this coordination is to be developed. All these factors point to some time in the second 6 months of life as the best time for an operation, but I believe that few hand surgeons have a sufficient volume of these cases, and therefore sufficient experience, to justify surgery much before 1 year of age.

The proponents of "later" surgery cite the technical difficulties related to the very small anatomical structures, the thick subcutaneous fat and thin skin, the risk of loss of the digit from vascular embarrassment, and the lack of patient cooperation. Although I am sympathetic to these objections, I do not believe them to be valid since most can be obviated by meticulous surgical technique and by obtaining the parents' understanding of the values of early surgery.

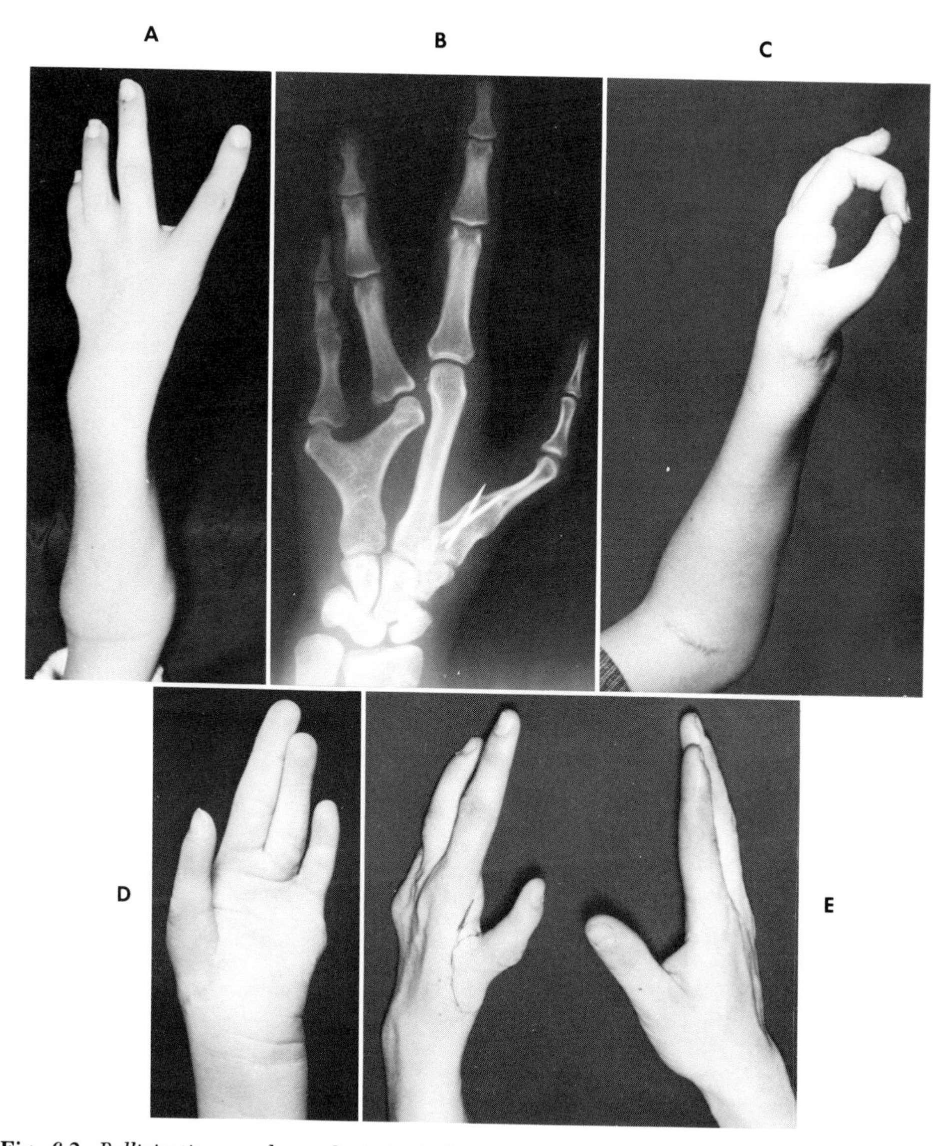

Fig. 6-2. *Pollicization—early method.* **A,** Before pollicization. **B,** The rigidly fixed union between the proximal and distal portions of the index metacarpal. **C** and **D,** Postoperative position and function. **E,** A comparison with the normal hand showing the small size of the new thumb in adult life.

Development of the operation

Pollicization has gone through three major stages of development, and for those of us who have worked throughout the period, it is a chastening experience to review one's early cases.

In patients with congenital absence of the thumb, the trapezium is usually absent and the scaphoid often malformed. Because of these deformities, in the 1950s one was proud to be able to obtain survival of a "thumb" that had to be rigidly fixed at its base by fusing its distal portion at the appropriate angle to the proximal metacarpal shaft (Figs. 6-2 and 6-3).

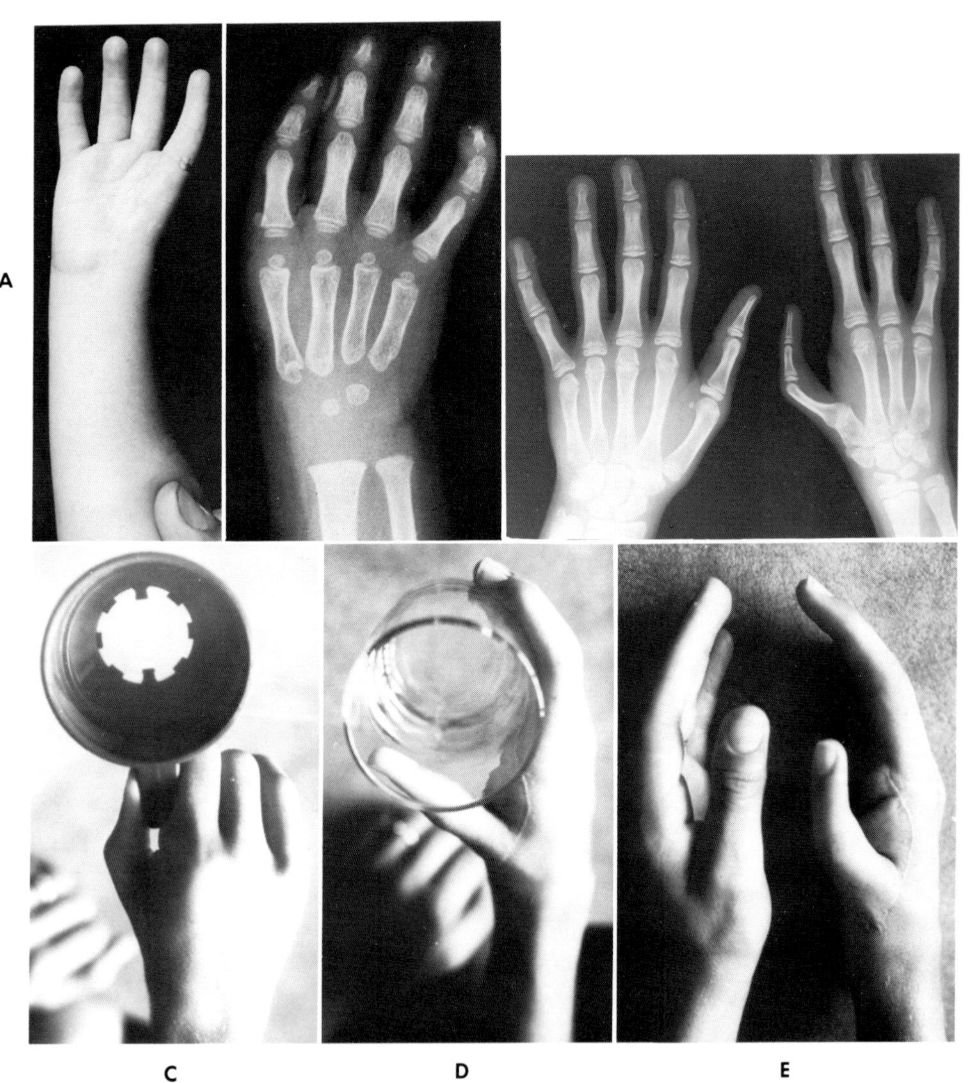

Fig. 6-3. *Pollicization—early method.* **A,** Before pollicization. **B,** The angulation and rotation of the index after pollicization. **C** and **D,** The span of grasp possible. **E,** The small size of the new thumb compared with the normal.

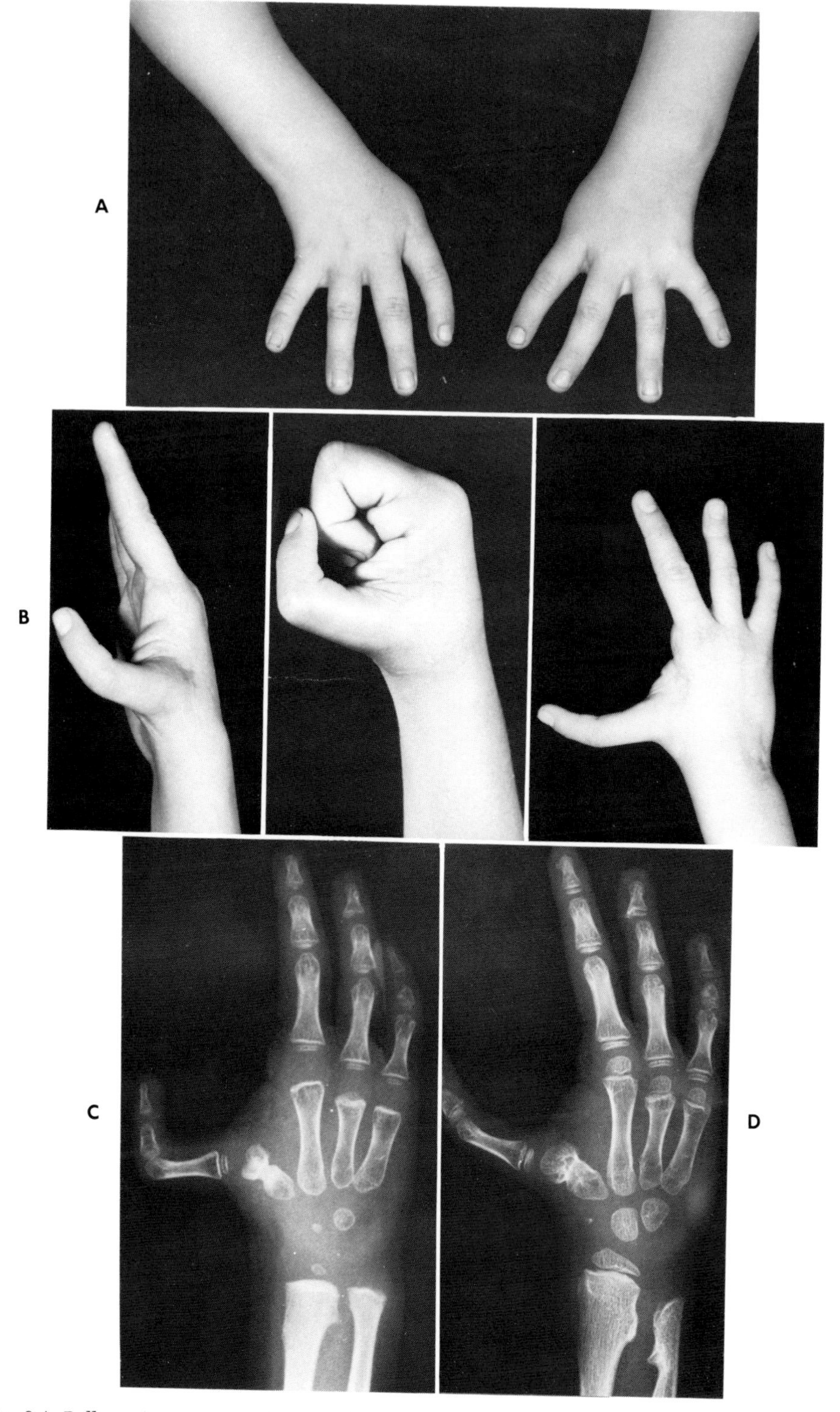

Fig. 6-4. *Pollicization—intermediate method.* **A,** Before pollicization. **B,** Function after pollicization. Note the excessively long, ugly thumb. **C** and **D,** Lengthening continuing over 5 years in the growth center of the head of the metacarpal.

In the 1960s attempts were made to fashion a carpometacarpal joint out of the finger metacarpophalangeal joint by retaining the joint and fusing the distal metacarpal shaft into its proximal stump. Unfortunately, growth continued in the metacarpal head, producing an increasingly grotesque thumb (Fig. 6-4).

In this decade all major problems appear to have been solved, and the thumbs now being made are cosmetically and functionally almost indistinguishable from normal.

OPERATIVE PROCEDURE

The conversion of an index finger to a thumb necessitates some shortening, since in the normal hand the tip of the thumb usually reaches just proximal to the line of the proximal interphalangeal joint of the index finger. This shortening is accomplished by removal of most of the metacarpal shaft, but the head is retained since it will become a new trapezium.

In effect there is a one joint recession of the finger so that its distal interphalangeal joint becomes the thumb interphalangeal joint; the proximal interphalangeal, the metacarpophalangeal joint; and the metacarpophalangeal, the carpometacarpal joint (Fig. 6-5). The distal phalanges in both are the same, but the index middle phalanx becomes the thumb proximal phalanx, the index proximal phalanx becomes the thumb metacarpal, and the head of the index metacarpal constitutes the new trapezium. The key to this shortening is destruction of the growth plate of the metacarpal head. This arrests longitudinal growth at the base of the new thumb and prevents the development of the grotesque thumbs made by earlier methods. When the new thumb is placed,

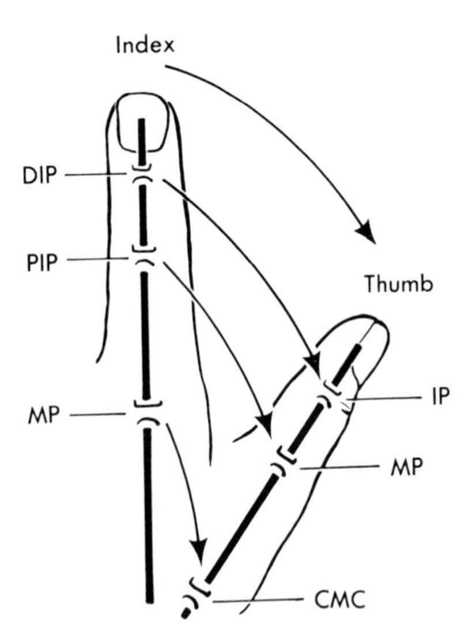

Fig. 6-5. *Pollicization—joint distribution.* The three joints of the index finger become the three joints of the thumb, but the index metacarpophalangeal joint becomes the carpometacarpal joint of the thumb. (Modified from Buck-Gramcko, D.: Pollicization of the index finger, J. Bone Joint Surg. [Am.] 53:1605-1617, 1971.)

it will have to be angled out in about 40 degrees of palmar abduction so that it can readily move into opposition. To obtain freely moving and proper opposition, one will have to rotate the index around its longitudinal axis to allow the proper pulp-to-pulp posture. It is important to bring the finger around far enough to easily reach the ring finger. Buck-Gramcko recommends a rotation of about 160 degrees since, he points out, a certain amount of derotation occurs during suturing of the muscles and skin and the final rotation desired is about 120 degrees, which is far in excess of the 90 degree rotation that has often been recommended.

Six major operative steps

There are six major steps to the operation:

1. Planning and mobilizing the skin flaps to ensure good thumb-web coverage
2. Identifying and extensively mobilizing the two neurovascular bundles and the dorsal venous network
3. Identifying both the extrinsic and intrinsic muscles and detaching the palmar and dorsal interossei from the index metacarpal so that they may become the new thenar muscles
4. Shortening the finger by cutting through the metacarpal neck epiphysis and removing the shaft and most of its proximal end
5. Placing the new thumb by suturing the cartilaginous metacarpal head anterior or palmar to the stump of the second metacarpal
6. Attaching all extrinsic and intrinsic muscles, rotating the skin flaps, and closing the skin wounds

The following detailed operative description is based upon the protocol I give my trainees, but this description lacks the final paragraph on how to prepare a cup of tea for the exhausted surgeon. This surgery is far from easy, and it is even more trying to lead a trainee through his first pollicization! I believe it should not be attempted for the first time without the active assistance of a surgeon experienced in pollicization.

The operation must be done under tourniquet, and the field can probably be safely kept bloodless for 2 hours. A tourniquet time longer than this may be cause for concern, but release of the tourniquet for, say, ten minutes and its reapplication is a mixed blessing because the anoxic tissues will swell so much that it may be extremely difficult to close the skin wounds. I keep the tourniquet at 200 mm of mercury until I have completed the operation.

Skin incisions

Opinions vary as to the best approach to the various tissues, but I have found, in total thumb absence, that the incisions recommended by Buck-Gramcko are useful since they allow good exposure with gentle skin retraction. All skin incisions are planned and drawn before the tourniquet is inflated. The first line is that of the palmar incision (Fig. 6-6, *A*, *B-A*). This is slightly S-shaped with the distal limit of the line being about the midpoint of the length of the

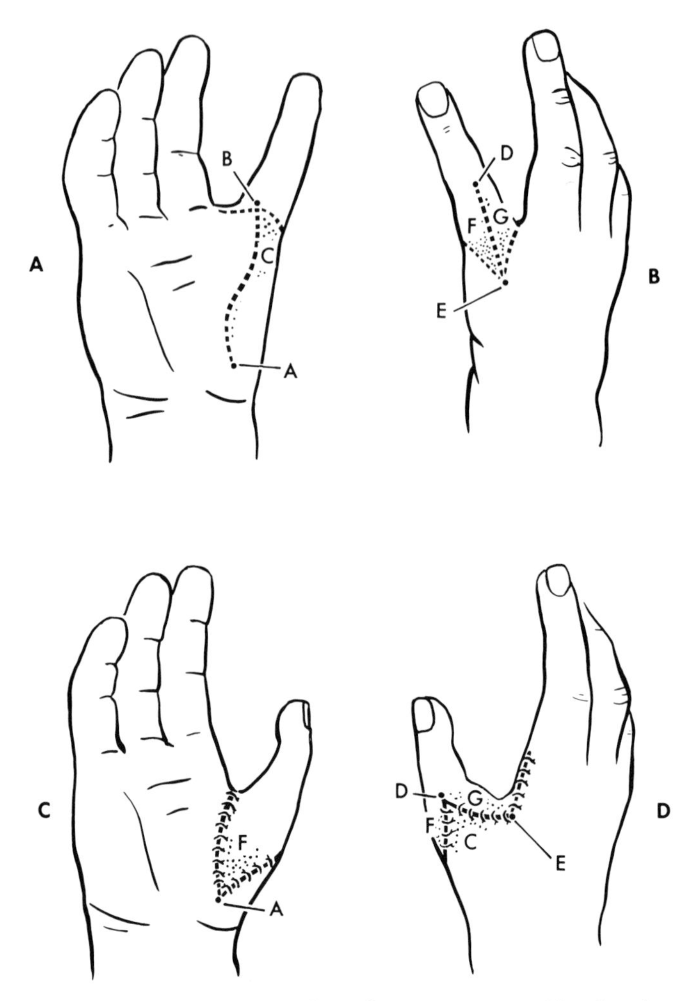

Fig. 6-6. *Pollicization—skin incision.* **A,** The palmar incisions. **B,** The dorsal incisions. **C** and **D,** After implantation of the new thumb. The two dorsal flaps on the index finger are separated by the flap *C.* (Modified from Buck-Gramcko, D.: Pollicization of the index finger, J. Bone Joint Surg. [Am.] **53:**1605-1617, 1971.)

proximal phalanx on its midpalmar aspect. From here the incision curves proximally with the first curve of the shallow S being to the radial side and the second, more proximal curve being to the ulnar side. The proximal limit of the incision is at point A, which should be about the level of the base of the second metacarpal. The second incision is on the dorsum of the proximal phalanx and covers its whole length from the proximal interphalangeal joint to the metacarpophalangeal joint (Fig. 6-6, *B,* D-E). The third line encircles the finger, passing from point E obliquely palmarward on either side of the metacarpophalangeal joint and transversely across the palmar aspect of the finger and intersecting with point B.

These incisions provide three major flaps, F and G on each dorsal-lateral side of the finger and a large radial-palmar flap C based on the preaxial border. During the closure this flap C will interdigitate with F and G (Fig. 6-6, *D*). I usually find that these three flaps are slightly large and trim excess skin from each; the end of the C flap almost always has to be reduced in length.

Neurovascular mobilization

Having planned all the incisions, I put on my magnifying loupes, inflate the tourniquet to 200 mm of mercury, and open all incisions into the subcutaneous fat. I then mobilize the two dorsal flaps, F and G, taking great care to preserve any dorsal veins draining the finger. The tips of these two flaps must be freed completely from the vein or veins so that later trimming will not inadvertently damage this vital venous return. With the hand still palm down, I explore the space between index and long fingers to find the strong intermetacarpal ligament. Immediately beneath this tough white sheet lie the common neurovascular pedicles to the index and long fingers. The nerve should be split back by gentle teasing and the arterial supply to the radial side of the long finger tied off so that the common artery will now support the ulnar side of the index-thumb pedicle. It is often possible to see the distal perforating artery joining the main trunk, and this must also be ligated between two ligatures of No. 4-0 catgut.

Having dissected the nerve and vessel as far as possible, I turn the hand over and freely mobilize the radial-palmar flap, C. This gives access to both neurovascular bundles of the index. Occasionally the radial artery is absent. Its presence is not vital to success; the digit can survive on the ulnar artery alone. Both neurovascular bundles must be freed as far proximally as possible. Retracting the flap also allows a good view of the dorsal venous drainage, which has to be dissected free from the subcutaneous tissues as far proximally as the base of the metacarpals.

Mobilization of extrinsic and intrinsic muscles

Having defined the two neurovascular pedicles and the dorsal venous drainage, I keep the hand palm up and free the flexor tendons both proximally and distally. The flexor tendon sheath should be cut away from the palmar plate so that the flexors enter the new thumb near the base of the true proximal phalanx. The intrinsic wing tendons can be seen edge-on on either side of the finger and should be dissected free proximally into the musculotendinous junction. Further mobilization of both the palmar and dorsal interossei must be done very carefully to preserve their nerve and blood supply. They should be stripped subperiosteally from the metacarpal shaft. To do this properly, one will need to pronate the hand to complete the mobilization of the muscles from the dorsum.

With the hand in this position the two wing tendons must now be freed from the central extensor tendon. The thickened free edges of these lateral bands are mobilized distally to the dorsum of the proximal interphalangeal joint. They are transected just distal to their origin from their respective intrinsic muscles, but they must be left attached to the confluence of the extensor mechanism over the joint. By this means two strong bands are left to be passed through the musculotendinous junctions of the two intrinsic muscles when they are later rejoined (Fig. 6-8).

The extensor digitorum communis and the extensor indicis proprius should now be identified and the former cut at the level of the metacarpophalangeal joint; its proximal end should be tagged so that it can later be reattached as the new abductor pollicis longus.

Positioning the new thumb

All soft tissue dissection and preparation has now been done and the metacarpal can be mobilized. With the hand still palm down, the flare at the base of the metacarpal is identified and the bone cut off at this level. Distally the epiphyseal line can be found with the point of the knife and the cartilaginous head sliced off at its junction with the growth plate. After removal of the shaft, the "thumb" is now free to be fixed in place.

This is the only really technically difficult part of the procedure. Riordan has pointed out that the key to proper placement is to site the metacarpal head—or new trapezium—in front of, or palmar to, the stump of the second metacarpal. This is because in the normal hand the basal joint of the thumb is not in line with, but palmar to, the line of the metacarpal bases.

To do this properly, one has to develop a small hole in the soft tissues with the points of scissors or with a periosteal stripper and test the fit of the metacarpal head. Once it can be accommodated, the proper rotation of the head must be obtained. This is a vital part of the operation, since it provides capsular stability for the new carpometacarpal joint and prevents its hyperextension (Fig. 6-7).

The normal metacarpophalangeal joint laxity is greatly reduced by bringing the proximal phalanx into a position of hyperextension in relation to the metacarpal head. This is best done by rotating the head nearly 90 degrees so that its palmar surface is now proximal. The raw cut surface will now face outward when the head is sewn in place, and the tight palmar capsule will prevent hyperextension of the joint (Fig. 6-7).

The head is anchored in place with one or two strong, absorbable synthetic sutures (size No. 4-0) passed right through the cartilaginous head and into the tissues of its bed near the second metacarpal base. Trial pulling on the two sutures will show whether the necessary 40 degrees abduction tilt and 160 degrees of rotation have been obtained. Occasionally a third suture may be used to firmly secure the head in the correct position.

Reattachment of muscles

The distribution of muscles to the new thumb is as follows:

Extensor digitorum communis becomes abductor pollicis longus

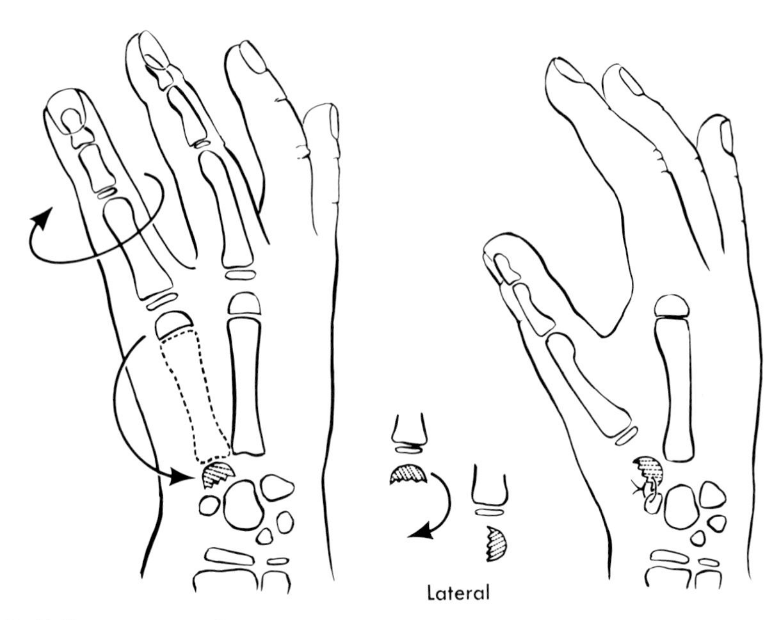

Fig. 6-7. *Pollicization—implantation.* The metacarpal head—or new trapezium—is placed palmar to the stump of the index metacarpal. Before being sewn in place, the cartilaginous head is rotated so that the palmar surface is proximal and the tight palmar capsule will prevent hyperextension of the joint. (Modified from Buck-Gramcko, D.: Pollicization of the index finger, J. Bone Joint Surg. [Am.] **53**:1605-1617, 1971.)

Extensor indicis proprius becomes extensor pollicis longus
First dorsal interosseus becomes abductor pollicis brevis
First palmar interosseus becomes adductor pollicis

First the proximal end of the extensor digitorum communis should be attached to the base of the old proximal phalanx or new metacarpal of the thumb to become a new abductor. The tension is hard to judge—it should be reasonably tight but not so great that the thumb resists a gentle passive adduction push (Fig. 6-8).

Next the two intrinsic muscles should be placed so they clasp the new metacarpal shaft with the dorsal interosseus becoming the new abductor pollicis brevis and the deeper palmar interosseus becoming the new adductor pollicis. The two lateral bands must be passed through a hole in the musculotendinous junction area of each muscle and looped back on themselves and the tension on the two adjusted so that the thumb stands out straight with a normal posture. The radial lateral band will go through the dorsal interosseus and the ulnar band goes to the future adductor pollicis. After these two bands have each been sewn to themselves, the excess lateral band tissue is excised.

The flexor tendons do not need to be shortened, since experience shows that they will always take up sufficiently for proper flexor power to develop. However, the extensor indicis proprius will need to be shortened sufficiently to bring the new metacarpophalangeal and interphalangeal joints into full extension against gravity.

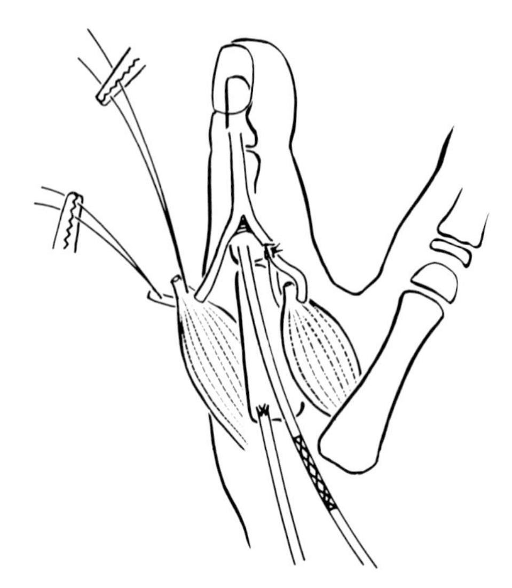

Fig. 6-8. *Pollicization—intrinsic muscles.* The two intrinsic muscles clasp the metacarpal shaft; the dorsal interosseus becomes the abductor pollicis brevis and the deeper palmar interosseus becomes the adductor pollicis. (Modified from Buck-Gramcko, D.: Pollicization of the index finger, J. Bone Joint Surg. [Am.] **53**:1605-1617, 1971.)

Closing the incisions

The thumb is now freestanding with good muscular balance, and all that remains to be done is to clothe the thenar area with skin. I use No. 6-0 ophthalmic catgut sutures for all skin closure so as to avoid the risk of struggling during the suture removal. I usually first place the tip of flap F into the V at point A and then judge how much skin I will need to excise from the tip of flap C to fit comfortably into the Λ at D (Fig. 6-9). Having got this loosely sutured into place, I close the incision around the base of the thumb on the palmar web side. I next close the incision along the radial side of the long finger from point E distally. This now leaves me the option of trimming excess skin from either flap G or flap C or both so that all skin lies in place without tension.

The great hazard in this operation for the less experienced surgeon is to be so impressed by the appearance of the new thumb that he or she spends an inordinate amount of time fussing over small cosmetic niceties in placing the flaps. Dog ears are acceptable and can always be removed in a later procedure if they have not absorbed. It is not good judgment to prolong tourniquet time and jeopardize the survival of the thumb for the sake of cosmesis. The parents will be sufficiently impressed by a working thumb to gladly accede to a later tidying-up operation.

I believe a drain is essential and usually place a twist drain deep down near the base of the second metacarpal and bring it out on the palmar aspect distal to point A. Occasionally I put a second drain in from the dorsum near point E.

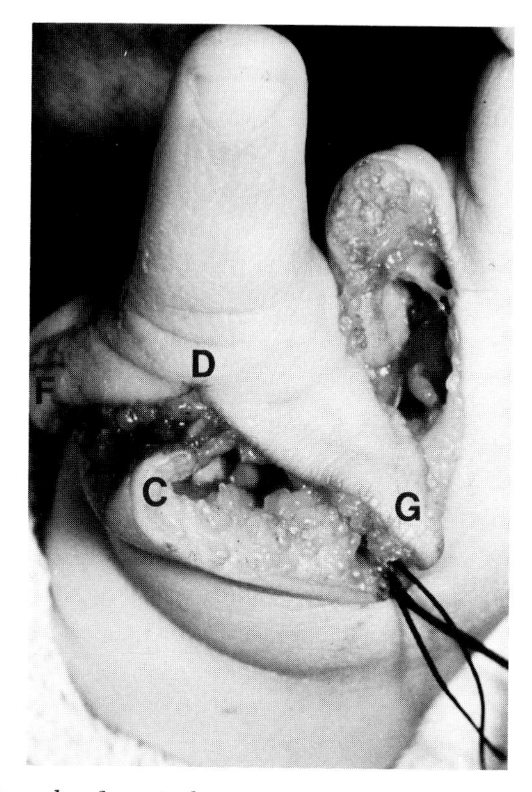

Fig. 6-9. *Pollicization—skin flaps.* A close up after implantation of the thumb and prior to suturing the skin flaps. Note how flap *C* separates flaps *F* and *G*.

Dressings

Adaptic, Telfa, or some other nonadherent dressings and voluminous Dacron fluff dressings bandaged into a compression dressing with a narrow Kling bandage should always be applied and the hand elevated before the tourniquet is released. It should immediately be removed from the arm to prevent venous engorgement in the wound. A plaster of Paris slab should be carefully molded to hold the thumb in the proper abduction and opposition position. The thumb's terminal phalanx should be left exposed, and the cast should extend above the elbow, which must be flexed to at least 90 degrees. The one or two drains should be wrapped in a little cotton and led out through the cast so that they can be removed in 24 hours without destroying the cast.

POSTOPERATIVE CARE

Complications are rare, but if vascular embarrassment is suspected, all dressings must be removed and any hematoma evacuated. The hand should always be suspended for at least 48 hours no matter how much the child objects (Fig. 6-10). With sensible parents it is quite possible to let them hold the child in their lap while the child's arm is safety-pinned to the clothing on their shoulder.

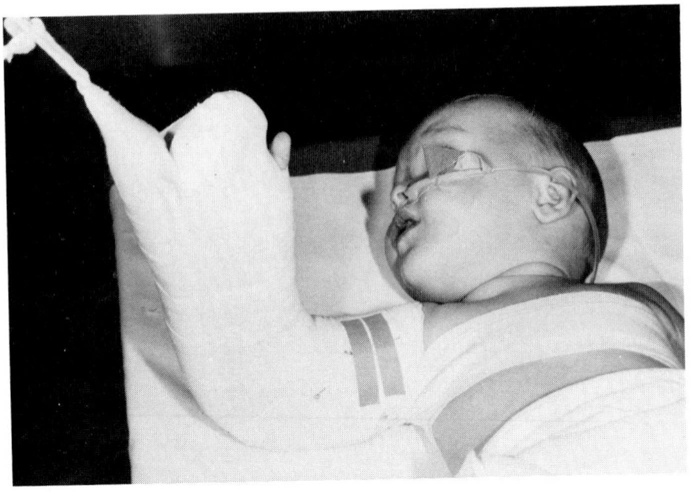

Fig. 6-10. *Pollicization—postoperative suspension.* The hand should always be suspended for the first 48 hours after the operation. The thumb must be left free for inspection, and care must be taken that the bandages are not tight across the proximal dorsal portion of the thumb. (Courtesy Dr. Dieter Buck-Gramcko.)

Long-term complications have been seen in the metacarpal head. Buck-Gramcko has had three instances of avascular necrosis and crumbling of the head, but subsequent follow-up has not shown any significant interference with the motion of the carpometacarpal joint.

It is my practice to discharge the child on the third or fourth postoperative day and inspect the cast at the third week. If it is not loose and remains sanitary I leave it in place for a further 3 weeks. Soggy, moist dressings can play havoc with the healing of the incisions, and I have been amazed at the variety of food that can be stuffed into the dressings by the unbandaged hand. If in doubt, I replace the cast.

After 6 weeks of immobilization I believe that the thumb is sufficiently healed to allow unrestricted motion. Dr. Buck-Gramcko tells me that he believes 3 weeks immobilization is adequate. He points out that he does not seek firm bony union between the metacarpal head and the carpal region. He prefers the slight extra mobility provided by a fibrous union. He does not provide any extra protection even for children in the crawling stage. If there is a tendency for an adduction contracture in the first few weeks after cast removal, he provides a foam rubber web spacer that is bandaged in place at night by the parents.

Development of full dexterity in the new thumb will take many months, and take-up in the flexor tendons will occur at varying speeds in different children (Fig. 6-11). I encourage the parents to place uninteresting objects in the unoperated hand and then place food or a favorite toy in front of the new thumb. The child will rapidly develop functional scooping by the fingers against the new thumb, and as flexor power develops, so will true prehension.

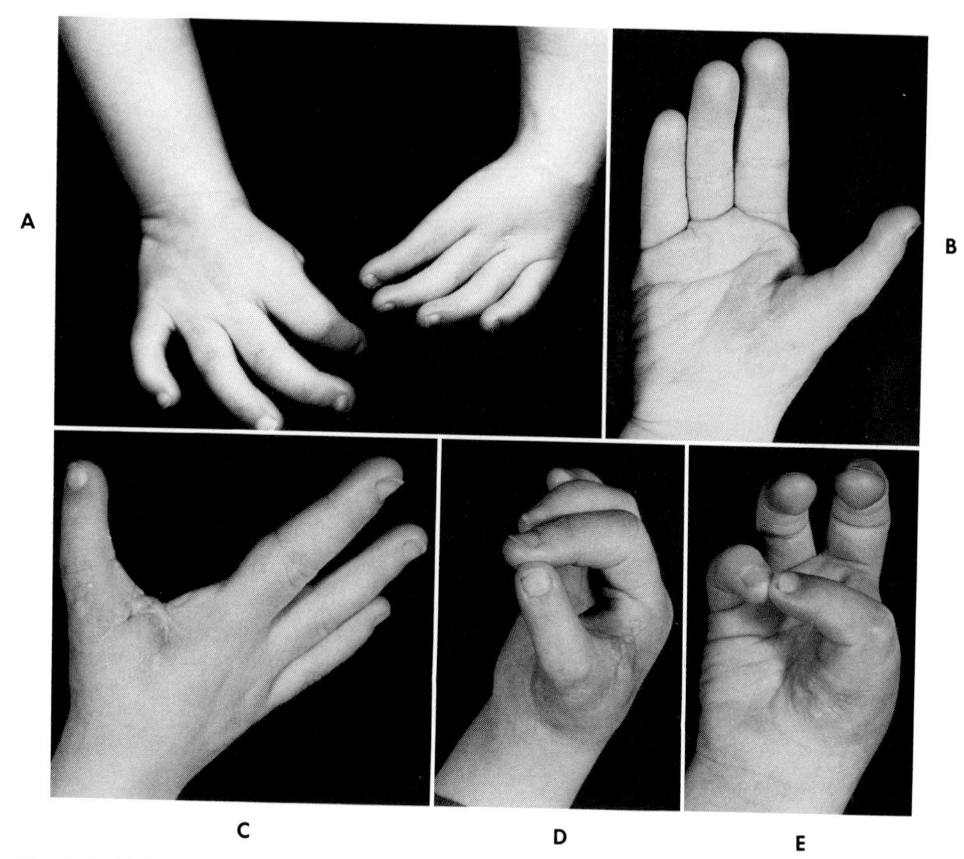

Fig. 6-11. *Pollicization—result.* **A,** This child had early stabilization of her left radial clubhand by ulna implantation at 18 months of age. She had no right thumb, and pollicization was carried out a year later. **B** and **C,** The posture and size of the new thumb is pleasing. **D** and **E,** Lateral pinch and full opposition are possible.

POLLICIZATION AND THE FLOATING THUMB

When planning pollicization in the presence of a floating thumb, one should recognize that there are a number of associated anatomical anomalies. These have been nicely collected in one diagram by Edgerton, Snyder, and Webb (Fig. 6-12). Probably the commonest is a neural ring in the common digital nerve, which encircles the common digital artery. When the ring is near the index metacarpal head, the nerve fibers have to be separated over a considerable distance proximally to allow mobilization of the index finger in pollicization. Another anomaly that causes problems in pollicization is an enlargement of the distal perforating artery, which replaces the common digital artery. This means that the index will have to be carried on the radial-sided artery, which may also be anomalous.

The operation can be done in one or two stages: the vestigial thumb can be amputated first and pollicization performed some months later or both

Radial proper volar digital artery may be missing

Small nail

Small cartilaginous phalanges

Small or missing thenar intrinsic muscle

Distal floppy attachment

Small wrist and distal radius

Anomalous neural ring

Abnormal position of radial artery

Absent trapezium (greater multangular and scaphoid fused)

Fig. 6-12. For legend see opposite page.

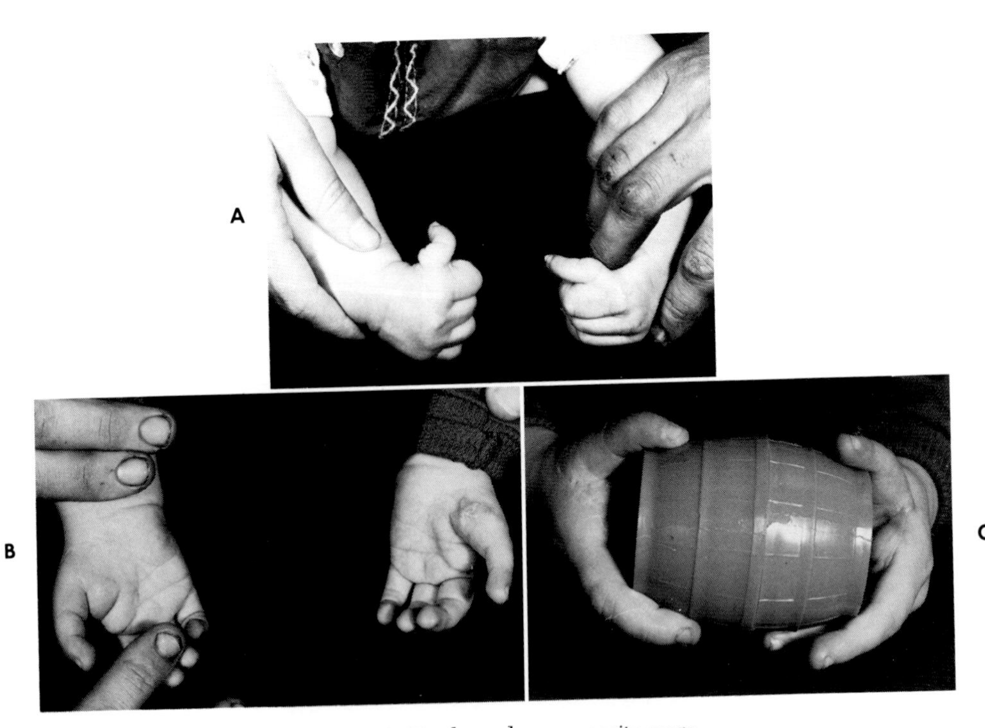

Fig. 6-13. For legend see opposite page.

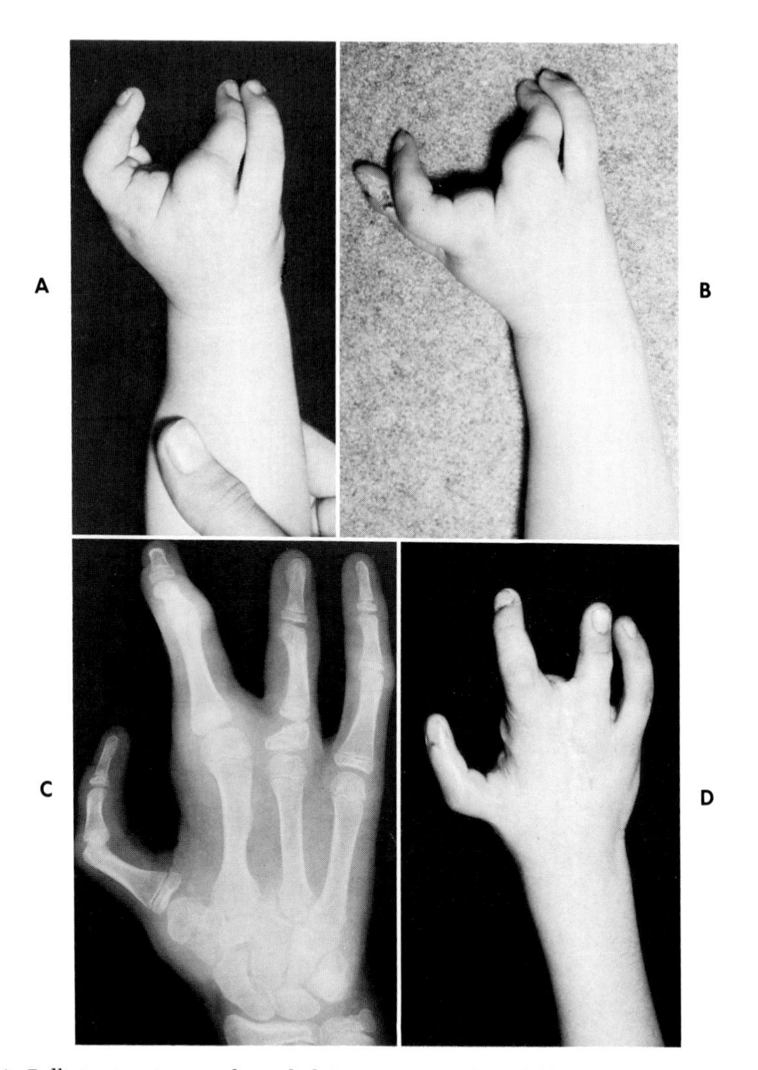

Fig. 6-14. *Pollicization in complicated deformities.* **A,** This child was born without a thumb, with a syndactyly of the index and long fingers, with a central cleft, and with a delta phalanx in the ring finger. The operative plan included: **B,** Separation of the syndactyly and **C** and **D,** closure of the cleft with osteotomy of the long finger metacarpal and pollicization of the index. The result is a far from normal, but very functional, hand. (NOTE: This is the small child that appears in Fig. 7-13.)

Fig. 6-12. *Floating thumb and anatomical anomalies.* Many variations can be encountered during pollicization. The commonest is the anomalous neural ring; the radial artery to the index may be missing, and the radial artery may be abnormally placed.

Fig. 6-13. *Floating thumb and pollicization.* **A,** Both these useless thumbs were replaced by pollicization. Amputation of the thumb and transfer of the index were carried out at the same time; this demands very careful planning of the skin flaps. **B** and **C,** One year after the first pollicization and 6 months after the second the patient has good span and good strength.

procedures can be done at the same time. Since the site of the pedicle will vary in the area of flap C, it is hard to give guidance if both procedures are done together. In general I fillet the thumb and leave its skin opened out as a flap until the final skin closure. I then either amputate it or extend the line of the fillet incision to the edge of flap C and thereby break it into two large flaps. These may or may not interdigitate with flap F and more often are useful on the palmar aspect where extra skin may be needed. Some ingenuity is often necessary to place the various flaps satisfactorily.

POLLICIZATION IN COMPLICATED DEFORMITIES

The possible combinations of deformities are infinite and this short section is included to show that it is possible to pollicize a finger even after several procedures have been done to a hand. The child shown in Fig. 6-14 was born without a thumb and with a syndactyly between the index and long fingers, a central cleft of the hand, and a distorted ring finger because of a delta-shaped proximal phalanx. The syndactyly was separated and the central cleft closed prior to pollicization of the liberated index finger. The end result is not pretty, but the hand can now function adequately.

CHAPTER 7

EXTRA THUMBS

Duplication of the thumb occurs in many forms ranging from varying degrees of longitudinal splitting to abnormal delta phalanges and triphalangeal thumbs. There may be no more than splaying of the nail with a central longitudinal furrow, which requires no treatment, or there may be any of the intermediate stages up to a complete duplication of the thumb with controlled flexion and extension motion. One rarely finds small fleshy nubbins on the radial border of the hand such as are commonly found on the ulnar, postaxial side.

When duplication occurs alone, it is invariably unilateral and sporadic. When there is a duplication of a triphalangeal thumb, it usually results from autosomal dominant inheritance. Thumb polydactyly is also part of a variety of syndromes, but none of these occurs with any great frequency. The commonest are probably the two types of acrocephalopolysyndactyly—the Noack type is transmitted by a dominant inheritance, while the Carpenter type is caused by a recessive trait. Much less common is its appearance with such syndromes as Fanconi and Holt-Oram. It can also be associated with syndactyly, brachydactyly, nail dystrophies, deafness, facial clefts, and other less common anomalies.

TYPES

For such a relatively common abnormality as thumb duplication, the literature contains very little detailed analysis of the different types or of their treatment. In a paper published in 1967, Millesi reported 14 cases of thumb polydactyly in 19 patients with polydactyly of the hands and feet, and he suggested a classification of thumb duplication into five different types.

In 1969 Wassel published an extensive review of the thumb duplications

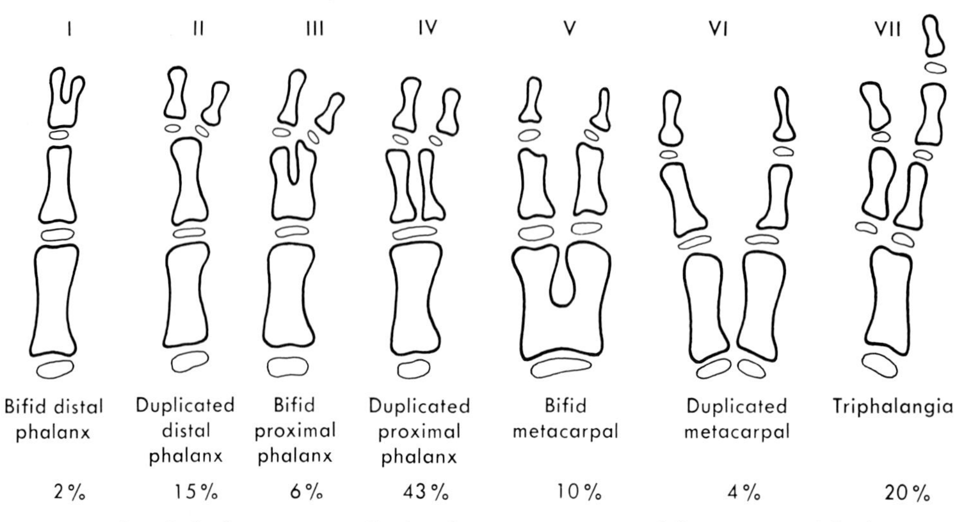

Fig. 7-1. *Thumb duplication.* Wassel's classification shows seven different types of duplication; Type IV is the commonest.

Table 7-1. Incidence and varieties of thumb duplication in 70 charts reviewed*

Types	*Number of thumbs*	*Percent*
I	2	2
II	12	15
III	5	6
IV	34	43
V	8	10
VI	3	4
VII	15	20
Total	79	100

*Nine patients had bilateral involvement making a total of 79 thumbs.

Table 7-2. Incidence of different types of thumb duplication in 18 patients reviewed

Types	*Number of thumbs*	*Percent*
I	1	4.5
II	5	22.7
III	2	9.1
IV	6	27.3
V	2	9.1
VI	0	0.0
VII	6	27.3
Total	22	100.0

seen at the University of Iowa. Seventy patients were found who had thumb duplications, and nine patients had bilateral involvement. His work showed that it would probably be better to classify thumb polydactyly into seven types (Fig. 7-1) rather than the five types of Millesi.

INCIDENCE

The incidence and varieties of thumb polydactyly in these 70 patients are shown in Tables 7-1 and 7-2. Of the 18 patients who returned for review, five were women, and four had bilateral involvement, for a total of 22 thumbs reviewed. In this small series, Types IV and VII polydactyly were equally common, but in the larger series of all 70 patients, it is seen that the Type IV involvement is twice as common as the Type VII.

COMPLICATIONS

This survey showed that significant complications occur either as a result of inadequate surgery or because of dynamic imbalance of the tendon forces.

Many patients were found to have a decreased range of motion both in the interphalangeal and metacarpophalangeal joints; they had radioulnar instability at the metacarpophalangeal joint, loss of tip and palmar pinch, a decrease in tip and palmar power, decreased or diminished opposition, radial or ulnar deviation at the interphalangeal and metacarpophalangeal joints, and in some cases, contracted thumb-web space.

Most impairment of function arose from the excessively radially deviated thumb. True precision handling was lost, and only a crude lateral grip utilizing the ulnar side of the thumb was possible. Decrease in the range of motion at the interphalangeal and metacarpophalangeal levels also interfered with both fine precision activity and the pinch mechanism. Radioulnar instability made the picking up of fine objects from a flat surface difficult and, in some cases, impossible. Wassel was unable to show that the decrease in opposition produced any measurable functional impairment. The cosmetic appearance of the hand was affected if the radial or ulnar deviation at the interphalangeal or metacarpophalangeal joint was greater than 25 degrees.

Review of the surgical approaches used showed that a linear incision should be avoided. In several cases such an incision produced a contracted scar that always exaggerated, and in some cases even caused, a deviation. If a linear approach is inevitable, then the primary incision must be broken up with multiple Z-flaps.

OPERATIVE PRINCIPLES

There is no such thing as simple ablation of a reproduced thumb. One must keep in mind the potential functional hazards of such surgery. Every effort should be made to restore the normal anatomical relationship of the remaining bones to the longitudinal axis. Operative treatment must be preceded by very careful planning, which in turn must be based on very careful examination of both thumbs. The range of motion in the interphalangeal and metacarpophalangeal joints of normal thumbs shows a considerable variation, and 90 degrees of flexion in both joints is not common. Because of this, there is no absolute pattern for a "normal" thumb, and occasionally there will be a great problem in selecting which thumb is in fact the functional one or, even more particularly, which portions of each thumb are the best to use in the reconstruction of a single thumb. I have on occasion found it necessary to transfer the distal portion of one duplicated thumb onto the proximal portion of the other and, in addition, to transfer flexor or extensor tendons from one to the other digit so that the final, composite result may yield the best possible function. The selection of which components should be incorporated in the final thumb can be extremely difficult. Observation of the child at play, often for prolonged periods, is the only satisfactory method of determining which tendons and which joints will yield the greatest function.

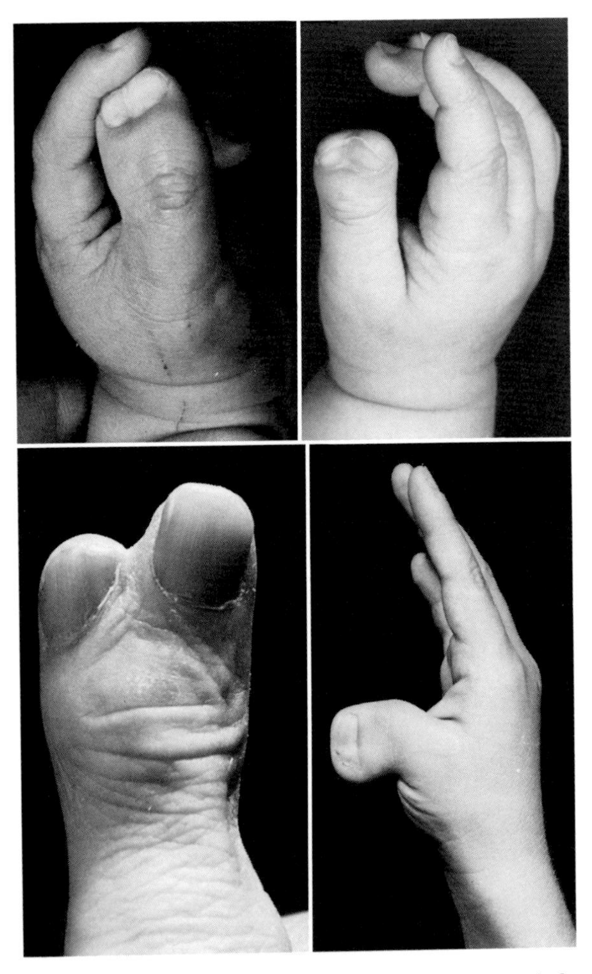

Fig. 7-2. *Type I and Type II duplications.* Examples of variations of the first two types of duplications. (Top two photographs are Type I; bottom two photographs are Type II.) In most cases one thumb is smaller than the other.

If both thumbs function well in all normal tasks, then size and cosmetic appearance can be included in the selection. Frequently the radial of the two digits is more hypoplastic, which is convenient since a stable pinch can be more easily obtained by retaining the better developed, more ulnar digit with its ulnar collateral ligament intact.

It must be emphasized that no matter what type of duplication is present, the remaining thumb will almost invariably be smaller than a normal thumb. The parents must be forewarned about this reduction in size and also about the possible functional limitations of the residual thumb.

OPERATIVE DETAILS

In dealing with the more distal duplications such as Types I and II (Fig. 7-2), I agree with Millesi that the preferred treatment is the Bilhaut-Cloquet

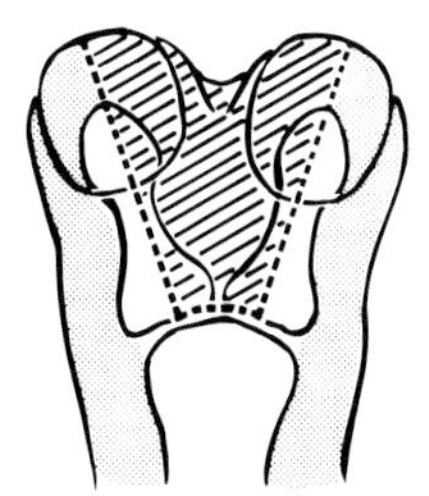

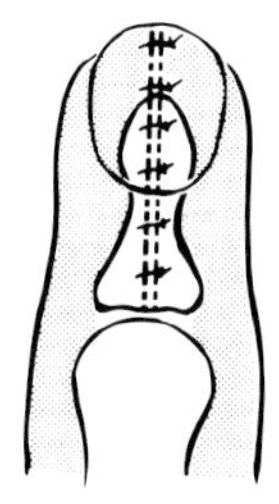

Fig. 7-3. *Bilhaut-Cloquet closure.* The preferred treatment for Types I and II is this method in which a central wedge is resected.

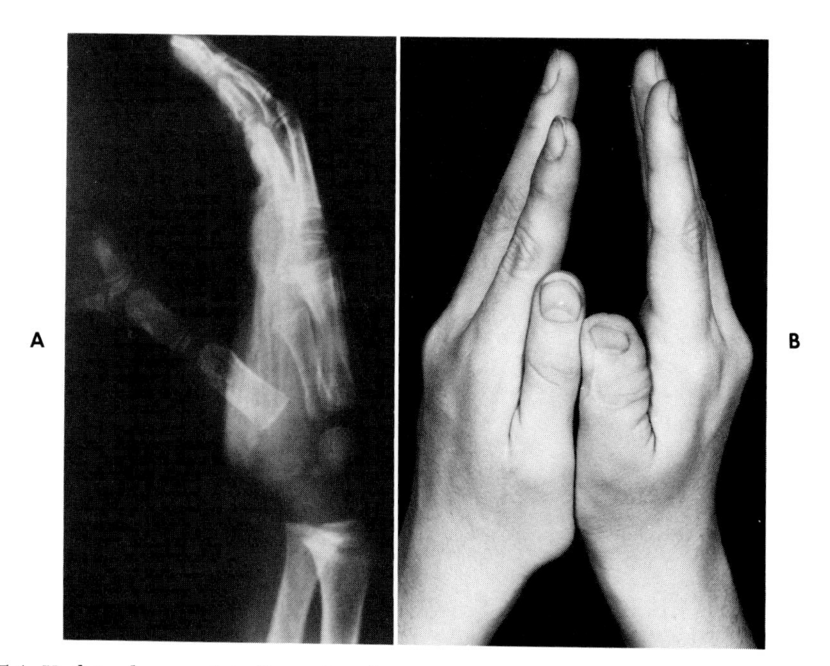

Fig. 7-4. *Unilateral removal in Types I and II.* **A,** If one of the two components is significantly smaller, then unilateral resection is justified. **B,** Long-term follow-up shows that the retained thumb is significantly smaller than the opposite normal thumb.

method of central wedge resection (Fig. 7-3), particularly when there is equal divergence of the two components from the longitudinal axis. Removal of one component might maintain the inherent deviation of the remaining component. This can be avoided by central fusion of the two duplicated portions rather than by the ablation of one. Care must be taken not to damage the basal epiphysis when removing the central wedge. If one of the two components is significantly smaller or if only one deviates from the midline, then unilateral removal can produce a satisfactory result (Fig. 7-4).

In Type III cases the involvement is more proximal and two separate interphalangeal joints are present (Fig. 7-5). This problem is usually too complicated to yield satisfactory results by a central fusion of the two longitudinal elements, and I believe amputation gives the best result. However, Karchinov has extended the Bilhaut-Cloquet principle more proximally and has reported

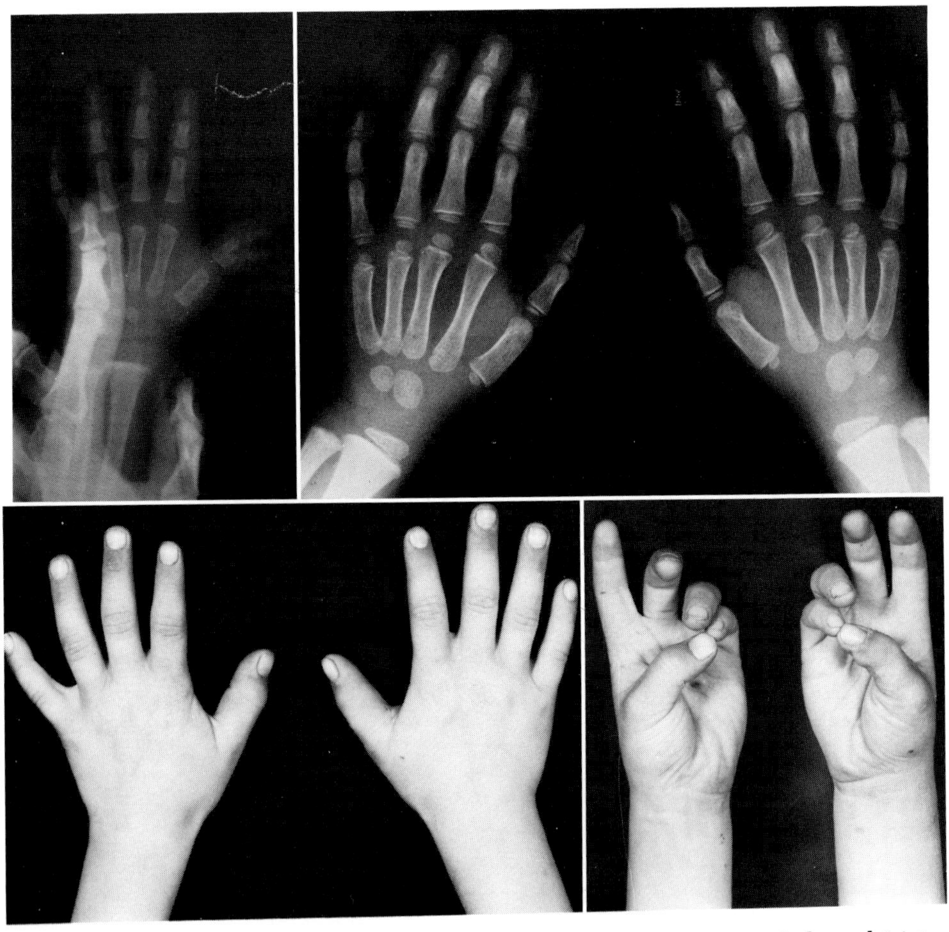

Fig. 7-5. *Type III duplication.* In these thumbs there are two separate interphalangeal joints, but the bases of the proximal phalanges are fused. I believe that amputation generally gives a better result than central fusion after wedge excision.

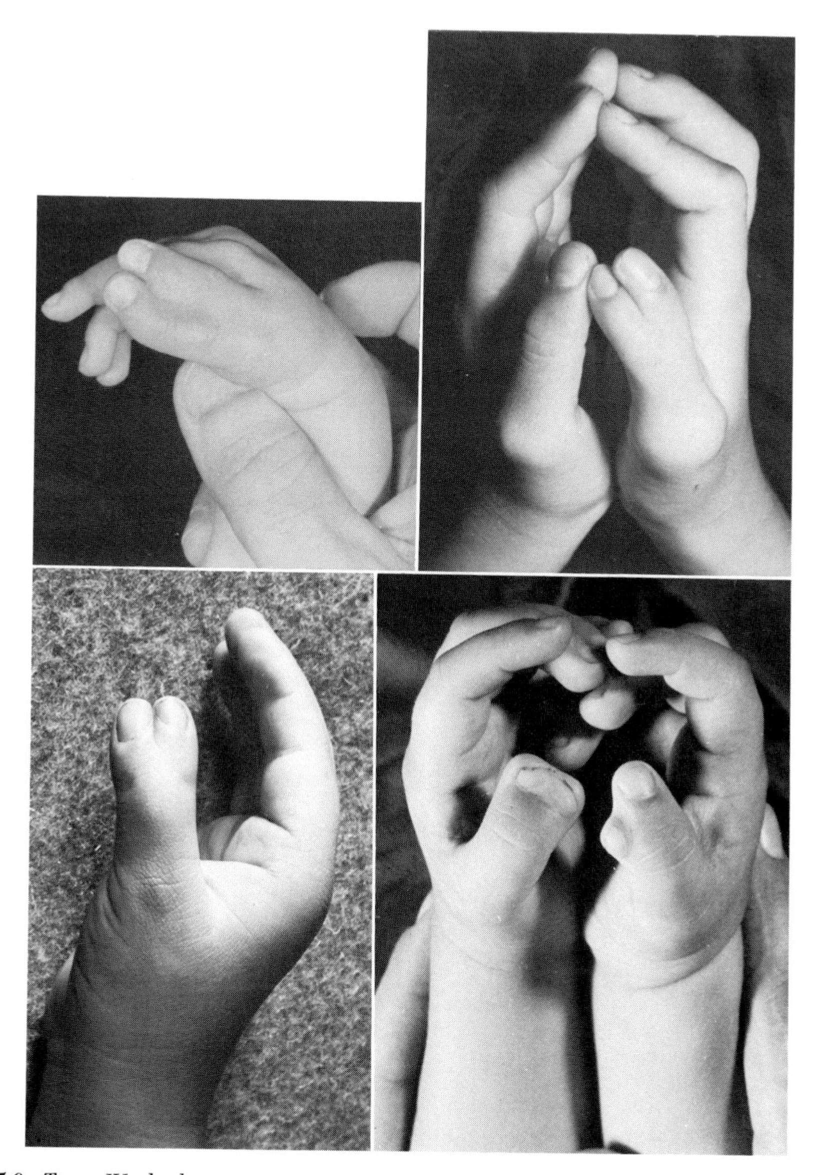

Fig. 7-6. *Type IV duplication.* In this, the commonest duplication, there is a complete reduplication of the proximal and distal phalanges. As in other types there is a tendency for one thumb of the pair to be larger.

two cases in which he constructed satisfactorily sized thumbs by longitudinal fusion over at least two thirds of their length.

The commonest duplication is the Type IV, in which there is a complete reproduction of the proximal and distal phalanges (Fig. 7-6). The technical problems in this case are to provide good stability at the metacarpophalangeal joint and a correctly aligned thumb. If the additional thumb is acutely angled to the principal thumb, it is usually possible to obtain a good result (Fig. 7-7). When the two thumbs are parallel, the base of the twinned proximal phalanges is wider than that of a single normal phalanx. The two components articulate in a common synovial cavity with the head of the metacarpal, which may be facetted rather than smooth and rounded. This angulation of the articular surface is one cause of deviation in the retained thumb. Another cause, which has been stressed by Palmieri, is eccentric insertions of the extrinsic flexor and extensor tendons. If there is a tendency for interphalangeal deviation in the thumb that is to be retained, then exploration of the insertions of the extrinsic tendons should be done. If the insertions are not balanced in the midline then the eccentric side must be dissected off, crossed over, and reinserted (Fig. 7-8).

In patients with two parallel thumbs I always plan generous skin flaps based on the retained thumb and raised at the expense of the skin cover on the thumb to be amputated. These flaps provide ample skin, which can be trimmed in a zigzag fashion so that the final closure does not have linear scars (Fig. 7-9).

When operating on these cases, one should identify the digital nerves and vessels early; they must be shown to be present in the chosen thumb. The tendons in this thumb must be found and their insertions corrected if there is any tendency for joint deviation. Only after I am satisfied that the neurovascular and tendon mechanisms are satisfactory in the selected thumb do I resect those in the rejected thumb well back into the proximal portion of the wound. Transfer of the tendons from the amputated to the retained thumb has been suggested; I very rarely do this if any reasonable mechanism is present in the retained thumb. Transfer from one to the other is certainly indicated if there is total absence of tendon mechanisms, but I do not regard hypoplasia of the tendons as an absolute reason for transfer.

An important transfer that is necessary is the reinsertion of any of the thenar muscles that may be attached to the base of the proximal phalanx that is being excised. Frequently this transfer can be combined with, and used as a reinforcement for, the reconstruction of the metacarpophalangeal collateral ligament on the side of the joint from which the extra phalanx has been excised.

Reconstruction of the collateral ligament of the metacarpophalangeal joint is a difficult problem in all types of thumb polydactyly, but it is particularly difficult in Types III and V, in which there is reproduction of the head and neck of the proximal bone. The extra head and neck will have to be excised in a sloping fashion so the shaft of the retained bone slopes up to the remaining head. The

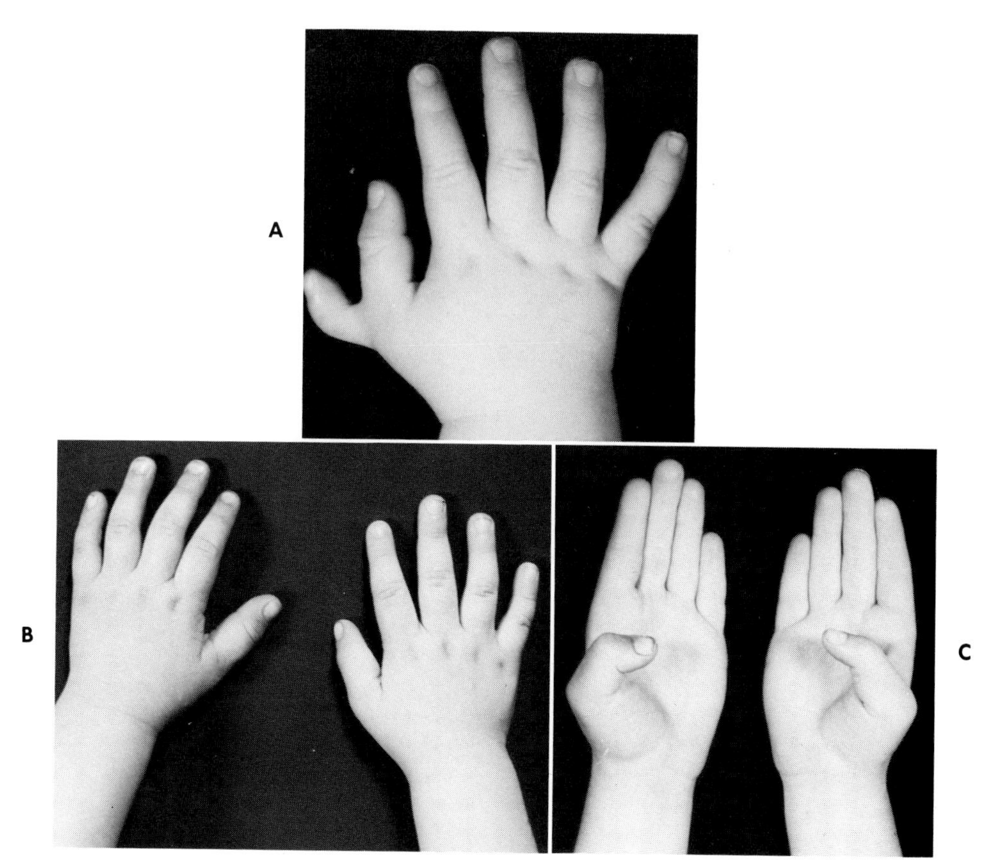

Fig. 7-7. *Type IV treatment—divergent thumbs.* **A,** When the additional thumb is acutely angled to the principal thumb, it is usually possible to get a good result (**B**). However, **C** shows the size and range of motion of the retained thumb may still not be the same as the normal side.

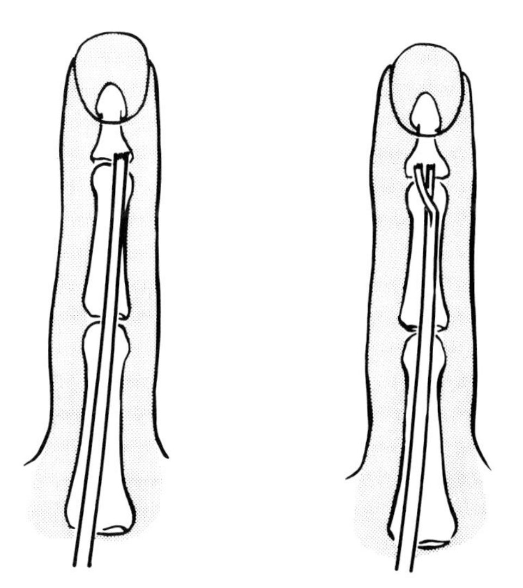

Fig. 7-8. *Extrinsic tendon centralization.* Eccentric insertion of the extrinsic extensor tendon or the flexor tendon or both can cause interphalangeal joint deviation in the thumb. When it is found, the eccentric side must be dissected off, crossed over, and reattached.

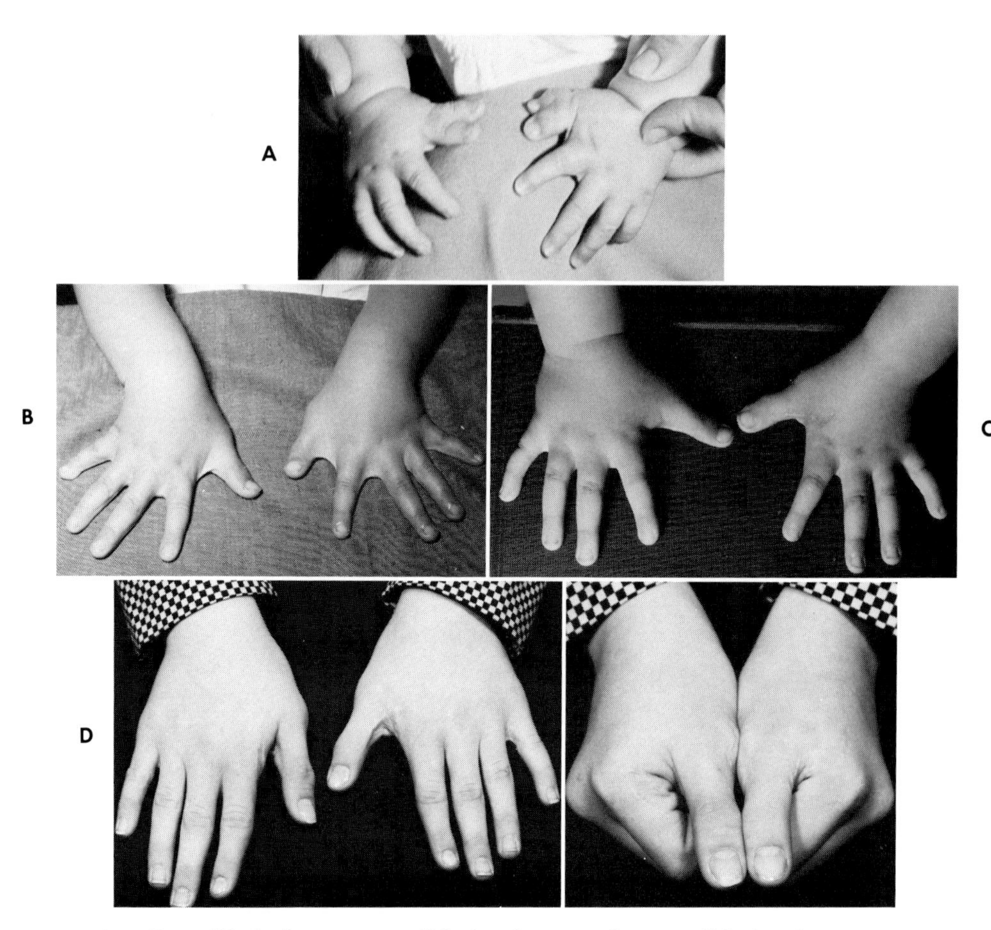

Fig. 7-9. *Type IV duplication—parallel thumbs.* **A,** When parallel thumbs are present, a choice has to be made as to which to retain; I always plan generous skin flaps at the expense of the discarded thumb. **B** and **C,** When the web space needs deepening, I usually do it at a later operation so as to avoid vascular embarrassment by operating on both sides of a thumb simultaneously. **D,** Thirteen years later, in teenage, the result remains satisfactory.

distal ligamentous attachment will have to be dissected free at the subperiosteal level, since it is connected to the phalanx that will be discarded. Great care must be taken during this subperiosteal dissection of the proximal origin of the ligament to leave it attached to the side of the proximal bone. If this can be done, then a relatively taut collateral ligament can be made; I do not hesitate to pass absorbable sutures through holes in the distal bone if by this means I can get a secure attachment.

Types V and VI duplications are uncommon and complete duplication of an entire thumb is rare. Almost always one is considerably smaller than the other, and it usually does not have a properly constituted carpometacarpal joint. Appropriate excision is the operation of choice, but occasionally one will come across an adult who has gone through life with such a digit and put it to good use (Fig. 7-10).

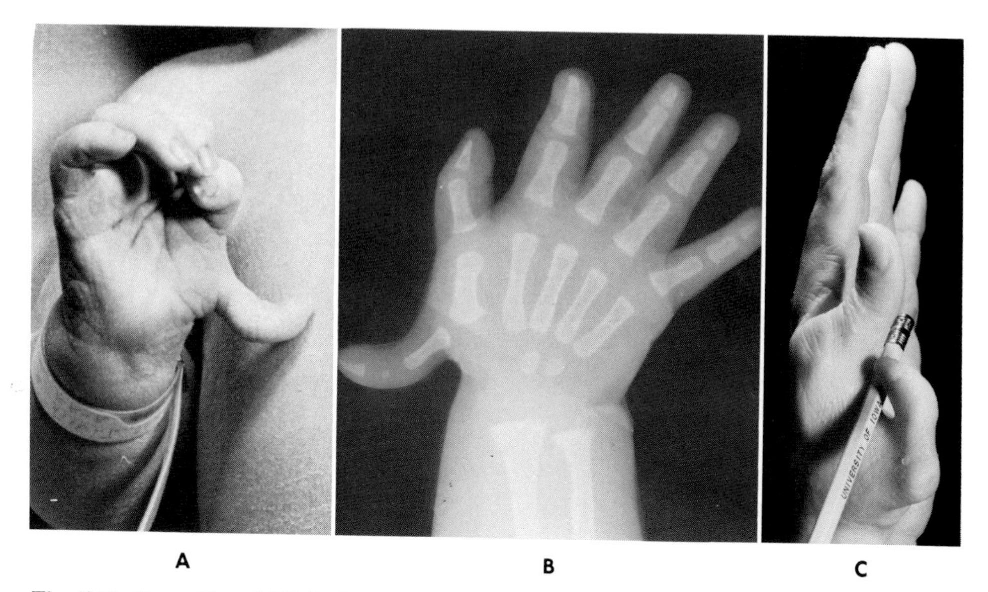

Fig. 7-10. *Types V and VI duplications.* These are uncommon. **A** and **B,** When seen in infants the extra thumbs should be removed. **C,** Occasionally they are seen in adults; this cast of a carpenter's hand shows how he uses his extra thumb to hold a pencil.

TRIPHALANGEAL THUMB

Type VII is the second most common type of thumb duplication (20% incidence in Wassel's series). In this group a "normal" thumb is accompanied by a triphalangeal digit (Fig. 7-11). A triphalangeal thumb can also be associated with a duplicated thumb or even absence of a thumb on the opposite hand (Fig. 7-12). It can also occur in isolation in an otherwise normal hand. Two different large surveys of mixed populations have established its incidence as about 1 per 25,000. It can be inherited as a dominant family trait, and Swanson has reported a large family in which 30 individuals spanning four generations were affected. It has been associated with a variety of anomalies such as duplication of the great toe or the little toe; cleft hands and feet; absence of the tibia, the scaphoid, and the trapezium; and absent pectoral muscles (Fig. 7-13). It takes part in several syndromes, such as Holt-Oram, Juberg-Haywood, and Blackfan-Diamond anemia, and has been coupled with a congenital hemopoietic disorder by Aase and Smith.

Types

Whatever the cause, there seems little doubt that there are two types of triphalangeal thumbs, both of which are simple autosomal dominant traits and either of which can be associated with duplication. The extra phalanx will vary in shape from a small delta-shaped nodule to an apparently normal phalanx (Fig. 7-14). In most cases the metacarpal is relatively long compared with a normal thumb metacarpal.

Text continued on p. 113.

A

B

C

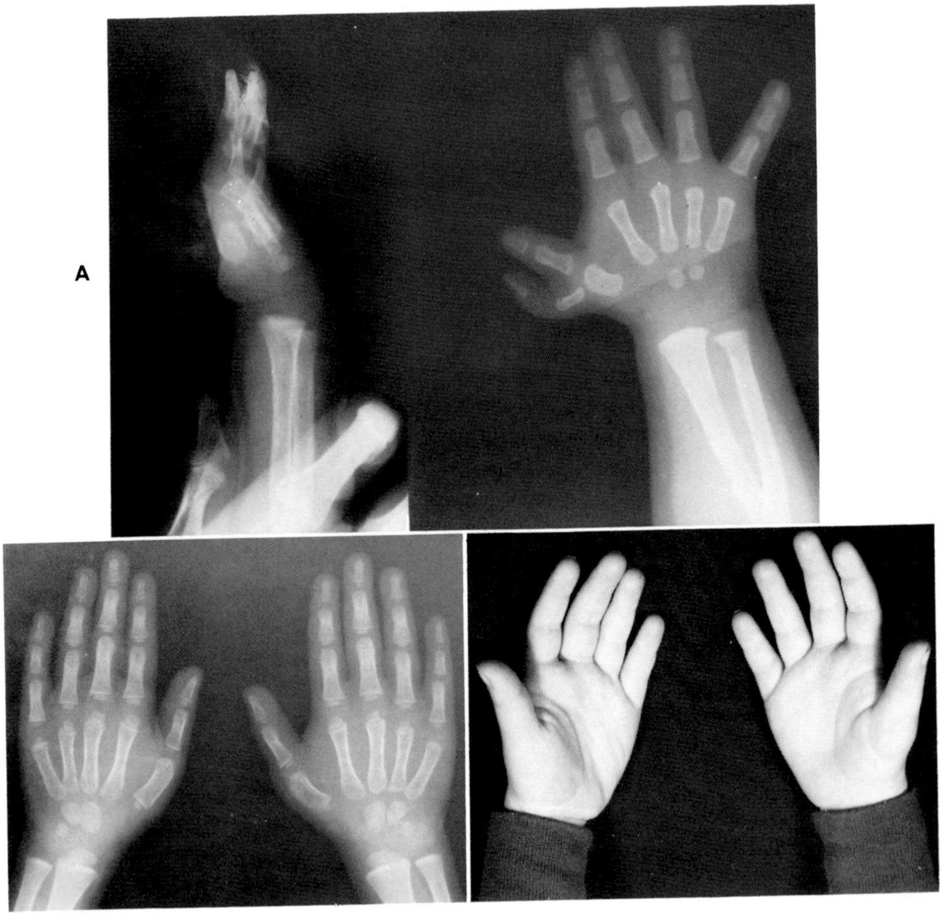

Fig. 7-11. ***Type VII duplication. A,*** In this second most common duplication the normal thumb is paired with a triphalangeal digit. **B,** Removal is relatively easy, but the enlarged metacarpal remains. **C,** The retained right thumb is thinner and more tapered than the unaffected left thumb.

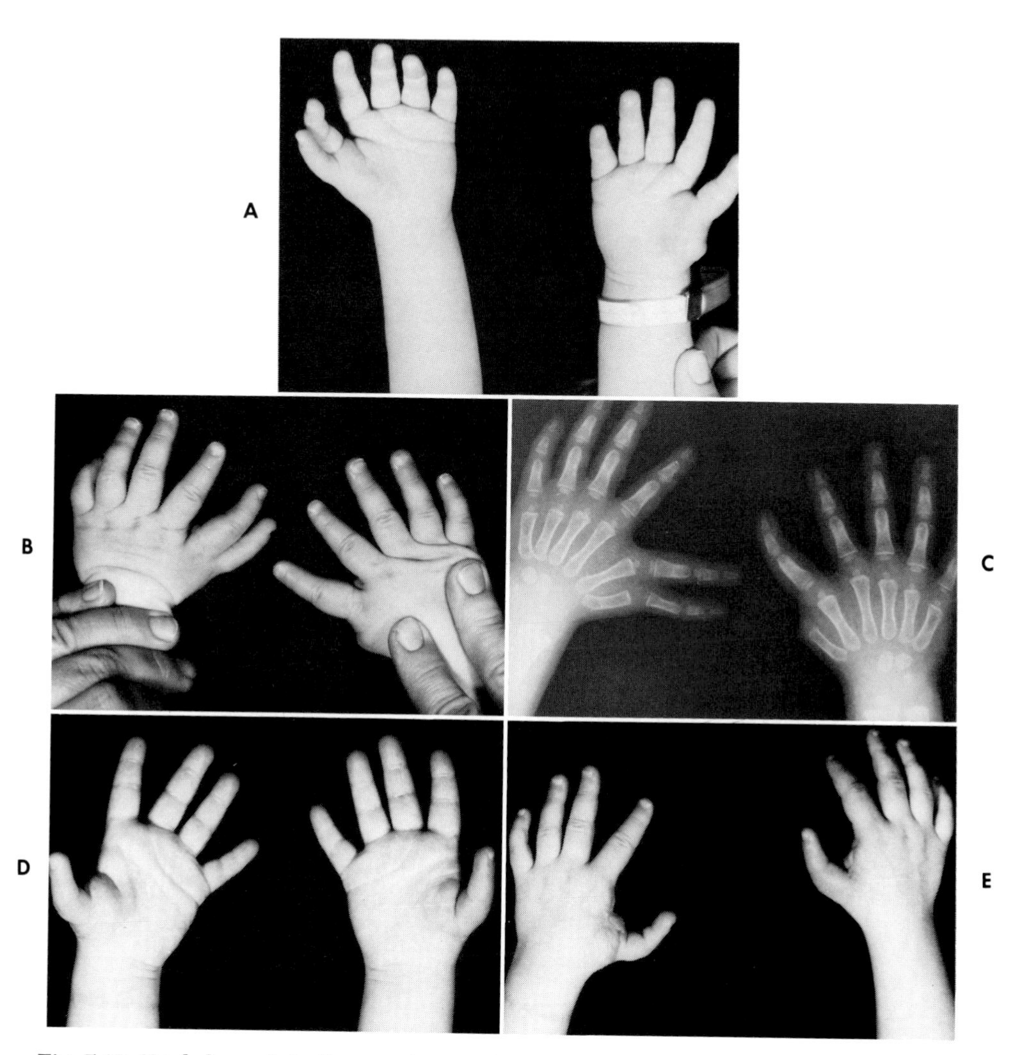

Fig. 7-12. *Triphalangeal duplication.* **A, B,** and **C,** This child with Holt-Oram syndrome had a triphalangeal duplication on the left side and an extra metacarpal on the right side. **D** and **E,** Pollicization was easier on the right side than the left because of the problems in designing appropriate skin flaps around the amputation site of the most radial digit.

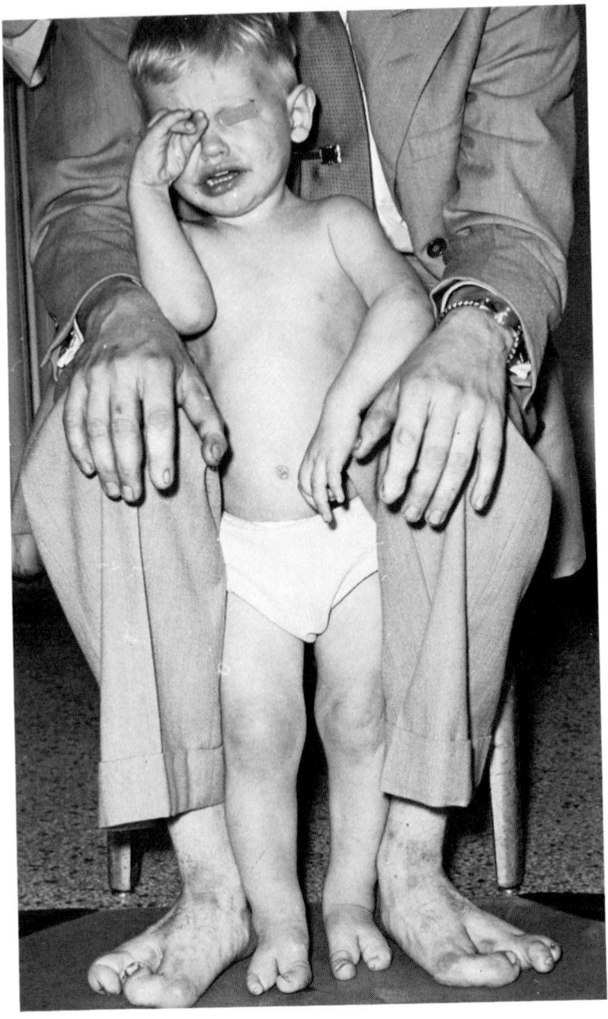

Fig. 7-13. *Anomalies associated with triphalangeal thumb.* This child, born to a father with cleft feet and triphalangeal thumbs, also had cleft feet, cleft hands, and syndactyly of triphalangeal radial digits. His right hand was operated on and is illustrated in Fig. 6-14.

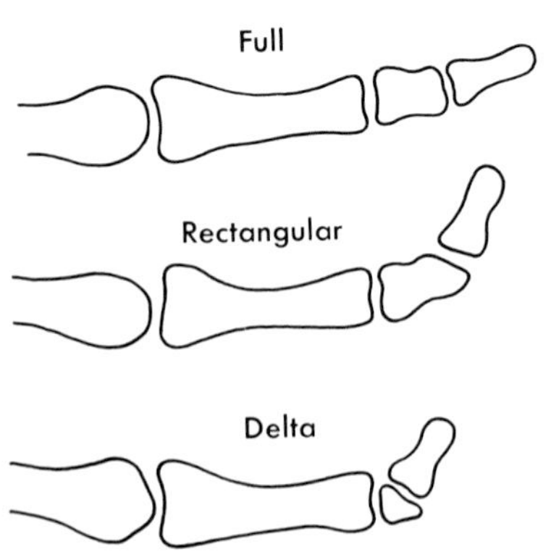

Fig. 7-14. *Types of extra phalanx.* In the triphalangeal thumb three different types of abnormal middle phalanx are found: full-sized, rectangular, and delta-shaped. (From Wood, V. E.: Treatment of the triphalangeal thumb, Clin. Orthop. **120**:188-199, 1976.)

Those thumbs that are not significantly increased in length can usually oppose and have small trapezoidal second phalanges that often cause an ulnarward angulation. The thumbs that have a full-length phalanx more closely resemble a finger and the hand is often called a five-fingered or a "thumbless" hand. In these patients the thenar muscles are often absent and opposition does not occur (Fig. 7-15). Treatment for these patients is discussed in Chapter 6.

Galen was the first to speculate on the proper names for the three bones of a normal thumb. Since his time, much has been written, but with total lack of agreement about whether or not the thumb metacarpal is a proximal phalanx because of its proximal epiphysis. All forms of life from the amphibians up have a

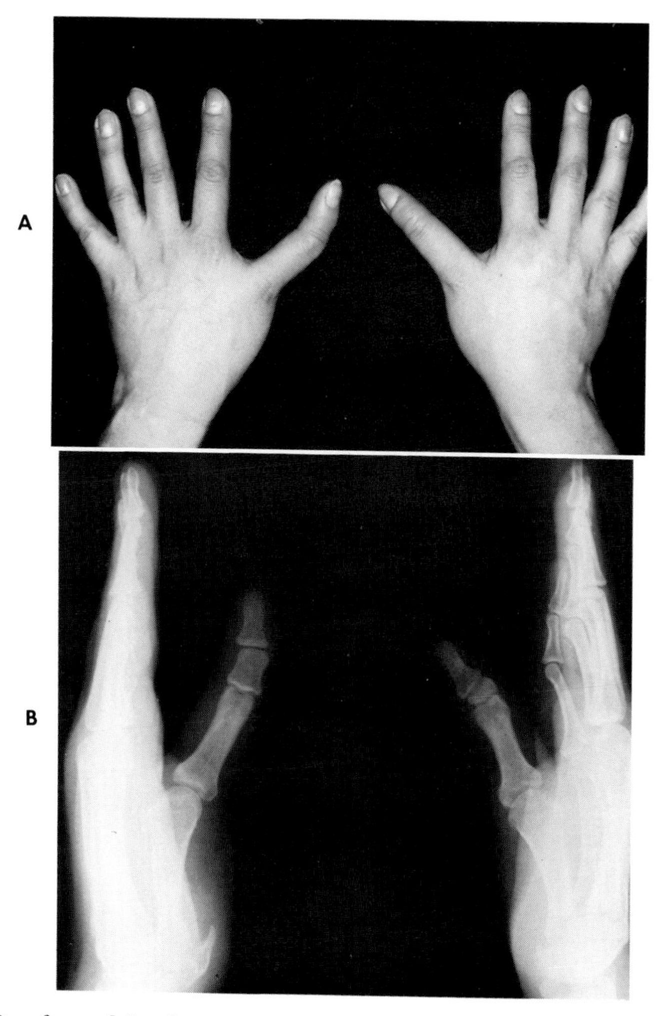

Fig. 7-15. *Five-fingered hand.* **A,** Triphalangeal thumbs resemble fingers, giving the appearance of a five-fingered hand. **B,** Their lengths will vary depending on the extent of development of the extra phalanx. In these hands the phalanx on the left is fully developed while that on the right is rectangular bordering on delta in shape. (From Wood, V. E.: Treatment of the triphalangeal thumb, Clin. Orthop. **120**:188-199, 1976.)

three-boned first or preaxial digit, and no adequate embryological explanation exists for the occurrence of a fourth bone or a "third" phalanx. It is of interest that in the thalidomide era of 1960 to 1962 most cases of triphalangeal thumb seen in Sweden and Hamburg were related to thalidomide taken during the forty-fifth to fiftieth day of gestation.

Individuals who have reached adult life with triphalangeal thumbs usually insist they are "normal" and that they can do everything they wish with their hands. In fact, the extra long thumb significantly alters their precision handling abilities and is a gross cosmetic defect.

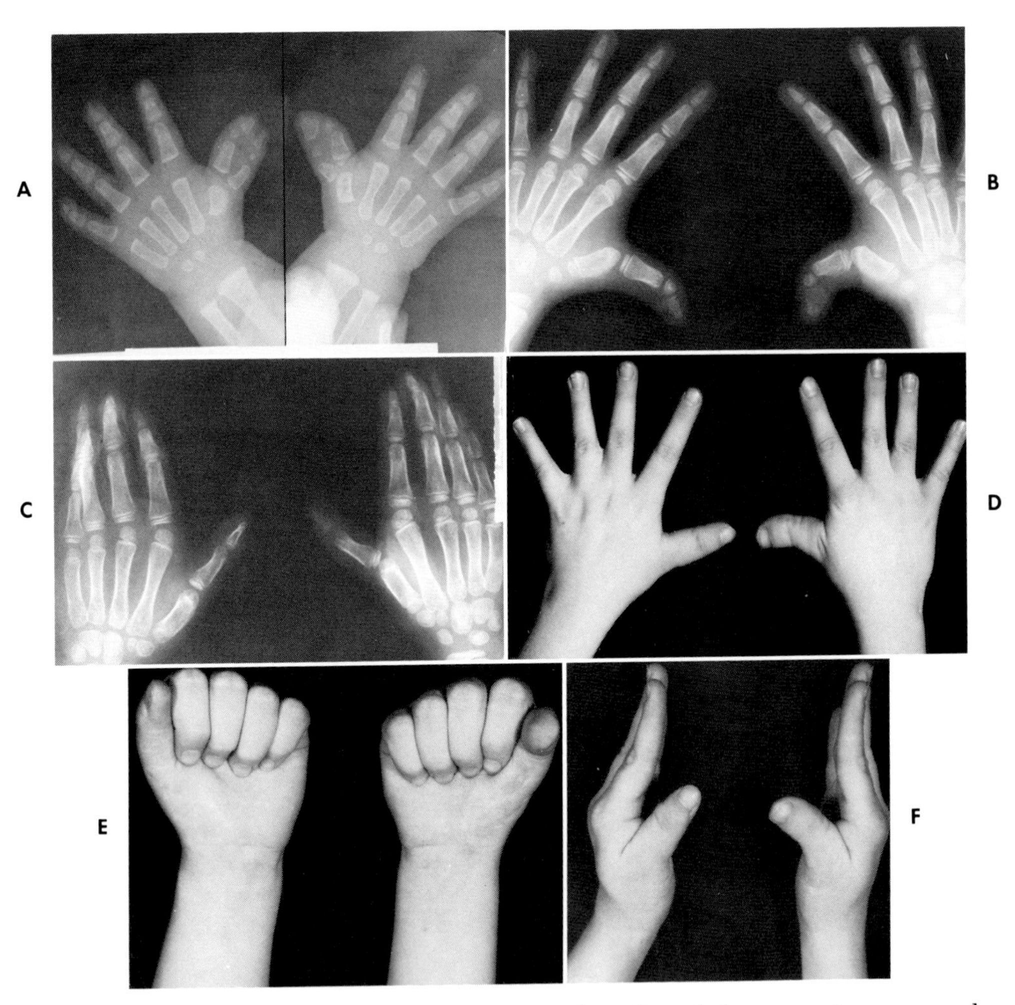

Fig. 7-16. *Triphalangeal thumbs—amputations.* **A** and **B,** The radial component was removed from both thumbs when the patient was 13 months old. Excessive radial deviation occurred around the supernumerary middle phalanx. **C,** Four years after bilateral removal of the middle phalanx. **D, E,** and **F,** Functional and cosmetic appearance 6 years after removal of the middle phalanges and 9 years after removal of the duplicated thumbs. (**A** to **C,** From Wassel, H. D.: The results of surgery for polydactyly of the thumb, Clin. Orthop. **64:**175-193, 1969.)

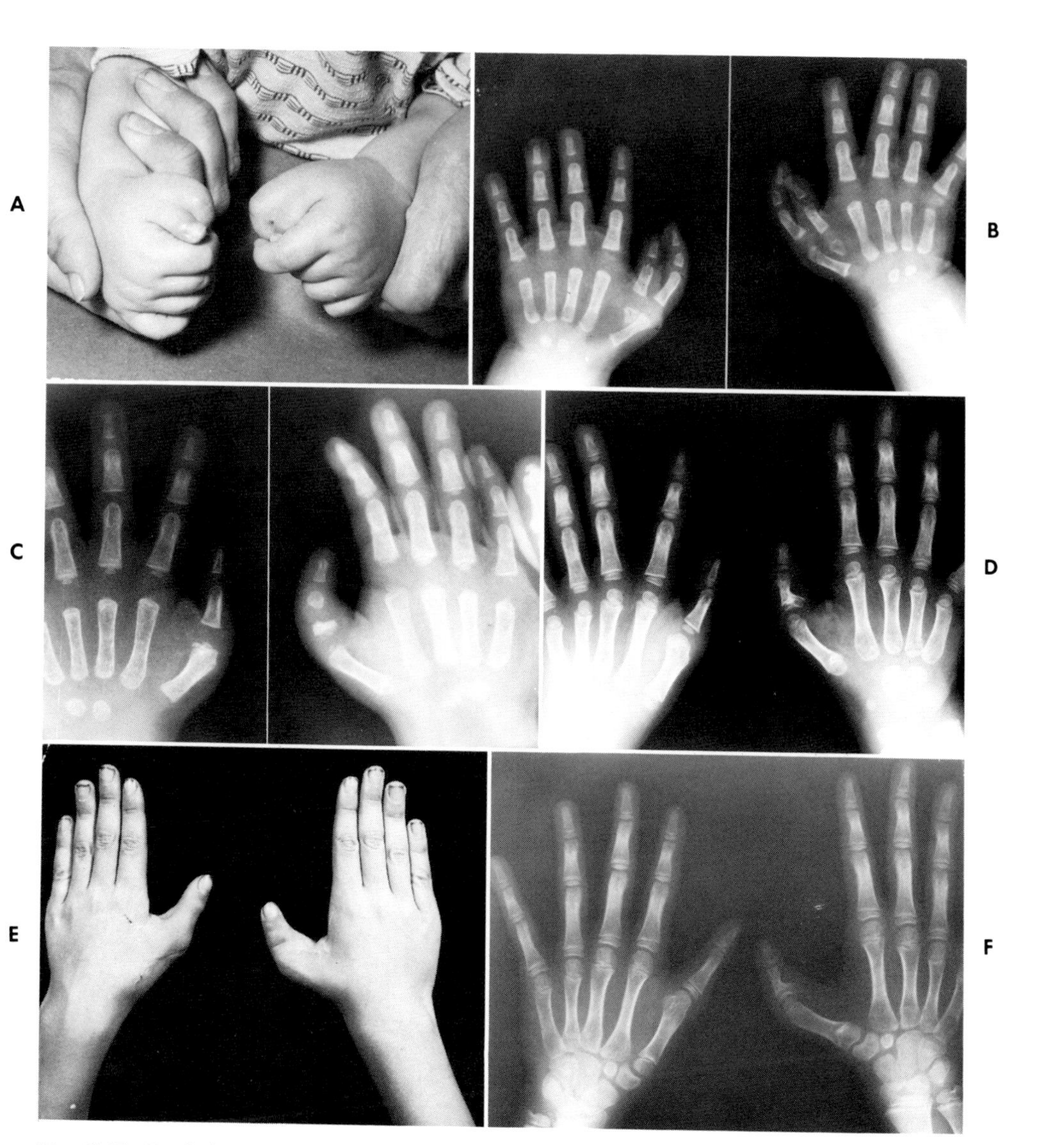

Fig. 7-17. *Triphalangeal thumbs—amputation and transposition.* **A** and **B,** Preoperative views of both thumbs. **C,** Right side, 2 years after removal of ulnar duplication and middle phalanx of radial component. A space has developed between the head of the proximal phalanx and its base. Left side, 3 years after removal of the radial phalanges and transposition of the ulnar digit onto the hollowed out metacarpal of the radial triphalangeal portion. **D,** Three years after bone grafting to the defect in the proximal phalanx of the right thumb. **E** and **F,** Eight years after the start of surgical treatment. There is 50 degrees of ulnar deviation of the right MP joint and 25 degrees of ulnar deviation of the left MP joint. (**B** to **F,** From Wassel, H. D.: The results of surgery for polydactyly of the thumb, Clin. Orthop. **64:**175-193, 1969.)

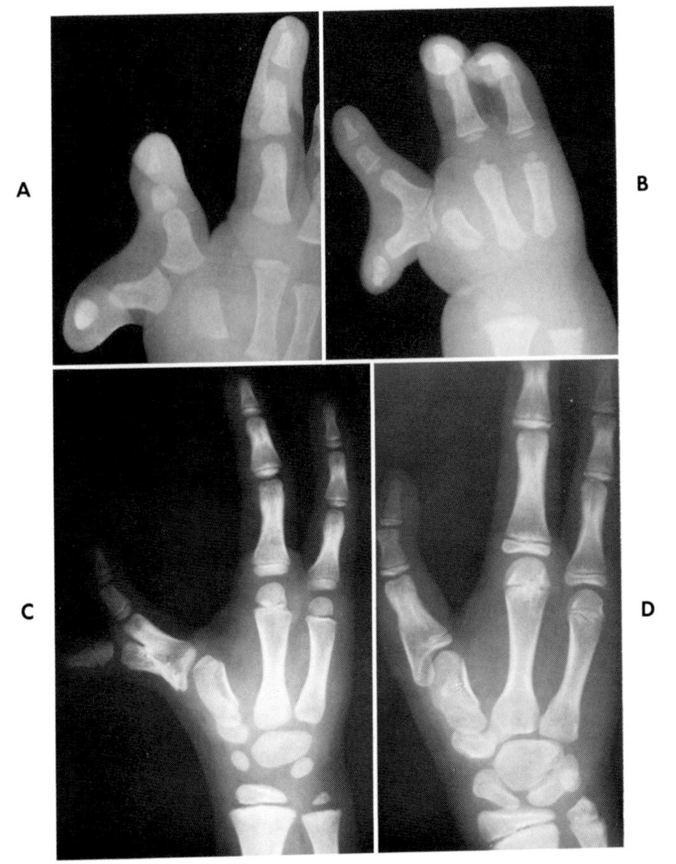

Fig. 7-18. *Triphalangeal thumb retention.* Stability in a thumb is of paramount importance. When the triphalangeal thumb is well aligned and stable, it may be wise to retain it. **A,** The two phalangeal thumb is badly deviated, and the triphalangeal thumb is in good position. **B** and **C,** As the child grows, the more radial thumb remains deformed. **D, E, F,** and **G,** Retention of the right triphalangeal thumb gives good function and a good appearance.

Treatment

In duplications involving triphalangeal thumbs, varying combinations of thumbs can occur. When both thumbs are triphalangeal, one should amputate the lesser and remove the extra phalanx in the retained thumb either at the same time or in a second operation shortly afterwards (Fig. 7-16). When there is a two-phalangeal normal thumb accompanying a triphalangeal thumb, the latter can be removed (Fig. 7-17). When the two-phalangeal thumb is hypoplastic, it is often difficult to decide what to remove. Stability in the joints is important, and if the extra phalanx is relatively large and the hypoplasia of the "normal" thumb mild, it may be wiser to retain the two-phalanx thumb because of potential instability at the site of excision of the accessory phalanx (Fig. 7-18).

Excision of an accessory phalanx should be done early in life and certainly by 1 year of age if the patient is seen early enough. Many of these phalanges

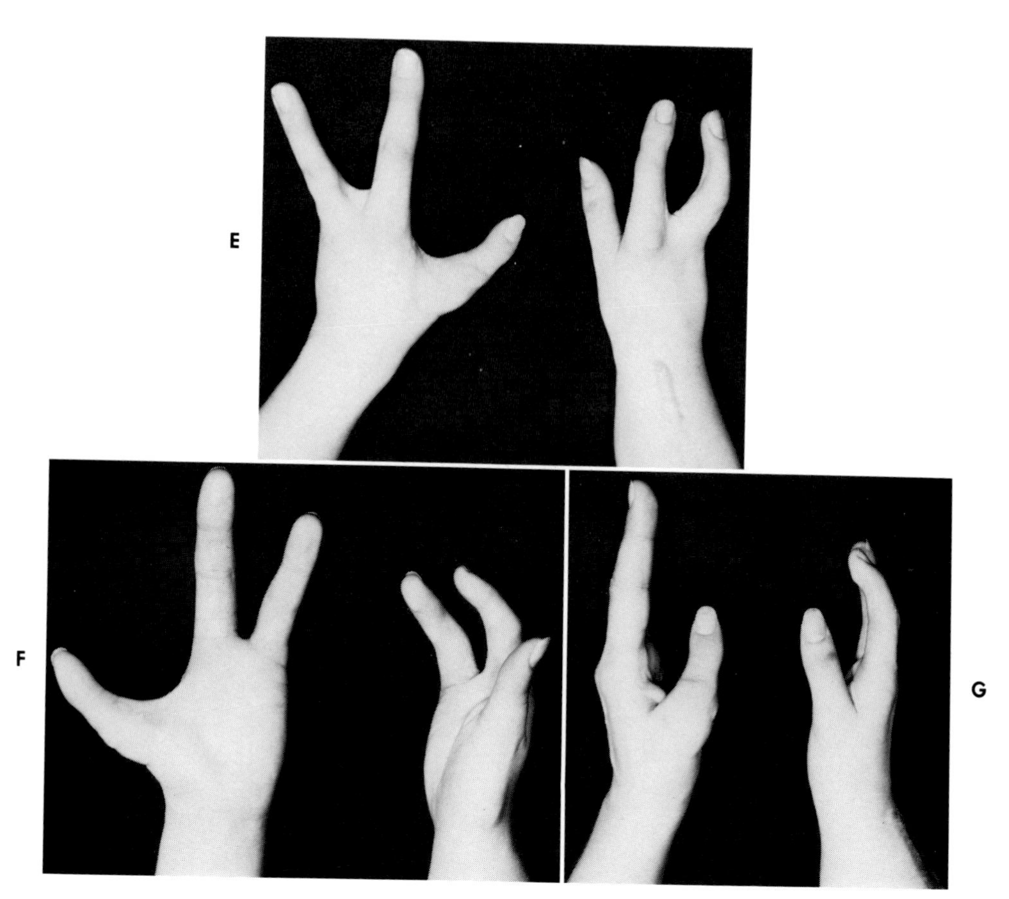

Fig. 7-18, cont'd. For legend see opposite page.

are of the delta type, and there is no value in operating upon their aberrant epiphysis. The whole bone should be removed, the proximal and distal phalanges impaled by a single, small Kirschner wire, and the soft tissues tightened around the new interphalangeal joint. I usually leave the Kirschner wire in place for 6 weeks and then support the thumb with a short, padded aluminum splint for another 6 weeks.

When length reduction is necessary in later life, probably the most satisfactory procedure is to retain the distal interphalangeal joint intact and to excise sufficient bone on either side of the extra or proximal interphalangeal joint to produce the correct length after a fusion between the head of the extra phalanx and the base of the proximal phalanx. The extrinsic flexor tendon will usually take up and good control return, but the extrinsic extensor tendons will need to be formally shortened at the time of bone removal. The site of shortening should be well proximal to the level of bony fusion.

SECTION THREE

THE FINGERS

CHAPTER 8

SMALL AND ABSENT FINGERS

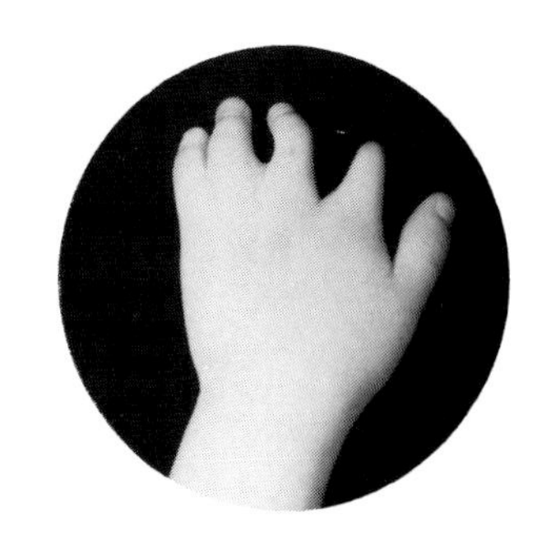

A great variety of growth disturbances occurs in the fingers. They can be hypoplastic but normal in appearance, short, stiff, stiff and short, or even absent. Surgery does not have a great deal to offer children with these disorders; although microsurgical techniques have now moved from the laboratory to the operating room, the use of whole toe or finger transfers is neither as easy nor as commonplace as lay publications would have one believe.

Hypoplastic fingers are often difficult problems to the surgeon because of a natural reluctance to amputate something that is normal in form although dwarfed in size, particularly since it is impossible to reconstruct anything as good. If there is good joint mobility and tendon control, I believe the finger should be left in place and only removed if its owner requests it after reaching the age of reason (whatever that may be nowadays).

It is important that the parents realize that hypoplastic digits will never catch up in growth and that the size disparity may increase throughout childhood (Fig. 8-1). A single, flail hypoplastic digit is probably not worth retaining since it will constantly get in the way during use of the hand. If it is consistently splinted or taped to an adjacent normal finger, the latter may be adversely affected. It is important to realize that short digits may be a clue to significant systemic disorders; an example is the association between brachymetacarpia and the Holt-Oram syndrome. A thorough general examination of patients with short digits is mandatory.

Attempts have been made to supply whole joints and tendon control to

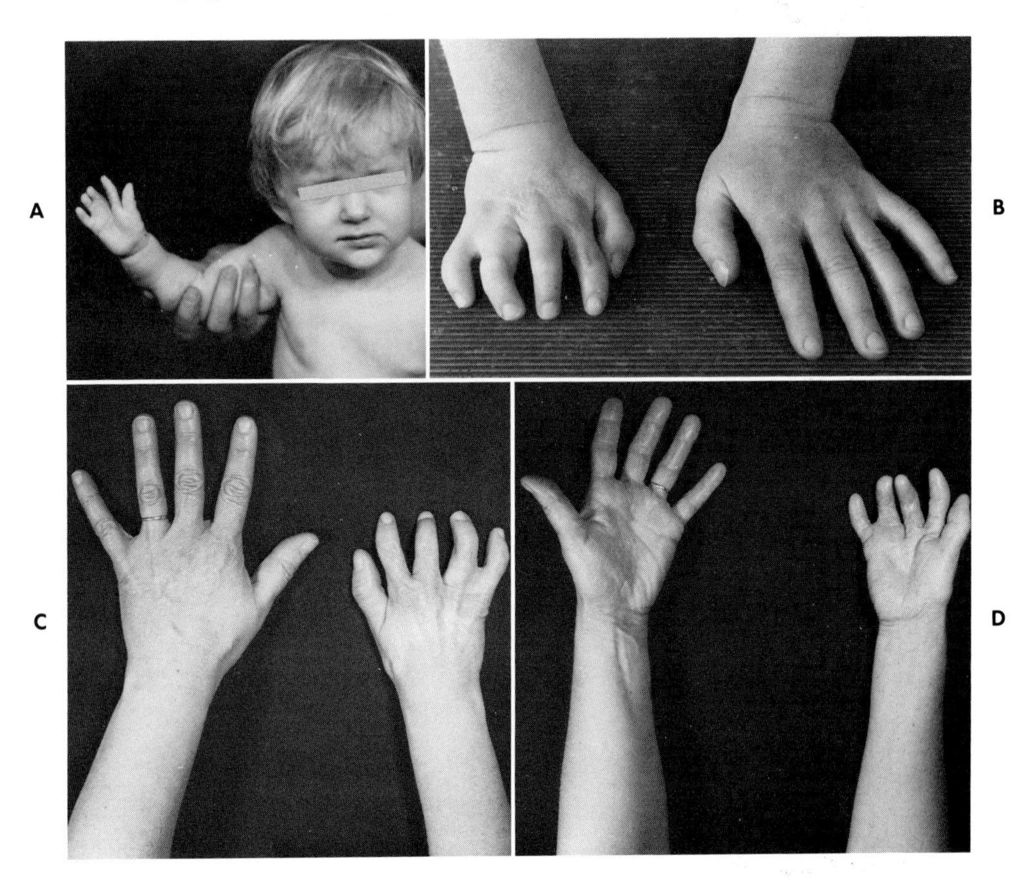

Fig. 8-1. *Hypoplastic digits and growth.* Parents should realize that hypoplastic digits never catch up in growth with normal digits. **A,** This 1-year-old child was born with a hypoplastic hand and syndactyly. **B,** At the age of 9 years, it was obvious the right hand was undersized. **C** and **D,** At 46 years of age, her hand is usable but still remains small.

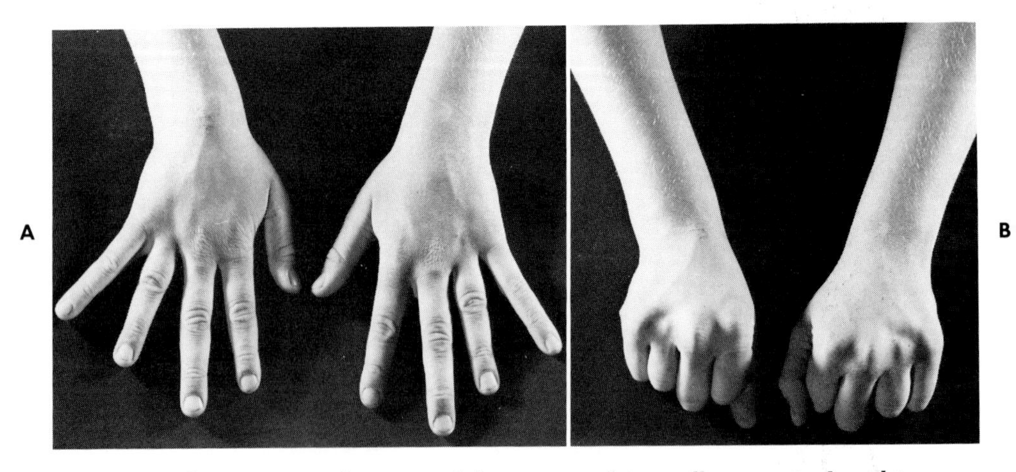

Fig. 8-2. *Brachymetacarpia.* Shortening of the metacarpal is usually not noticed until teenage. **A,** The ring fingers are somewhat short. **B,** The reason is apparent on fist-making, when the gap or insetting at the site of the metacarpal head appears.

hypoplastic fingers, but the ultimate functional results of these heroic operations are not usually profitable to the patient. I do not stress early amputation in these patients but do emphasize to the parents that the size disparity may increase and recommend they seriously consider amputation before their child enters school.

SHORT METACARPALS

Short fingers resulting from small metacarpals are uncommon and are rarely detected in early childhood. The "failure to grow" is often not recognized until around the teenage growth spurt and is thought to probably be caused by premature closure of the epiphysis (Fig. 8-2). It is usually first noticed by the gap or insetting that appears in the line of the metacarpal heads when a fist is made. Although the thumb metacarpal is often shortened in association with

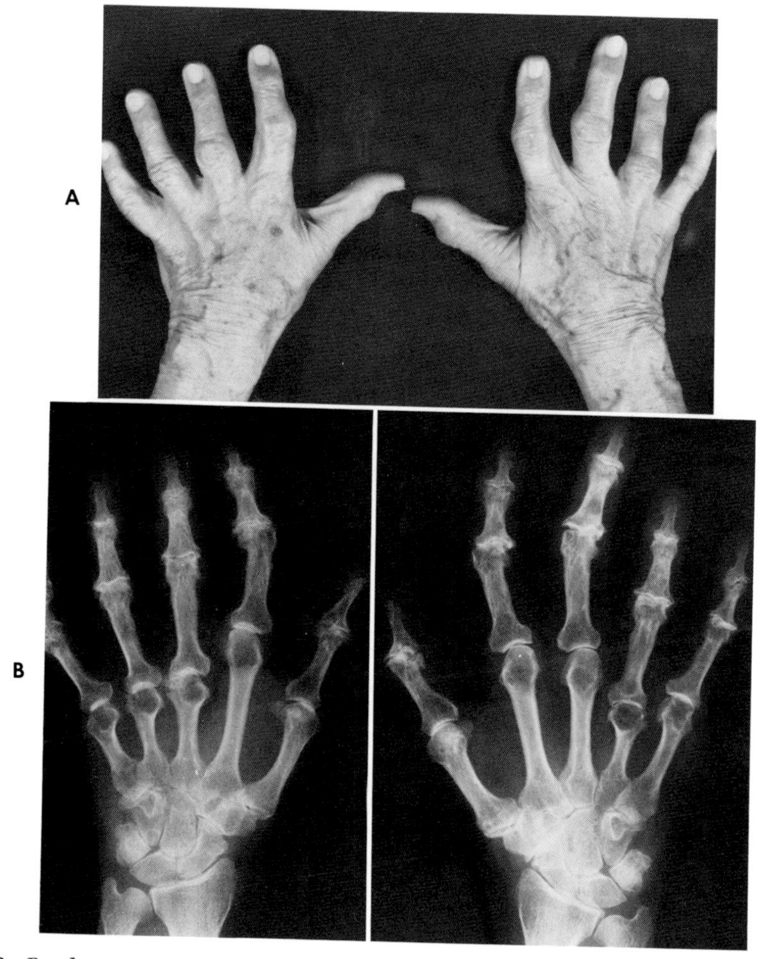

Fig. 8-3. *Brachymetacarpia in an adult.* **A,** This middle-age lady was not aware that her hands were abnormal. **B,** The metacarpals of the long, ring, and small fingers on the left hand and the ring finger on the right hand are all short.

radial aplasia, it is the ulnar two fingers' metacarpals that are most commonly affected.

When the fourth metacarpal is affected, it may be associated with mild pseudohypoparathyroidism. In cases of well developed pseudohypoparathyroidism it is said that the fifth and even the third metacarpal may also show shortening. A number of syndromes, including Turner, Biemond, and Silver, are also associated with short metacarpals, usually the fourth and fifth, with or without any other associated abnormalities. Isolated shortening of the fifth metacarpal has been described in association with the cri-du-chat syndrome.

Treatment is usually unnecessary, particularly if the patient is first seen as an adult (Fig. 8-3). The functional disability is small, and most of those affected adapt to the restriction in palmar grasp caused by the recession of the metacarpal head.

When the ring finger alone is affected or even when both ring and small finger metacarpals are affected, length can be gained by inserting a bone graft. Lengthening of these ulnar metacarpals increases the palm size and thereby aids in grasp. Tajima has described a method of inserting a bone graft with each end shaped into a V so that it will fit into a recipient chevron-like slot in the proximal and distal portions of the short metacarpal (Fig. 8-4). The

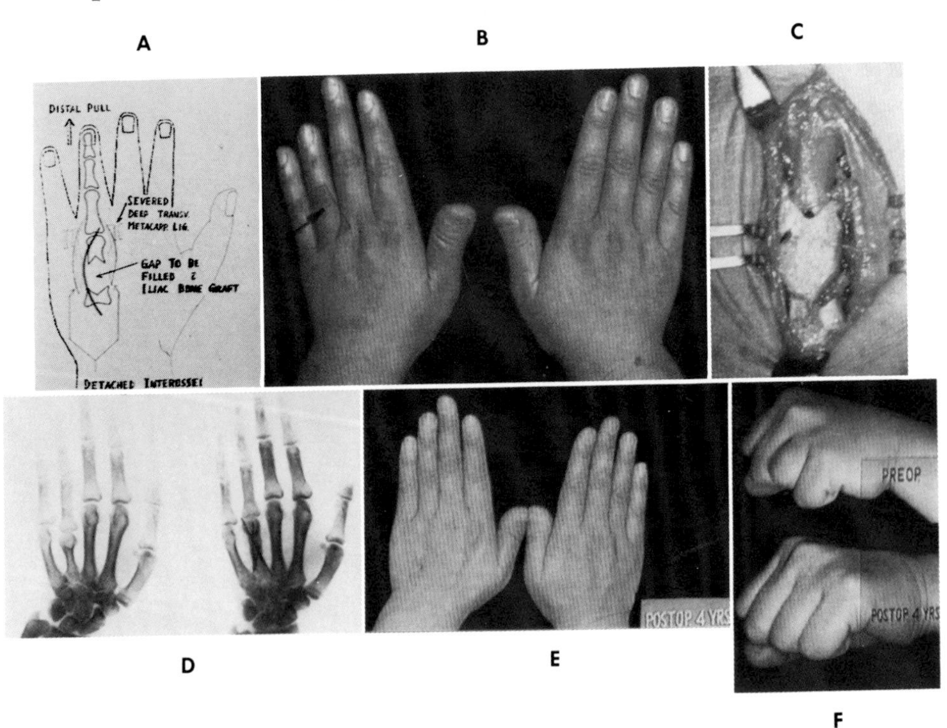

Fig. 8-4. *Brachymetacarpia and bone grafting.* Professor Tajima's method of lengthening a short metacarpal by bone grafting. A chevron-shaped bone graft is used to lengthen the affected bone. (From Tajima, T.: Operative treatment of congenital hand anomalies, Rinshô Seikei-geka [Clin. Orthop. Surg.] **11**:475, 1976.)

osteotomy should be done about midshaft in the affected bone but should not be done until the interossei have been detached from the shaft and the intermetacarpal ligaments on either side of the metacarpal neck have been severed. This is not an easy technical task, and the operation demands a lot of skill and judgment if length is to be obtained and motion retained in the affected finger. I believe it should be used only if a significant functional deficit has been demonstrated.

SHORT PHALANGES

Short fingers are the delight of the geneticist and the despair of the surgeon. Shortening of the phalanges is easy to see and easy to measure and was the first example of mendelian inheritance demonstrated in man. Surgical lengthening of these short digits is a formidable and frequently impossible task.

The genetic literature goes into extensive and detailed classifications of the various types of abnormal shortening of the phalanges, but these classifications are of little value to the surgeon in his work. Warkany has pointed out that no unitary concept has evolved from the many classifications that have been proposed. Anatomists, radiologists, geneticists, and surgeons all look at the anomalies from different points of view. Many quasiclassical terms have been devised for these classifications; the term brachydactyly is most commonly used and does mean short finger. Common usage restricts the term to digits in which the loss of length is caused by a shortness of the metacarpal or a phalanx. A short metacarpal has been named brachymetacarpia and a short phalanx, brachyphalangia. For those who wish to distinguish which phalanx is short, brachytelephalangia refers to the distal phalanx and brachymesophalangia to the middle, while the proximal phalangeal shortening has been termed brachybasophalangia. The essential use of brachydactyly is that it refers to a short digit in which the normal number of bones is present but there is a size reduction. The term ectrodactyly has been loosely used as a substitute for brachydactyly; this is incorrect since ectrodactyly is properly used to describe complete absence of one or more phalanges or the metacarpal.

Shortening of the fingers is a frequent dominant malformation. It is relatively easy to follow up over many generations, and Bell has done extensive work investigating and classifying the genetic aspects of families with short fingers (Fig. 8-5). The possible variations are so great that there is a sizable group of individuals who do not fit into the extensive classifications that have been developed. New papers are constantly describing further genetic variations, and in a recent paper Bass described the situation accurately when he wrote, "At present we must be content to have added another score to the growing repertoire of human genetics."*

It is said that brachydactylous women are compactly built and petite and

*Bass, H. N.: Familial absence of middle phalanges with nail dysplasia: a new syndrome, Pediatrics **42**:318-323, 1968.

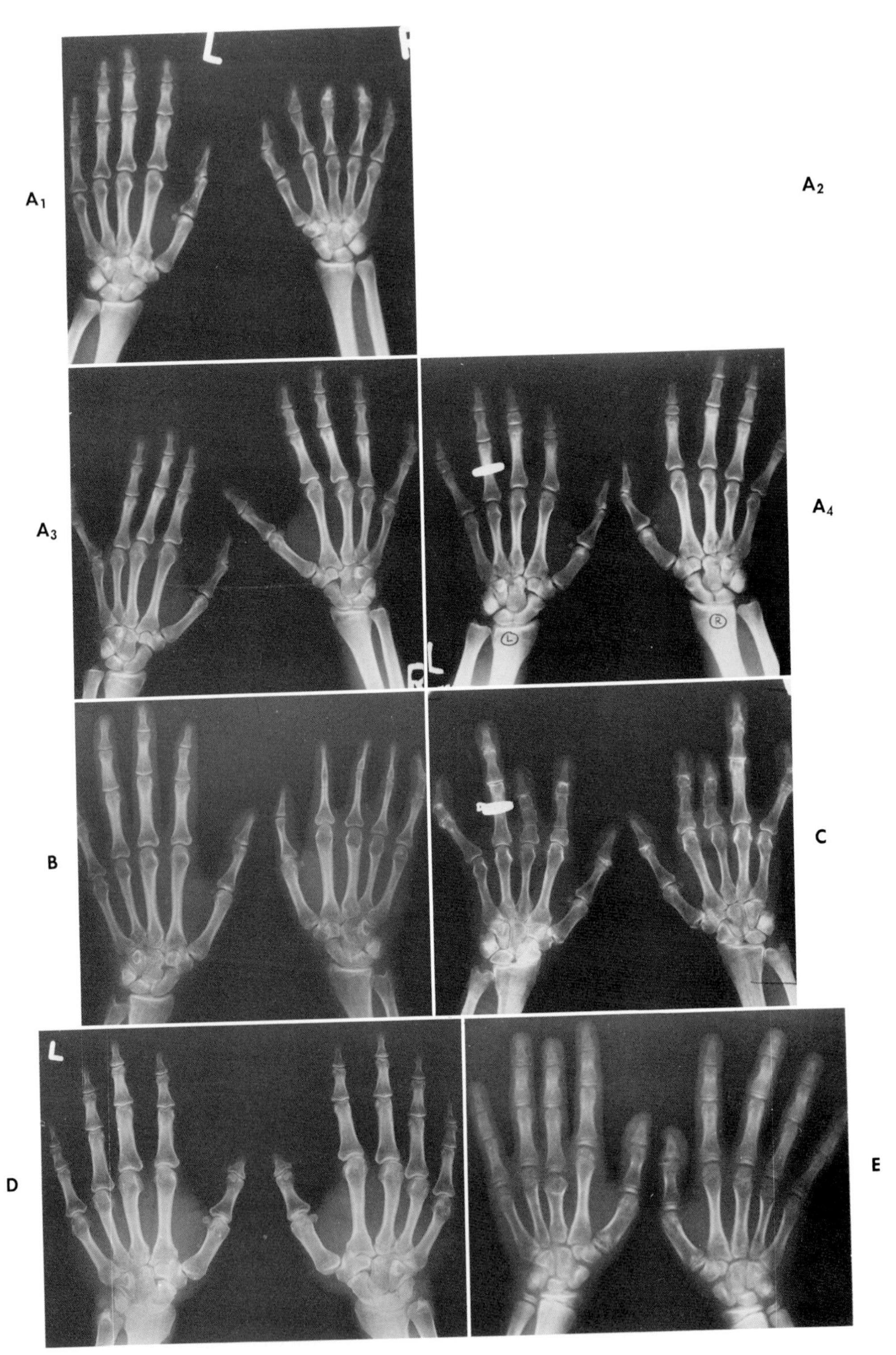

Fig. 8-5. For legend see opposite page.

are preferentially selected by normal men. Brachydactylous men are often short in stature, but their preference for, or as, partners is not recorded. The chances are good that a brachydactylous individual will marry a normal partner and that about half of their children will bear the same defect as the abnormal parent. Pedigrees have been published in which the anomaly was transmitted as an irregular dominant trait. In some, brachydactyly was transmitted with great uniformity, but in others, its occurrence was so variable that it could only be recorded as an inherited irregularity of digital development. Sporadic brachydactyly does occur, and not all cases of short fingers are genetically determined. Surgeons can only admire the devotion of geneticists to their monumental task, and most of us should not enter a guessing contest as to the likelihood of further children being born with abnormal hands.

The phalanges of the small and index fingers are the most frequently affected. There is a broad spectrum of abnormalities starting with a minimal shortening of the middle phalanx of the small finger through shortening of several fingers to complete absence of all digits. These deficiencies can occur alone, or they may be associated with similar deformities in the feet or abnormalities of other parts of the body.

Shortening of all the middle phalanges is often associated with shortening of the proximal phalanx of the thumb. It was classified by Bell as Type A-1 brachydactyly. The extent of shortening may vary in the fingers with the border fingers, the index and small, usually being most affected. Shortening of the middle phalanges also occurs in other types of brachydactyly, and it is seen in a large number of malformation syndromes such as Treacher Collins, Bloom, Cornelia de Lange, Holt-Oram, Silver, and Poland. In Poland syndrome the shortening affects only one hand (Fig. 8-6).

The middle phalanx of the small finger is the most variable in length of any of the hand bones in females, and Garn has stated, in Garn, Fels, and Israel, that it is almost the most variable in males. He has also demonstrated that there is a significant variation in shortening in different populations, from 0.6% in the adults of southwest Ohio to 5% in Peru and Hong Kong. Many syndromes have been associated with shortening of the middle phalanx of the small finger, the best known being trisomy 21. The other syndromes associated with shortening of

---

Fig. 8-5. *Brachydactyly—Bell's classification.* Bell has classified short digits into the following groups. **A1,** Middle phalanges, which may be rudimentary or fused to the distal phalanges. The proximal phalanx of the thumb is also short. **A2,** Extremely rare (no photographs). All digits are normal except the index in which the middle phalanx is delta shaped. **A3,** The middle phalanx of the small finger is short and may be delta shaped. **A4,** The index and small finger middle phalanges are short. Occasionally the ring finger is affected, but in this instance the long finger in the left hand has a short middle phalanx. **B,** The middle phalanges are short, and the distal phalanges are rudimentary or absent. The thumb may also be affected. **C,** The middle phalanges of the index and long fingers are short; the middle phalanx of the small finger is short and may be delta shaped. As in this patient, there may be hypersegmentation of the proximal phalanges of the index and long fingers. **D,** Short, broad terminal phalanx of the thumb. **E,** Brachymetacarpia.

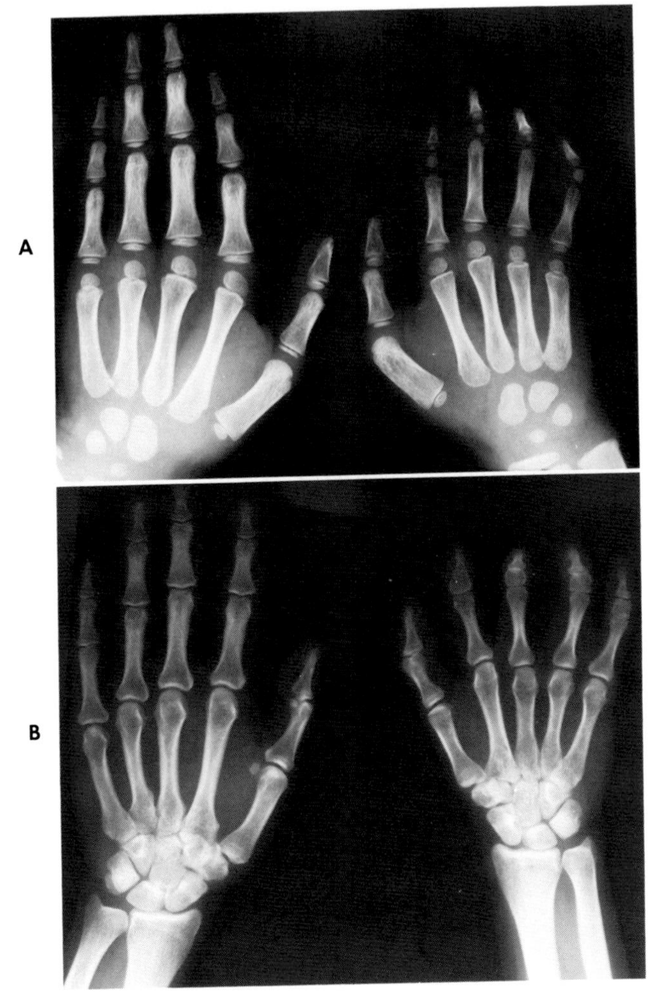

Fig. 8-6. *Short middle phalanges.* **A** and **B,** When a hand with hypoplastic middle phalanges matures, the thumb, which in childhood appears normal, is often found to have a short proximal phalanx.

the middle phalanx of several fingers can also be associated with isolated shortening of the small finger.

Single digit shortening, particularly in the small finger, needs no treatment (Fig. 8-7). If there is a difference in shortening on the two sides of a phalanx, a clinodactyly occurs (Fig. 8-8). For some unknown reason this shortening is more common on the radial than on the ulnar side of the phalanx, producing a radial deviation of the finger. The majority of these short, bent fingers are not a functional handicap; their treatment is discussed in Chapter 9, pages 154 to 158.

Following the middle phalanx of the small finger in frequency of size variations are the distal phalanges. They may be shorter or narrower than would be considered normal. The tufts to which the septa of the pulp of the terminal por-

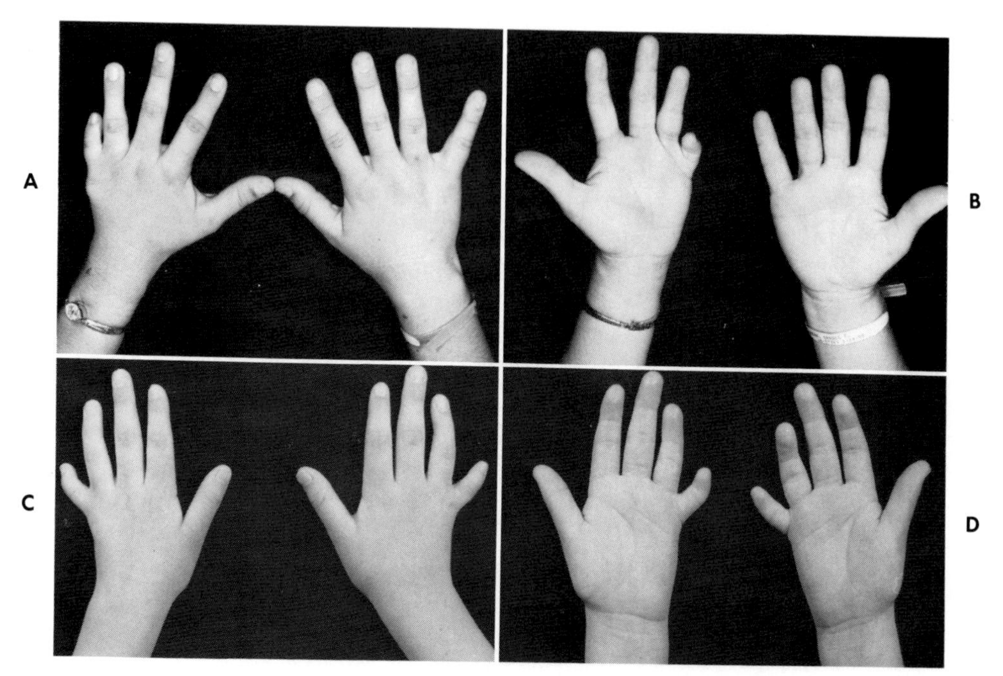

Fig. 8-7. *Small finger shortening.* Two patients with typical shortening of their small fingers. No treatment is needed for this type of shortening.

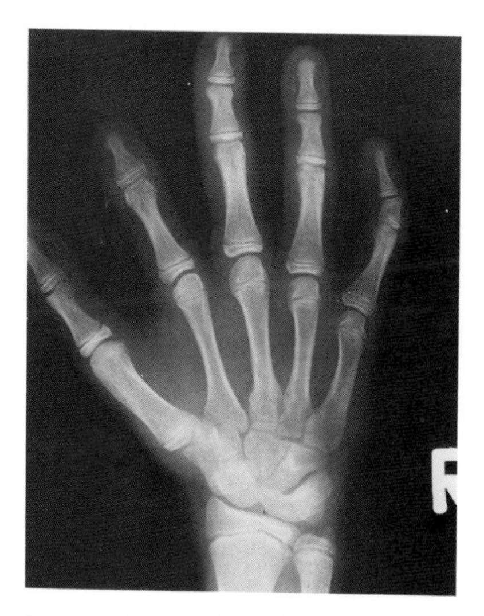

Fig. 8-8. *Small finger clinodactyly.* When a digit shortens unevenly, it is more common for it to be short on its radial side, thereby producing radial deviation of the finger. In this patient the index middle phalanx failed to develop.

tion of the finger are attached may vary considerably in size. Shortening of the terminal phalanx may be associated with hypoplasia or absence of the fingernails, but the converse is not necessarily true (Fig. 8-9); hypoplasia of the nails may occur over perfectly normal distal phalanges. When normal nails grow over shortened phalanges, there is sometimes a tendency for the nail end to grow over and palmarwards producing an unattractive almost club-like end to the finger. The differential diagnosis should always include the possibility of cold injury to the epiphysis producing premature epiphyseal closure. There are also some syndromes and a number of acquired diseases, particularly neurological, that are associated with reduction in size of the distal phalanges.

Shortening of the proximal phalanges is rare and does not seem to occur in isolation. It has been reported in the index and long fingers and has been associated with other hand abnormalities that Bell has grouped as Type C. It is seen in short thumbs, and in many conditions the proximal phalanx of the thumb is found to be reduced in size in proportion to the middle phalanges of the fingers in the same hand.

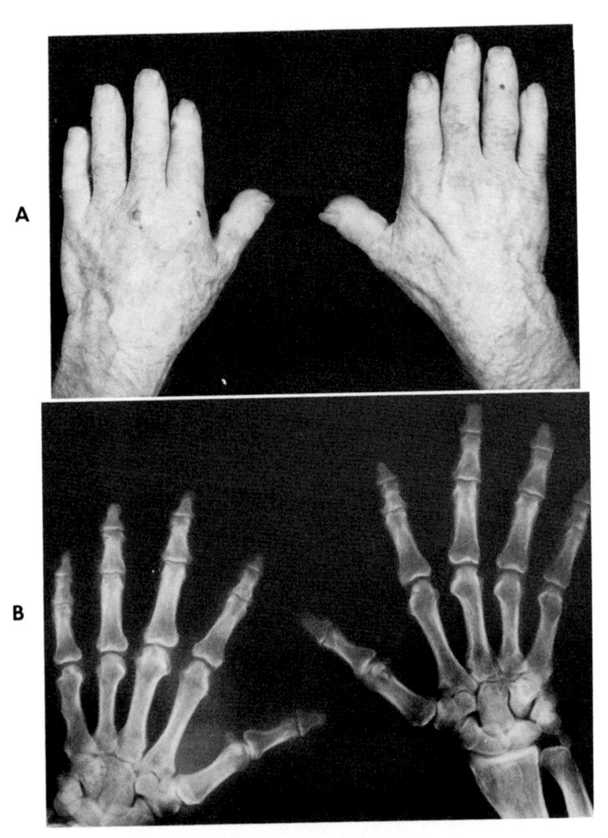

Fig. 8-9. *Distal phalanx hypoplasia.* **A,** When the terminal phalanges fail to fully develop, the fingernails may also be affected. **B,** The terminal phalanges are shorter and narrower than usual, and the terminal tufts are underdeveloped.

Most patients adapt to their shortened fingers and admit to little or no functional handicap. Although unsightly, the presence of a short finger enclosed between normal fingers does not seem to greatly interfere with function. I believe phalangeal lengthening procedures have little or no place in the treatment of short fingers because the lengthening rarely, if ever, produces a demonstrable increase in function.

SHORT, STIFF FINGERS

Strict usage of the term symphalangia should be restricted to a condition of complete continuity of the middle with the proximal phalanx and absence of the proximal interphalangeal joint. It is a rare condition, and even more rare is fusion of the distal and middle phalanges. Both types are well documented as being of autosomal dominant inheritance; however, differences in the number of joints affected and location of joints affected and variability in the degree of fusion of the affected joints all point to genetic heterogenicity.

In 1906 Harvey Cushing saw a young woman who, in addition to her cerebral glioma, had "an unusual condition of the fingers which could not be bent at the proximal interphalangeal joints."* During the next 10 years Cushing followed up this girl's family tree until in 1916 he published his findings and plunged us deep into a semantic jungle by coining the term "symphalangism." He used the term to mean hereditary clinical stiffness of the proximal interphalangeal joints with normal or near normal length of all phalanges. Unfortunately subsequent authors have seldom adhered to this definition, and most reported pedigrees include individuals with greatly shortened or even absent middle phalanges. In 1976 the *Journal of Bone and Joint Surgery* published a paper that opened with the statement "Symphalangism is a congenital anomaly of the digits manifested clinically by shortness. . . ."† To add to the confusion the condition has also been reported under such names as synostosis, anarthrosis, hereditary ankylosis, hereditary aplasia of joints, and even lateral fusion of finger joints.

Cushing traced the family back to a Scottish immigrant who settled in Virginia in the 1740s. Drinkwater claimed to have traced a similarly afflicted family back to John Talbot, the first Earl of Shrewsbury, who lived in the late 1300s. Recently Elkington and Huntsman concluded that adequate proof for transmission of the symphalangia through 14 generations of the Talbot family is lacking.

Stiff fingers are of great interest to geneticists; but not all stiff fingers are inherited. Failure of separation of phalanges can be seen in thalidomide children and in various congenital anomalies such as complex syndactyly; Apert, Poland, Mobius, and Marchesani syndromes; diastrophic dwarfism; Bell's brachydactyly

---

*Cushing, H.: Hereditary anchylosis of proximal phalangeal joints (symphalangism), Genetics **1**:90-106, 1916.

†Dellon, A. L., and Gaylor, R.: Bilateral symphalangism of the index finger, J. Bone Joint Surg. (Am.) **58**:270-271, 1976.

Types A and C. We at the University recently reported its association with gargoylism and dysplasia epiphysealis multiplex.

The great majority of congenitally stiff fingers are caused by interphalangeal fusions in fingers in which the middle phalanx is either shortened or absent. Cushing's paper established the general rule that the middle phalanges and the proximal interphalangeal joints are most severely affected. In less severely affected hands, he observed that the radial side of the hand was usually much less involved than the ulnar side.

Clinically the characteristic finding of a stiff finger is smooth skin overlying the affected joint in which no motion occurs (Fig. 8-10). Thus the condition can be diagnosed at birth. X-ray examinations are not helpful because a joint space may appear to be present between the phalanges when in reality it is a solid bar of cartilage. The epiphysis at the base of the middle phalanx is often abnormal when compared to its neighbors. In children the differential diagnosis between ankylosis of a finger joint and symphalangia is simple. Symphalangia is present from birth. Several fingers are usually affected; each finger is extended, and the proximal joints are usually involved. Ankylosis is acquired; the finger is usually flexed, and the distal joints are commonly involved.

Individuals with rigid digits are unable to make a fist (Fig. 8-11) and have difficulty in picking up small objects. Their fingers have been described by Bell as

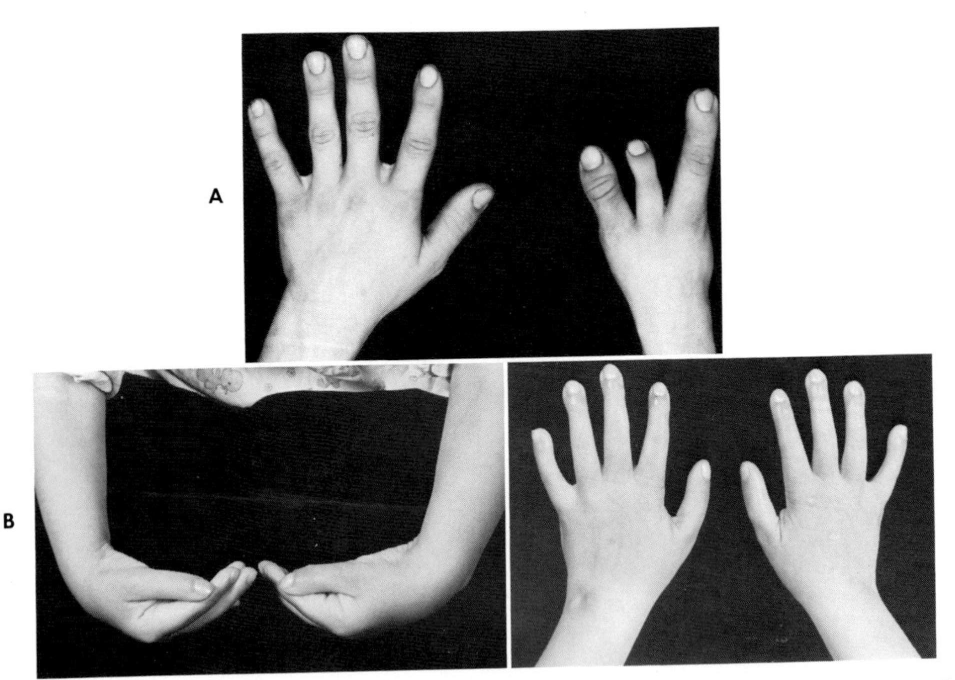

Fig. 8-10. *Short stiff fingers.* When digits cannot bend, they do not maintain joint creases and the skin is smooth. **A,** A stiff "middle" finger in the right hand. **B,** A child with stiff interphalangeal joints attempting to make fists. **C,** The dorsal view of her hands showing the smooth, shiny skin on her digits.

"shuffle fingers." It is extremely uncommon to find a congenitally stiff digit that is of normal size. This hypoplasia is caused by both inherent developmental defects and atrophy from lack of use.

Until 1935 all previously recorded examples of the condition had been observed in whites. In that year Cole reported an occurrence of the condition in a family that initially was thought to be black. It was, however, learned that one of the progenitors of the family was an American Indian. There have been no other reported cases of involvement in the black race.

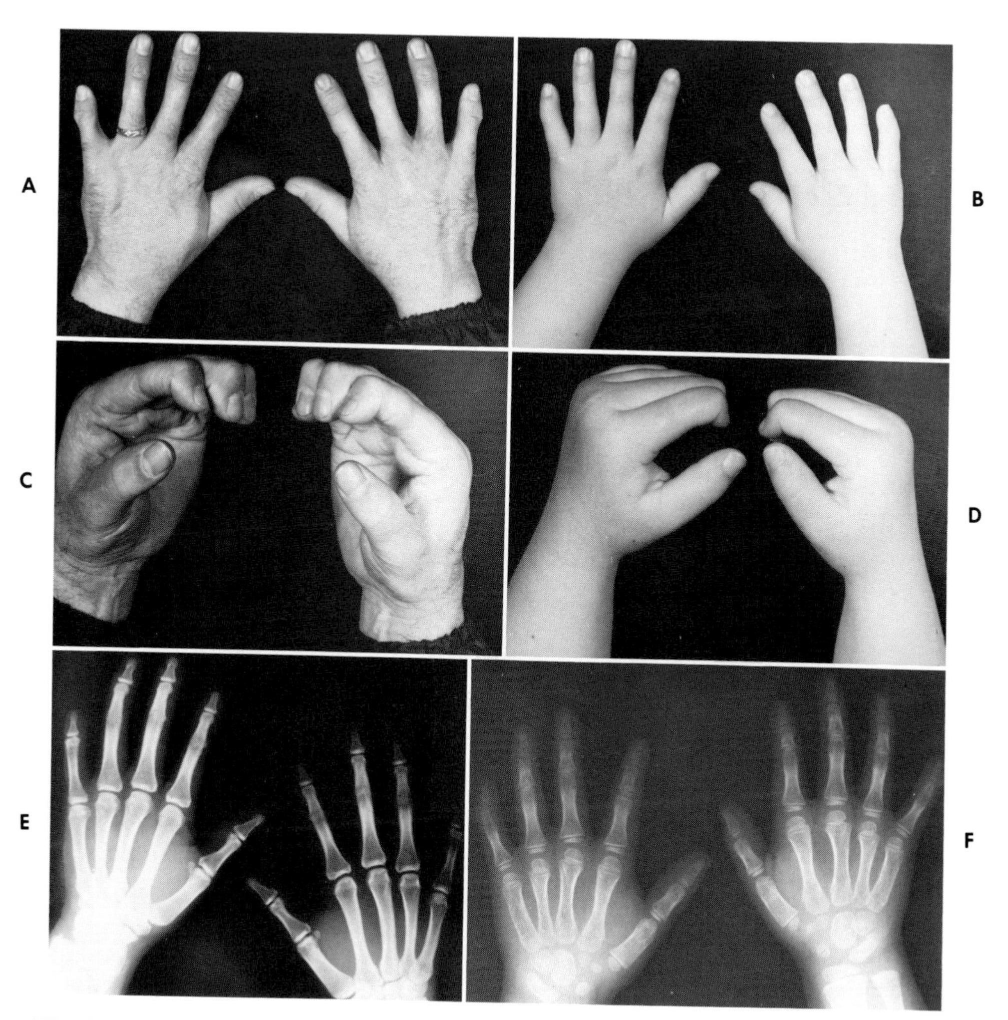

Fig. 8-11. *Hereditary stiff fingers.* Illustrated here are the hands of a father (**A, C,** and **E**) and his daughter (**B, D,** and **F**) showing multiple true symphalangia involving all the fingers of both hands. They attempt to compensate for their stiff proximal interphalangeal joints by fully flexing the distal joints. The father's joints are fused and the apparent joint spaces in his daughter's hands are deceptive. The child's grandfather also had a similar deformity. (From Flatt, A. E., and Wood, V. E.: Rigid digits or symphalangism, The Hand **7**:197-214, 1975.)

In a paper published in 1975 we reported the incidence of stiff fingers in both the world literature and in our patients in Iowa. As has been true in other similar studies we have reported, we found that the relative incidence of the target condition was higher than would be implied by reports in the literature. In this study we found a greater incidence of distal interphalangeal joint fusion in our 65 Iowa patients than we could establish in our survey of the literature (Tables 8-1 and 8-2).

These rigid digits fall into three main groups: Group I is composed of single and multiple true symphalangia, Group II comprises the most common type, symbrachydactyly, and Group III involves a large group in which symphalangia and other anomalies are associated.

Group I—true symphalangia. The single stiff digit is not common, and in our two cases the long finger was also involved (Fig. 8-12). The affected digit is usually shorter and more slender than adjacent fingers or its twin of the opposite hand. The skin of the digit is usually tighter and thinner than that of the adjacent digits, and there is an absence of transverse skin creases over the stiff

Table 8-1. Survey of literature (649 fingers)

	Index	*Long*	*Ring*	*Small*
Proximal interphalangeal fusion	112	135	168	177
Distal interphalangeal fusion	19	9	7	19
Metacarpal phalangeal fusion	0	0	1	2

Table 8-2. Iowa patients (557 fingers)

	Index	*Long*	*Ring*	*Small*
Proximal interphalangeal fusion	77	79	82	74
Distal interphalangeal fusion	58	54	60	53
Metacarpal phalangeal fusion	5	4	5	6

Table 8-3. Incidence of associated anomalies

Associated anomalies	*Number of patients*	*Percentage*
Syndactyly	54	83
Apert syndrome	19	24
Poland syndrome	15	23
Foot anomalies	14	21.5
Twins	4	6
Cleft palate	3	4
Facial paralysis	3	4
Bilateral dislocated hips	3	4
Hearing loss	2	3
Mobius syndrome	2	3
Dysplasia epiphysealis multiplex	2	3
Gargoylism	2	1.5

Data from Flatt, A. E., and Wood, V. E.: Rigid digits or symphalangism, The Hand 7:197-214, 1975.

joint. Growth of the finger is usually slower than normal, and inevitably, it is subjected to an excessive amount of trauma.

Several stiff digits of normal length in one hand were also uncommon in our patients (Fig. 8-11). When they occur most of these digits show fusion of the proximal interphalangeal joints. The distal interphalangeal joints compensate and can frequently flex to more than 90 degrees. This type of deformity of the distal interphalangeal joint was clearly illustrated in Cushing's original paper.

Group II—symbrachydactyly. Shortened, stiff digits were by far the most common of the three types found in our Iowa patients. The degree of shortening depends upon the extent of growth of the middle phalanx. All degrees can be found, and in our patients the whole spectrum of possibilities was represented. Our patients showed varying degrees of stiffness of both proximal and distal interphalangeal joints.

Group III—symphalangia with associated anomalies. A large group of patients falls under this general heading. They are included because they have stiff digits that are associated with other anomalies; the combination often constitutes a major problem of total care. The incidence of the associated anomalies we detected is shown in Table 8-3. Eighty-three percent of these 65 patients showed an associated syndactyly. Also commonly associated were different degrees of hypoplasia of the hand in which there were varying patterns of acrophalangia or brachyphalangia. I consider arthrogryposis to be primarily a soft tissue anomaly, and all arthrogrypotic patients were excluded from the study of our Iowa patients.

Treatment

Group I—true symphalangia. Occasionally one sees adults with this condition. In comparison with their normal peers, there is no question that they have

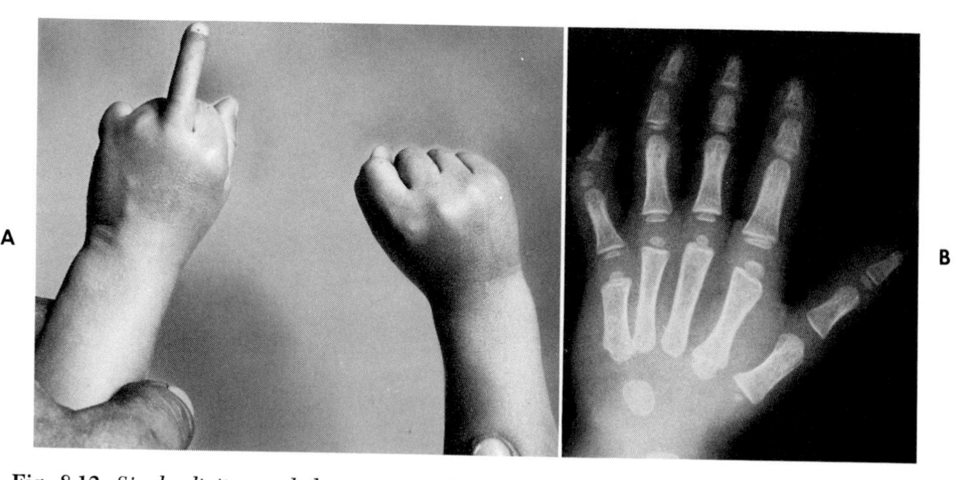

Fig. 8-12. *Single digit symphalangia.* A single stiff digit is rare. **A,** Attempting fist-making. **B,** The x-ray film shows apparent joint spaces in this rigid digit. (From Flatt, A. E., and Wood, V. E.: Rigid digits or symphalangism, The Hand **7:**197-214, 1975.)

impaired function. Most have adapted so well that they do not admit to being greatly handicapped even though the majority are unable to grasp many objects in the normal single-handed manner. The literature is very scant on this aspect of the problem. It is sad to note that the very extensive publications of geneticists pay little or no attention to the functional abilities of these deformed hands. One paper did comment that many different occupations were represented in the family studied and that the majority felt that "they could easily do any kind of work with their hand." One woman had learned to milk cows effectively with her rigid digits. Another paper commented that these patients appeared to be engaged in less skilled or less remunerative occupations than their normal contemporaries and that they found it very difficult to play most musical instruments.

I have never attempted to restore motion to the stiff digits of an adult by surgery. Stiffness of the distal interphalangeal joint is not a significant functional handicap. None of our patients with middle joint stiffness was interested in attempts to improve function by fusion of the proximal interphalangeal joint into the so-called functional position. One patient had had a fusion of the proximal interphalangeal joints done elsewhere and was extremely unhappy with his resulting "hook." He stated he was much happier and had better function when his proximal interphalangeal joints were straight. I believe therefore that adults who have adapted to their deformity should not be offered surgery. Great caution is urged in recommending surgery for those who do seek advice.

In teenage patients with single stiff proximal interphalangeal joints in otherwise normal hands, my attempts to restore motion by substitution with articulated metal or silicone prostheses have been uniformly unsatisfactory. These prostheses can withstand the relatively weak forces passing through a rheumatoid hand but should not be expected to resist the major stresses within an adult hand. An additional adverse factor is the state of the extensor mechanism over the joint. In the few fingers that I have explored, I found the central slip to be smaller than normal and abnormally adherent to the joint capsule and dorsum of the proximal phalanx.

I believe it significant that most of our adult patients state they would not want their children to go through life with stiff fingers. Therefore, despite the fact that rigid digits are apparently minor handicaps, I hope methods will be devised to restore motion to the proximal interphalangeal joints of such stiff digits in the hands of children.

In children, the problem of providing movement for stiff digits cannot be solved by prostheses with intermedullary prongs or silicone stems because of the risk of damage to growing epiphyses. Although the radiographs appear to show a joint space (Fig. 8-12), at operation one encounters a solid bar of cartilage connecting the two phalanges, and no joint cavity can be found. For such cases it is useless to excise a portion of cartilage and expect joint movement to be restored. Hematoma fills the space and becomes rapidly organized, and the digit stiffens once again. Repeated failures have also resulted from attempts at various types of excisional arthroplasty with or without the use of interpositional autogenous tissue. Despite the fact that both flexor and extensor tendons were found in these digits, the patients were unable to maintain their early voluntary postoperative motion. Within 6 months the steadily increasing fibrosis prevented even passive motion.

Because of repeated dissatisfaction with conventional arthroplasty procedures, I have inserted a variety of different shaped inert materials between the bone ends (Fig. 8-13). The results have been uniformly unsuccessful because little motion has been retained in these artificial articulations. Despite the great satisfaction expressed by my patients and by their parents, I am not entirely convinced that the motion I have been able to provide is of any functional value despite its obvious psychological benefit.

Group II—symbrachydactyly. In these patients there may be significant shortening of the fingers, commonly because of shortening or absence of the middle phalanges (Fig. 8-14). The fingers are usually slim, stiff, and spike-shaped in the sense that they taper toward the tips. Relative lengthening can be gained by recessing the interdigital webs into the spaces between the metacarpal heads. Attempts to provide interphalangeal motion for these slender fingers will meet

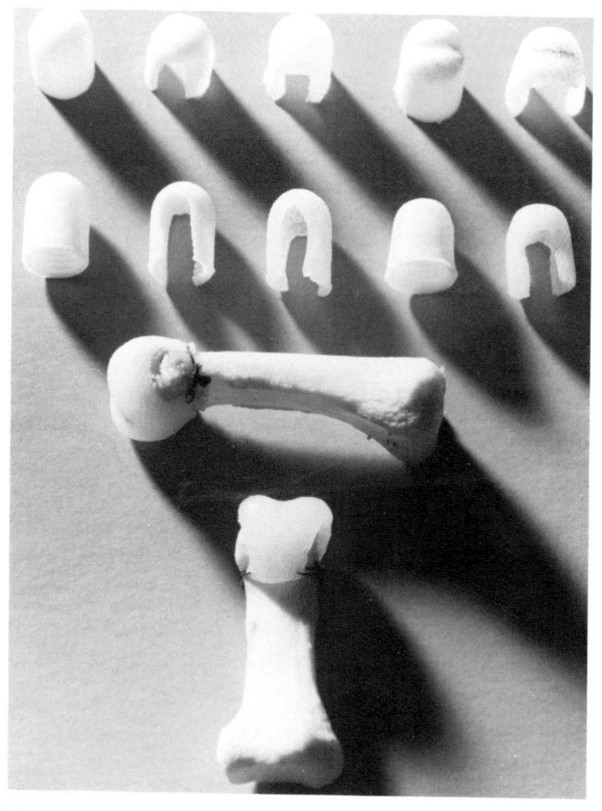

Fig. 8-13. *Rigid digit arthroplasty.* Restoration of controlled motion to these rigid digits is extremely difficult. I have tried a number of interpositioned forms of varied shapes, but the results have been uniformly poor.

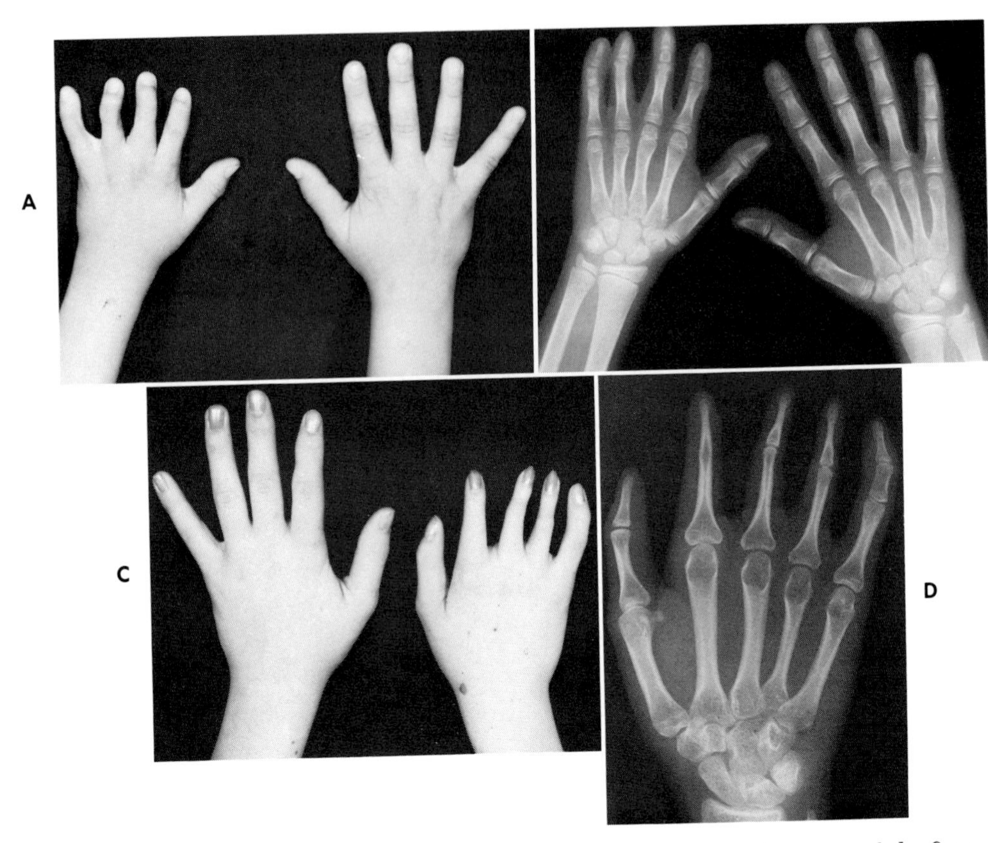

Fig. 8-14. *Symbrachydactyly.* These patients usually show significant shortening of the fingers because of shortening or absence of the middle phalanges. **A** and **B,** The left hand of a child. **C** and **D,** The right hand of an adult.

all the troubles encountered in the true symphalangia group; therefore, I encourage my patients to accept their handicap.

Group III—symphalangia with associated anomalies. Syndactyly is by far the commonest anomaly associated with hypoplastic, stiff fingers. The association is the rule in Poland syndrome, but webbing does occur alone. It can be partial or can involve all the fingers. These webbed, stiff, hypoplastic digits should certainly be separated early because the more movement provided for the hand, the greater the likelihood of growth. Although the hand will always remain relatively small, its use from an early age will allow as much growth as possible. Often a poorly developed, hypoplastic finger proves to be nothing more than a functional liability, and it may be wiser to plan its amputation at the primary operation. A properly functioning three-fingered hand is infinitely preferable to a poorly acting four-fingered hand. Occasionally one sees teenagers or adults who have not had their fingers separated, and often in their treatment it is wiser to establish a three-fingered hand using the extra skin to give good closure of flaps (Fig. 8-15).

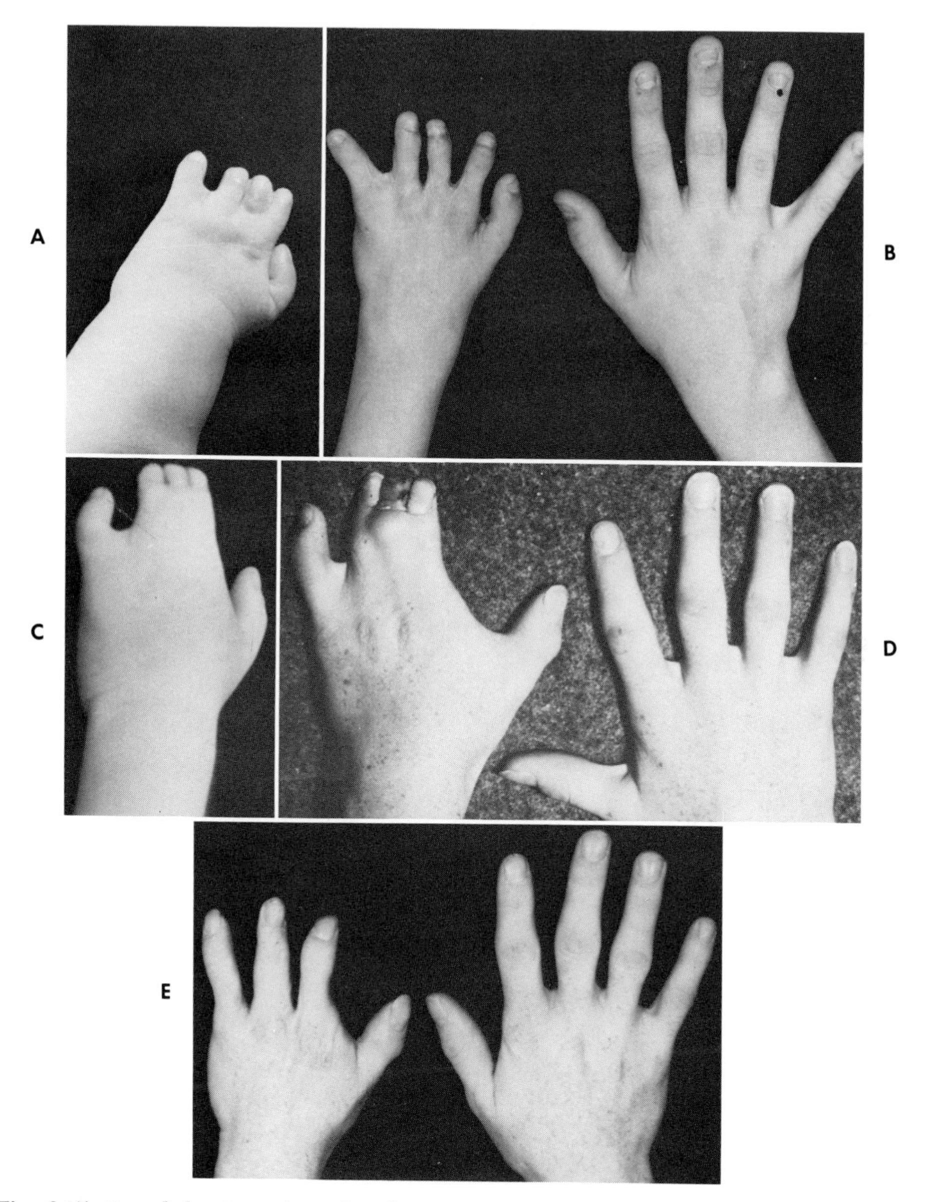

Fig. 8-15. *Symphalangia and syndactyly.* **A** and **B,** When these small hands are seen on children, separation at an early age will allow the maximum possible development. **C,** This patient was seen as a child, but the suggestion of surgery was not accepted until he was an adult, **D. E,** For this problem, a well functioning three-fingered hand was considered the best solution. (**A** and **E,** From Flatt, A. E.: Practical factors in the treatment of syndactyly. In Littler, J. W., Cramer, L. M., Smith, J. W., editors: Symposium on reconstructive hand surgery, vol. 9, St. Louis, 1974, The C. V. Mosby Co., pp. 144-156.)

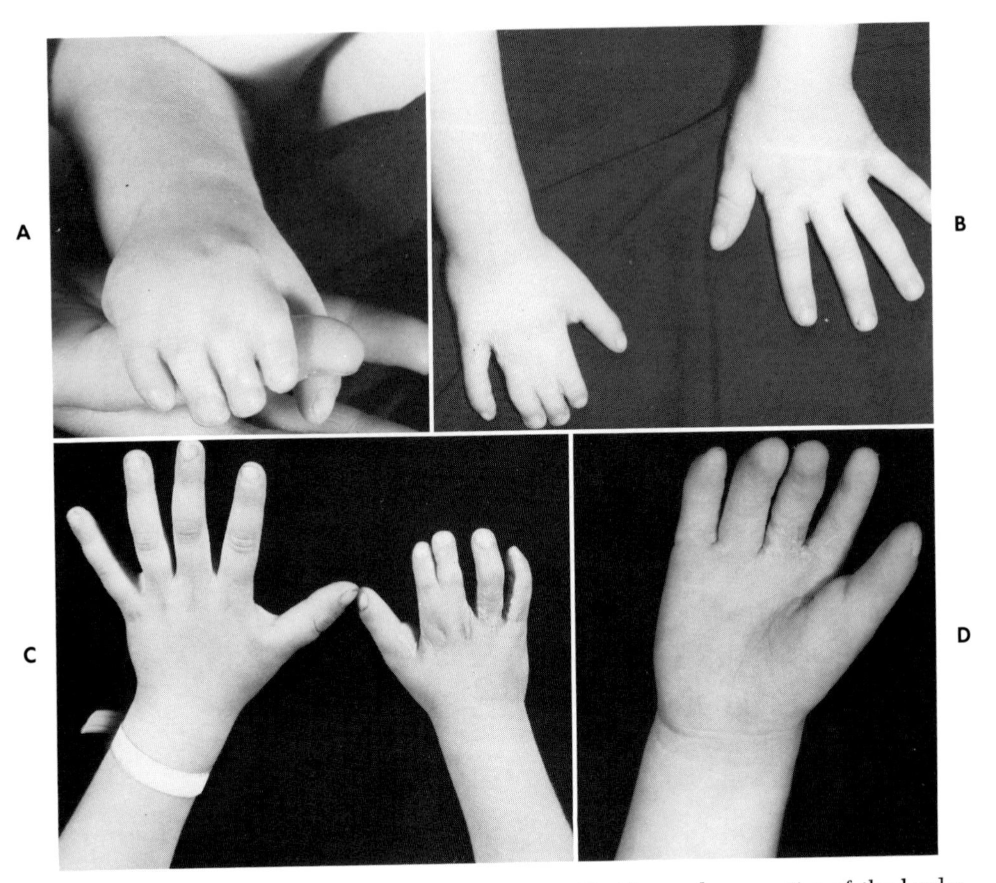

Fig. 8-16. *Hypoplasia in an infant.* **A,** Before surgery. **B,** After early separation of the border digits (thumb and small finger) to encourage use of the hand. **C,** Separation between long and ring fingers. **D,** Final state—all separations done before school entry age. Note the poor thenar muscles, tapered thumb, and lack of an interphalangeal flexion crease.

When the patient is seen as an infant or toddler, separation of the digits should certainly be done, applying the basic principles used in the separation of multiple syndactyly. Whenever possible the border digits should be separated early (Fig. 8-16). Division between the central mass can be delayed until later. The delay may have to be considerable if there is inadequate skeletal maturation. Failure of development or malformation of the middle phalanx is common in hypoplasia of the hand. Separation of all fingers in the presence of such problems might produce a floppy and even useless digit. Careful planning is therefore essential, and on several occasions I have decided to leave an inadequate finger adherent to a more normal neighbor thereby producing a broader but considerably more functional finger (Fig. 8-17).

STUBBY FINGERS

When the fingers are represented by short stumps containing at least part of the proximal phalanges and working metacarpophalangeal joints, the functional

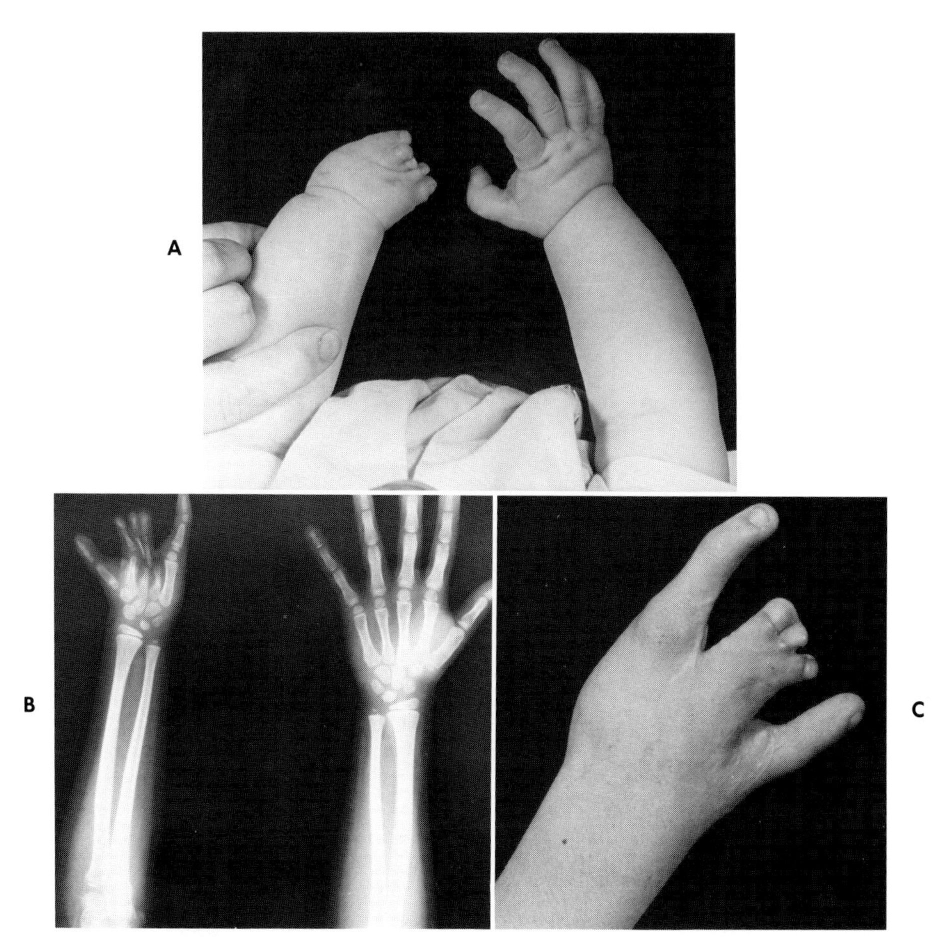

Fig. 8-17. *Hypoplasia and inadequate fingers.* The central three digits of this child's hypoplastic left hand were grossly underdeveloped. **A** and **B** show the state of the hand before surgery. **C,** The border digits were developed as far proximal as possible, but the central mass was left undisturbed to provide a broad but stable "middle" finger.

potential can still be encouraging. If, as is often the case, there is adjacent syndactyly between the fingers, then a good mitten hand is present. This type of hand has good sensory feedback and is an excellent scoop. The child uses it by pressing objects between the mitten and the body or other rigid structures (Fig. 8-18).

The combinations and permutations of shortening and webbing between the four fingers is almost infinite. Whatever the combination, it is probable that the functional potential can be increased by early separation of the fingers. This release of the short digits is helpful both functionally and psychologically; I almost invariably suggest that it be done (Fig. 8-19).

It is tempting to consider transposing one short stump on top of another, thereby lengthening selected rays. This somewhat dramatic surgery is pleasing to plan but hard to execute. I have seen a number of patients in whom a variety

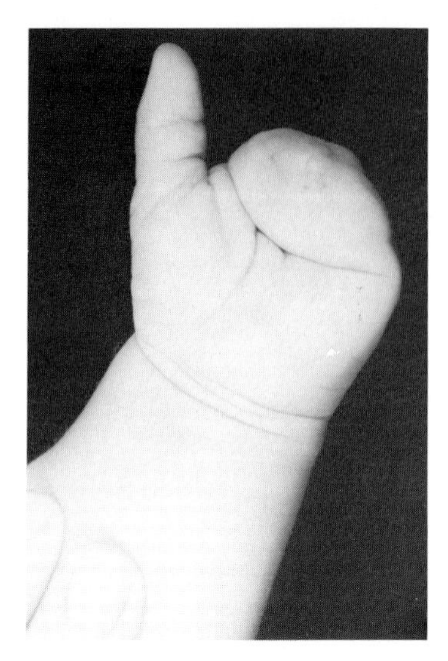

Fig. 8-18. *The mitten-scoop hand.* This type of hand, with only the basal portion of the proximal phalanges present, has good sensory feedback and is an excellent scoop.

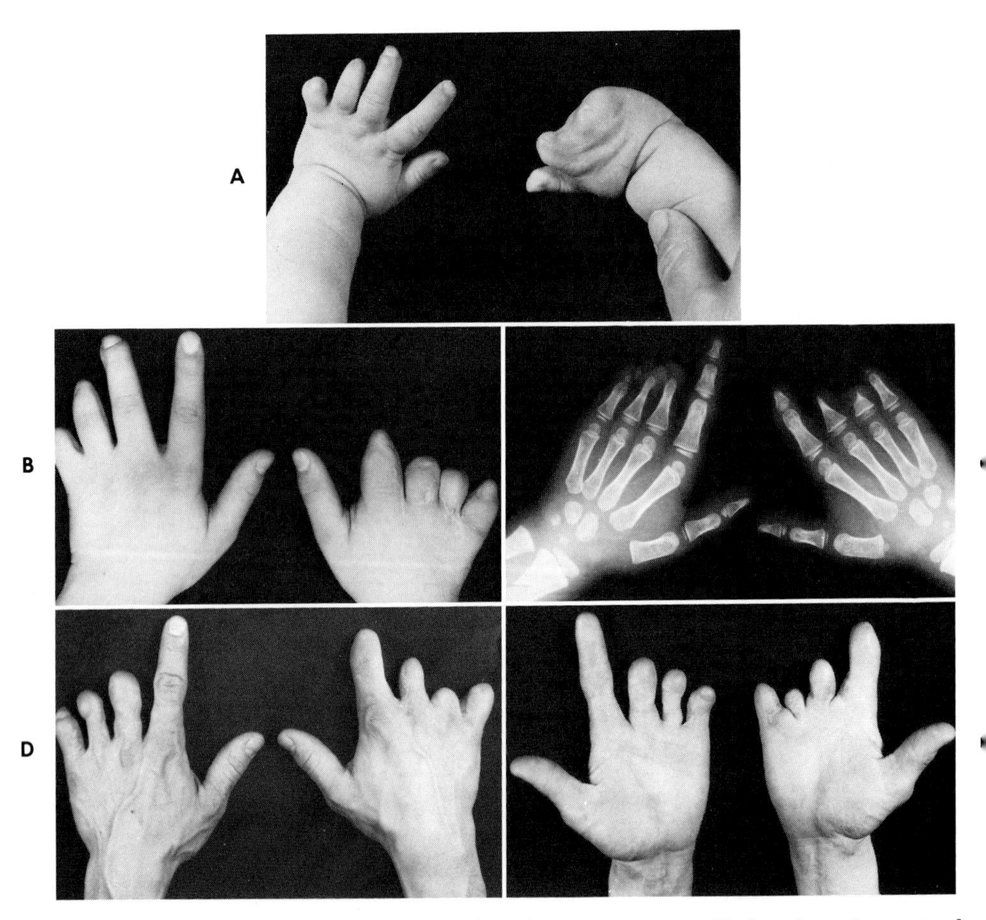

Fig. 8-19. *Stubby fingers—separation.* Providing there is any reasonable length to the proximal phalanges, early separation of the fingers will increase function. **A,** Before surgery. **B,** After separation of all the fingers. **C,** Another patient in whom all fingers were separated at an early age. **D** and **E,** The separations have allowed maximum independent growth of all the fingers into adult life.

of these heroic procedures have been done, but very few had a demonstrable increase in function.

It is certainly worthwhile to consider separation of the border digits to increase the span of the hand, thereby improving the ability to handle spherical objects. I doubt very much that the average surgeon should contemplate some type of "on-top plasty" in which, say, the ring finger stump is placed on the distal end of the small finger. Technically this can be done by way of a neurovascular pedicle or a skin subcutaneous neurovascular flap. Dobyns believes that the concept should be considered in any hand in which there are two or more short but partial digital segments. He believes it is particularly useful in rebuilding the functioning struts of the first web space. If the index and long fingers are of only partial length, a transfer of the index stump onto the end of the long finger would create a wider first web space and a longer ulnar component of grasp. The fundamental technical problem in these transfers is the maintenance of good circulation. To relieve tension on the neurovascular bundles it is customary to place the metacarpophalangeal joint in flexion. More often than not, this leads to a significant flexion contracture that may never be improved even by prolonged splinting. To avoid the risks of such flexion contractures one must do a very extensive dissection and mobilization of the neurovascular bundles in the palm. This demands considerable experience and excellent judgment; however, the incidence of these deformities is low and few surgeons can gain the necessary expertise.

An attractive alternative method for gaining length would be to insert bone into the nubbins or short stumps and thereby push the skin distally. Like many attractive propositions, this concept has liabilities. Elongation of the skin cylinder by inserting bone must mean that internal pressure will be applied to the distal tip of the skin by the end of the bone graft. The dilemma is that the shorter the insert, the better will be the circulation at the tip, but the functional gain will be correspondingly reduced. Again I can only counsel caution in the use of this concept; I have seen a depressing number of "lengthened" digits with a quiet necrosis at their tip or with a dried tip of bone protruding out of a reddened, granulating ring of skin. There is no doubt that the operation is feasible, and a number of surgeons have shown good results. The problem is that good judgment is vital in estimating the length of the graft to be inserted. The only way a surgeon can gain good judgment is by learning from a lot of bad judgments.

To obtain maximum length at minimum risk some surgeons prefer to repeat the operation at intervals, thereby gaining a greater ultimate length than would be possible at one operation. Iliac crest and toe phalanges have been used as bony inserts, but iliac crest tends to be absorbed over the years. The toes make good donor sites, and proper removal of phalanges creates little or no disturbance in the foot. Carroll has stressed the importance of careful technique in removing toe phalanges. He recommends taking the proximal phalanges from the toes in the sequence 5, 4, and 3. To avoid gait disturbance the medial two toes should not be used as donors. Usually the proximal phalanx is of adequate size any time after the age of 1 year. The toe is approached from the dorsum, the extensor tendon and the periosteum are divided longitudinally, and the periosteum is carefully peeled off the phalanx. The phalanx is removed together with its epiphysis and implanted in the finger. The periosteal tube should be closed with No. 5-0 catgut and the skin with interrupted No. 6-0 catgut sutures. Long-term follow-up shows that the hematoma in the periosteal tube organizes and occasionally bone is formed again. The functional and cosmetic results with donor toes are very satisfactory.

The recipient finger is opened through a dorsal approach, usually curved or zigzag in shape. The incision can be carried proximally over the bone, but distally it must not encroach on the end of the finger. The end or cap of skin must be left intact to receive the distal end of the toe graft. The phalanx should be held in place with a fine Kirschner wire passed through the tip of the finger and into the proximal bone. Carroll reports that in his patients the epiphysis usually shows growth for up to 2 years before it slows down and stops.

The most satisfactory way to judge the correct length of bone graft in relation to circulation in the skin cylinder is to deflate the tourniquet after the graft has been pinned in place. If blanching of the skin tip occurs and does not recede over several minutes, the graft must be shortened until good pinking occurs and remains present around the shaft of the protruding Kirschner wire. I usually bend the tip of the Kirschner wire into a rounded hook to prevent it migrating proximally and leave it in place 6 to 8 weeks.

ABSENT FINGERS

Any part of a hand may be missing. Ectrodactyly is used as the descriptive term; however, Barsky prefers the term partial ectrodactyly when only parts of the metacarpal or phalanges are missing. All the phalanges may be absent as well as the metacarpal of a single digit, or several fingers may be partially or completely affected. When all the fingers are absent, the hand looks like a little paw with hypoplastic metacarpals supplying the skeletal support. Sometimes the fingers are represented by small nubbins that may even carry fingernails (Fig. 8-20).

X-ray films of infants' hands do not always show the true skeletal potential because the bony elements, which may be felt on clinical examination, are still cartilaginous. Ossification of these structures is often delayed, and reconstructive or ablative surgical plans should be deferred until the full bony pattern has revealed itself. Absence of the border digits, thumb and ring and small fingers is usually associated with radial or ulnar absence or hypoplasia. Absence of the central rays also occurs and is usually considered a type of cleft hand, which is discussed in Chapter 14, page 265.

Complete absence of a digit or even the whole hand has been wrongly termed uterine amputation. Acceptable evidence that a digit or limb existed in utero and was subsequently amputated is extremely rare and has been reported more in the lower than the upper limbs. I believe that these cases usually should be con-

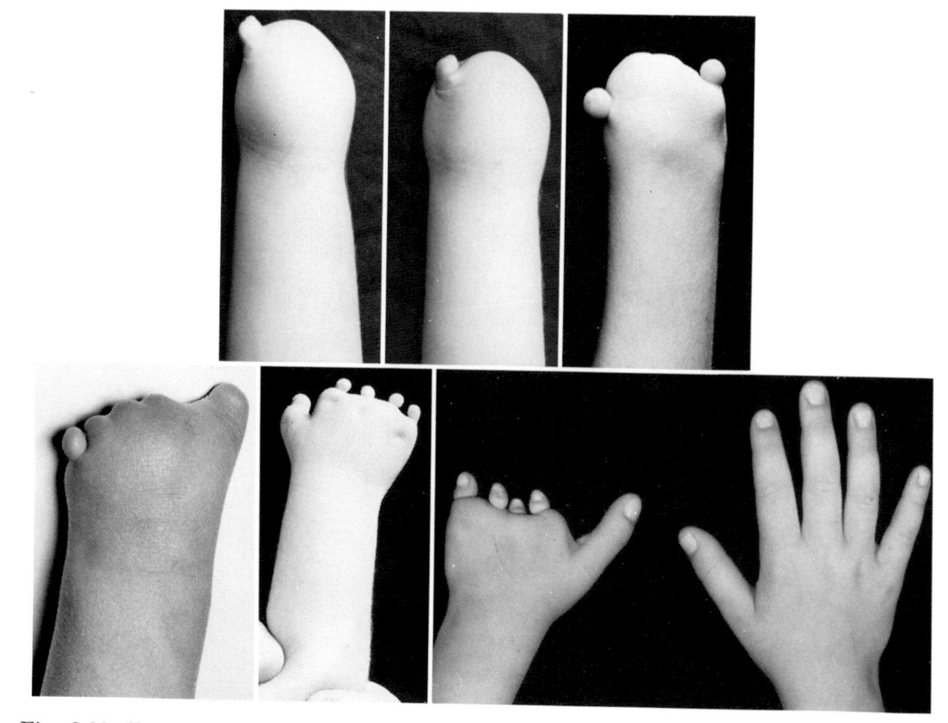

Fig. 8-20. *Finger nubbins.* A variety of functionally useless and cosmetically offensive finger nubbins. Removal of functionless rudimentary digital buds to produce what appears to be a simple amputation stump is, I believe, proper treatment.

sidered examples of failure to develop, or transverse deficiencies, of the affected parts.

There is virtually no treatment that can replace these missing parts. Finger nubbins that have no skeletal support are not functional and are probably better removed for cosmesis. The flexor tendons and occasionally the extensor tendons may reach into these nubbins so the child can develop a "party trick" of moving or retracting the nubbins proximally. This activity sometimes deludes the parents into unjustified hopes that purposeful function may yet develop.

I believe that in the present day atmosphere of greater scientific awareness, obvious congenital malformations are becoming less socially acceptable. I agree with Littler's recommendation that it is wise to consider altering "a stigma of congenitalism"* to make it appear to be the result of external trauma. Not all deformities lend themselves to this correction, but removal of functionless, rudimentary digital buds to produce what appears to be a simple amputation stump is often a kindness. Certainly the public will accept an amputation stump much more readily than a bizarre deformity.

*Littler, J. W.: Introduction to surgery of the hand, Reconstr. Plast. Surg. **4**:1543-1546, 1964.

CHAPTER 9

CROOKED FINGERS

Any of the digits of the hand of a newborn may show deviations away from their normal alignment to the palm. In the palmar plane, digits may be malaligned because of bone or joint abnormalities. Extrinsic tendon imbalance also leads to deviations at the metacarpophalangeal joint including frank ulnar drift. Hyperextension and flexion deformities also occur on the radial side of the hand. Flexion contractures of the thumb can result from congenital clasped thumb or trigger thumb. Such causes are not the subject of this chapter. We are concerned here with skeletal and soft tissue changes that occur largely on the ulnar side of the hand and particularly in the small finger.

The deformity usually occurs in a palmar, radial, or radiopalmar direction. Many descriptive terms have been used—camptodactyly, campylodactyly, clinarthrosis, clinodactyly, delta phalanx, dystelephalangy, dropped finger, familial finger contracture, hammer finger, Kirner deformity, and streblomicrodactyly. From all these choices, a few terms have found general acceptance. A palmar curvature is usually called camptodactyly and a radial curvature, clinodactyly (Fig. 9-1). Clinodactyly may be caused by a delta or longitudinally bracketed phalanx. The combined palmar and radial curving of the terminal

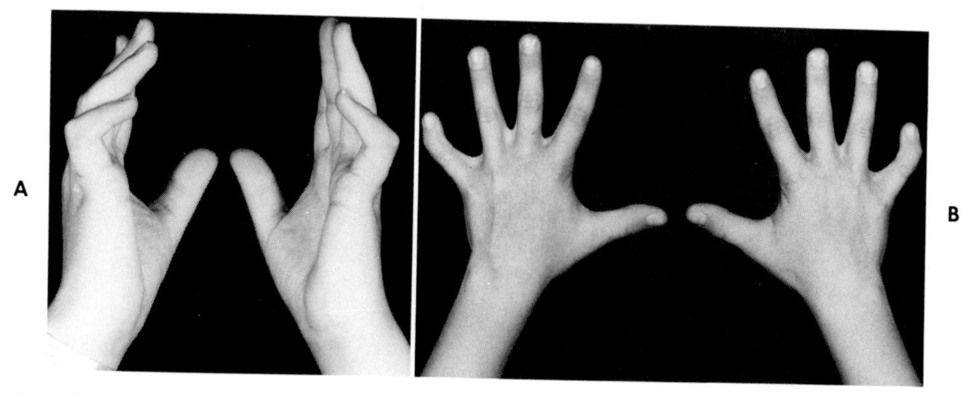

Fig. 9-1. *Camptodactyly and clinodactyly.* **A,** The typical flexion deformity of the proximal interphalangeal joint seen in camptodactyly. **B,** Radially deviated small fingers typical of clinodactyly.

phalanx was described by Kirner in 1927, and this deformity usually carries his name.

Many of these deformities are minor and need no treatment, but the importance of the condition is that its presence may indicate more serious, generalized, congenital deformities.

Camptodactyly is frequently associated with skeletal dysplasias and malformation syndromes. It is found in a variety of syndromes such as orofaciodigital, focal dermal hypoplasia, cerebrohepatorenal, Marfan, Holt-Oram, Poland, popliteal pterygium, and trisomy 13 and in craniocarpotarsal dystrophy and oculodentodigital dysplasia.

Clinodactyly is associated with a large number of conditions; one publication lists over 30 associated syndromes and 11 chromosomal disorders, mainly trisomy defects. The more commonly known conditions are Silver, Treacher Collins, Laurence-Moon-Bardet-Biedl, and orofaciodigital syndromes and oculodentodigital dysplasia.

Kirner deformity is occasionally seen in association with Silver and Cornelia de Lange syndromes. Delta phalanx does not have any regular associations with other abnormalities although it has been occasionally observed in Carpenter, Holt-Oram, and Rubinstein-Taybi syndromes.

CAMPTODACTYLY

The Greek word camptodactyly means "bent finger," and the deformity it describes characteristically occurs at the proximal interphalangeal joint of the small finger (Fig. 9-2). Other fingers may be affected, but the incidence of their involvement and the extent of the flexion contracture rapidly decrease toward the radial side of the hand (Fig. 9-3). The metacarpophalangeal joint occasionally shows hyperextension deformities, more particularly when the flexion of the proximal interphalangeal joint is severe. The degree of flexion of this joint varies from very mild to about 90 degrees of flexion, and the extent of flexion may

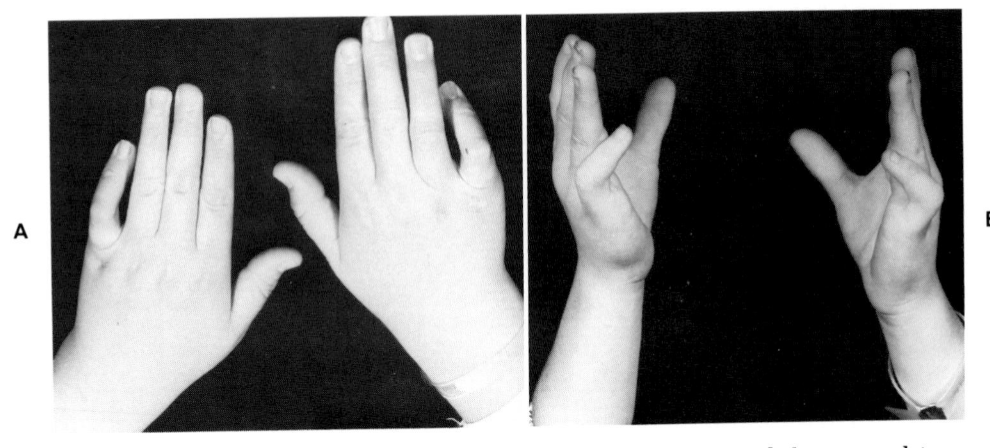

Fig. 9-2. *Camptodactyly.* **A,** In this deformity there is inability to extend the proximal interphalangeal joint although flexion beyond the deformity is possible. **B,** There may be accompanying hyperextension of the metacarpophalangeal joint as in this patient's left hand.

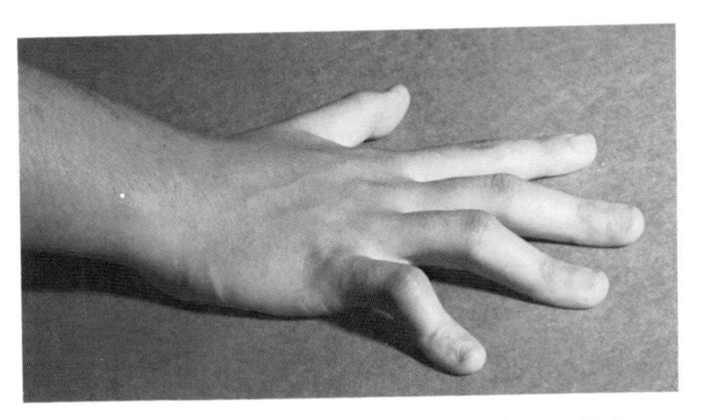

Fig. 9-3. *Camptodactyly.* Occasionally fingers other than the small fingers are involved; the extent of the flexion contracture decreases towards the radial side of the hand.

steadily increase as the child gets older (Fig. 9-4). It is rare for the distal interphalangeal joints to be affected, but if these joints are involved, the deformity progresses with age (Fig. 9-5).

There is controversy in the literature as to whether there are two types of camptodactyly. We at the University of Iowa have recently reviewed 110 hands of 66 affected patients, and the results of this survey lead me to side with those who believe that there are two general types, one that appears in infancy and affects males and females equally and another, which is probably less common, that usually first appears in adolescent females. If left untreated, the natural history of the condition is one of no improvement or of gradual progressive worsening in over 80% of cases.

The deformity often accelerates during the teenage growth spurt, but it ceases in young adulthood. Even in severe deformities there is surprisingly little func-

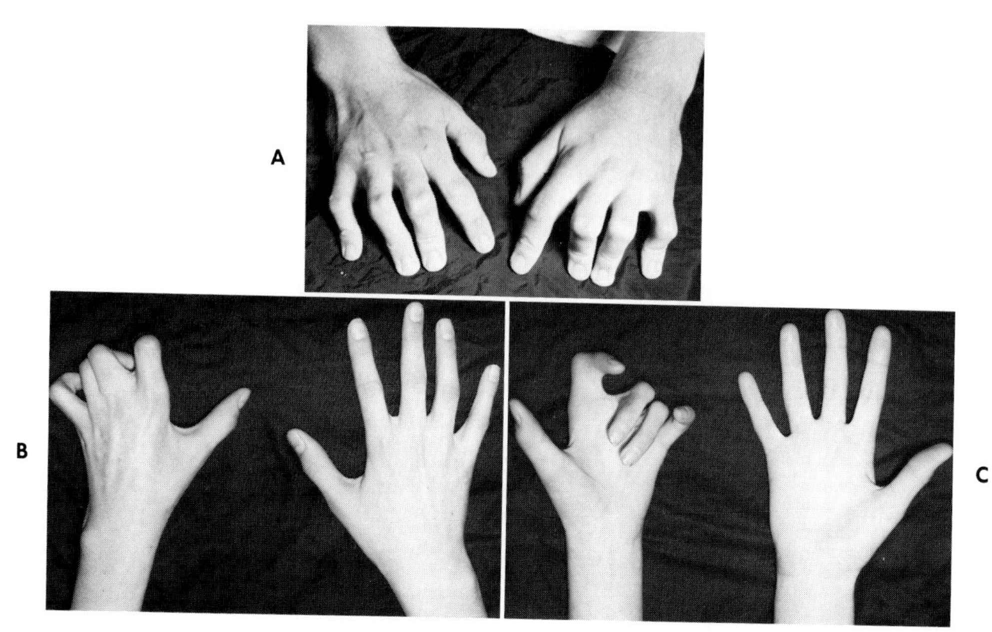

Fig. 9-4. *Severe camptodactyly.* **A,** Marked involvement of the proximal interphalangeal joints of all fingers but with the index being the least involved. **B** and **C,** Extreme involvement of the left hand with a normal right hand.

tional impairment (Fig. 9-6). The deformity is said to occur in less than 1% of the population. In familial cases there is a simple autosomal dominant pattern of inheritance, but many other cases are sporadic in occurrence.

Etiology

The pathogenesis of the condition is unknown, and a great variety of abnormalities have been thought to produce the deformity. It has been said to be caused by an imbalance between flexor and extensor tendon forces, by an abnormal insertion of the lumbrical muscle, by circulatory disturbances, by shortening and thickening of the collateral ligaments, by a vestigial or anomalous flexor digitorum superficialis or flexor digitorum profundus, or by faulty development of the dorsal extensor aponeurosis over the joint. In fact, as Smith and Kaplan have pointed out in their excellent paper, "virtually every structure about the base of the finger has been implicated as a determining factor."*

I do not believe there is one single etiological factor and have in fact found examples of several of the suggested causes at different operations.

Camptodactyly is really a clinical sign, possibly part of a syndrome, but is never a disease in its own right. Mild degrees are usually ignored by the child

*Smith, R. J., and Kaplan, E. B.: Camptodactyly and similar atraumatic flexion deformities of the proximal interphalangeal joints of the fingers, J. Bone Joint Surg. (Am.) **50:**1187-1204, 1968.

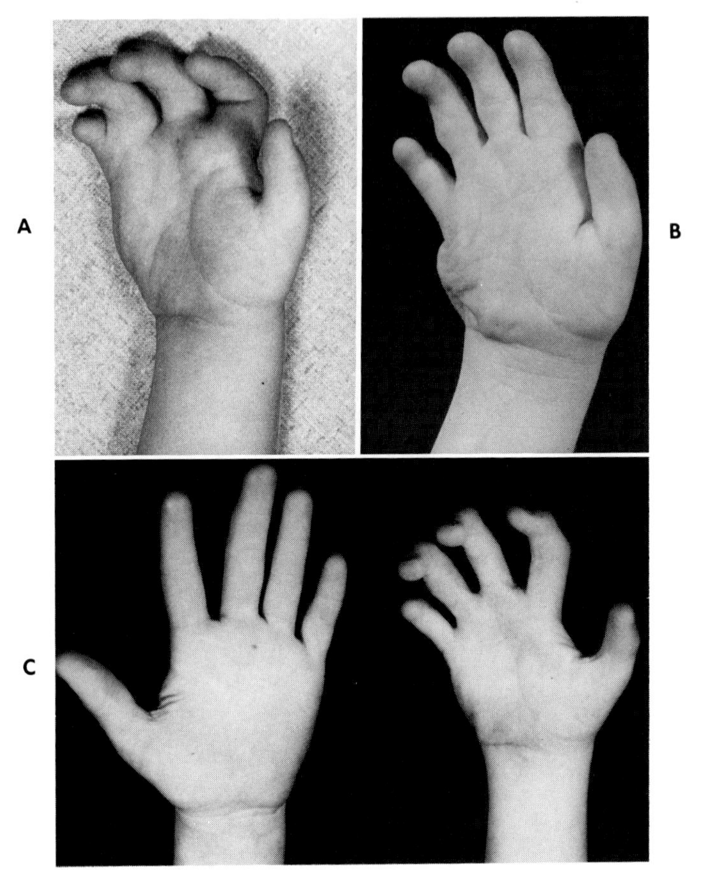

Fig. 9-5. *Camptodactyly—distal joints.* The distal interphalangeal joints are rarely involved, but when they are, the deformity increases with age. **A,** Age 4 years. **B,** Age 8 years. **C,** Age 12 years.

and often by the parents. More often than not the reason for consultation in the mild case is aesthetic rather than functional. In the more severe cases the flexion deformity can be a definite hindrance both in sports and in occupations such as typing or playing musical instruments.

Usually the date of onset is not known because the contracture is symptomless and its progression is insidious. The more severe of the cases appearing in infancy usually are seen in preschool children and are commonly bilateral. About two thirds of all patients show bilateral involvement, but the degree of contracture is not necessarily symmetrical.

Differential diagnosis

Diagnosis is made, to a certain extent, by exclusion. The differential diagnosis is limited, and the possible conditions are unlikely in children. Dupuytren contracture should be considered, but it is virtually unknown in a child and usually carries a strong family history. Boutonniere deformity is a possibility, but the lack of a

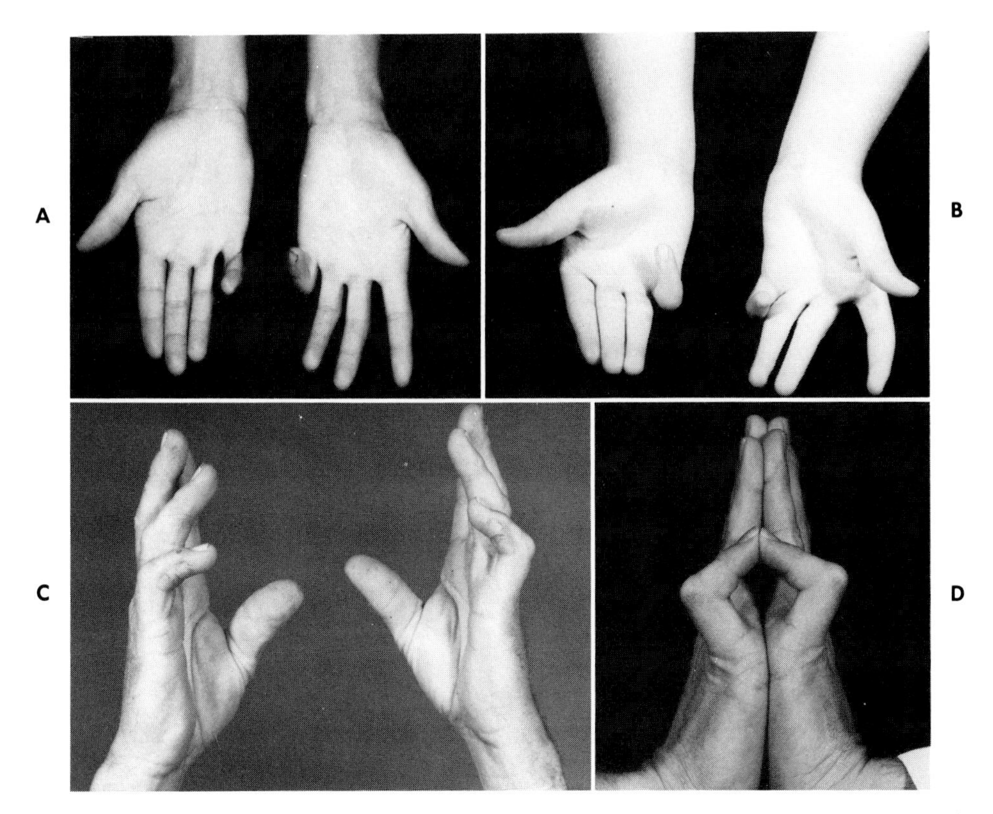

Fig. 9-6. *Camptodactyly and function.* Even in severe deformity there is little functional deficit. **A,** A teenage boy who "had a little trouble in baseball." **B,** A teenage girl who had difficulty learning to type. **C,** A farmer and, **D,** a professor of surgery, neither of whom would admit to any difficulty in using their hands.

history of trauma and lack of a swollen and possibly painful joint should exclude it.

Conditions such as trigger digit or congenital absence of the extrinsic extensor mechanism primarily affect the metacarpophalangeal joint rather than the proximal interphalangeal joint. More generalized conditions, such as mucopolysaccharidosis, juvenile rheumatoid disease, collagen disorders, and arthrogryposis, can be excluded by a thorough physical examination.

When examining the hand, flexion of the wrist or wrist and metacarpophalangeal joints together usually relieves the proximal interphalangeal joint flexion in younger patients. However, by the time they are teenage or older, their flexion contractures do not usually decrease with this maneuver.

When the joint cannot be fully straightened passively but does show active extension against resistance, then the primary pathology probably lies on the flexor aspect. However, recent literature stresses that some cases are really in an intrinsic minus posture and are not directly caused by abnormalities of the lumbrical or flexor superficialis.

A lateral x-ray film may show characteristic changes on either side of the joint (Fig. 9-7). The head of the proximal phalanx will often be flattened or narrowed and the base of the middle phalanx will be subluxed so that it lies beneath and proximal to the tapered head. In long-standing cases the palmar aspect of the base of the middle phalanx seems to create a pressure groove on the neck of the proximal phalanx.

Treatment

There is no consistently successful treatment because there seems to be no one single cause of the condition. For every paper that reports success with a particular procedure, another can usually be found reporting gross failure with the same method.

Conservative. My own approach is to try to persuade the patient to live with the deformity if at all possible. There are two age groups in which this approach is impractical: the young child with a strong family history and marked contracture and the teenager with a rapidly increasing contracture during the growth spurt years. Both need more than pious advice. Neither will get permanent benefit from splinting—either passive or dynamic; the problem is that in my experience they also do not greatly benefit from surgery.

Operative. In patients who show significant changes on lateral view x-ray films, soft tissue surgery will yield little or no improvement. If the deformity is marked, fixed, and creates a functional hazard, the only operation that can give

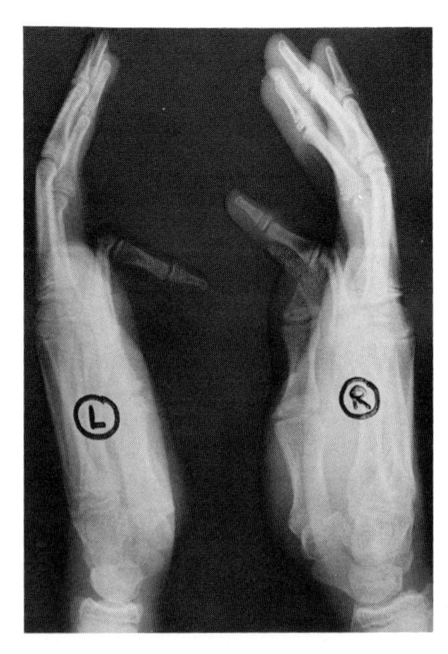

Fig. 9-7. *Camptodactyly and joint changes.* The lateral x-ray film will show the characteristic joint changes. In the left hand, the base of the middle phalanx has created a groove on the neck of the proximal phalanx. In the right hand, the head of the proximal phalanx has been flattened and narrowed.

any postural improvement is an angulation osteotomy of the neck of the proximal phalanx (Fig. 9-8). Oldfield originally recommended this in 1956, and Moberg has recently advocated the same concept for the acutely flexed finger in Dupuytren contracture so that amputation can be avoided.

The operation is far from curative; by dorsally angulating the distal portion of the proximal phalanx, the surgeon can make the finger appear straighter and retain the motion previously present in the proximal interphalangeal joint. The motion is now in a different arc, and some tightening of the ulnar grasp is sacrificed.

I regard this operation as a last resort but better than amputation. Technically, it is not easy to control the osteotomy site, and I recommend excising the dorsal wedge by defining it with multiple drill holes and obtaining correction by fracturing the palmar cortex and leaving its periosteal hinge intact. The correction will have to be maintained by fine Kirschner wires, which I usually bring out percutaneously and leave in place 4 to 6 weeks.

In patients in whom there are no obvious changes in the joint on x-ray examination, soft tissue operations can be attempted to obtain improvement. A preliminary treatment of several months of splinting can be tried, and every attempt should be made to obtain passive correction before the operation is performed. Splinting must be carefully applied, and it is difficult to do this well. Safety pin spring splints, ratchet type splints, or even plaster of Paris casts can be used, but with all these types great care must be taken to obtain the three points of application: the dorsum of the joint and the palmar aspects of the proximal and middle phalanges. The terminal phalanx must not be used for a purchase point because this will force the distal interphalangeal joint into hyperextension.

It is not always possible to get full passive extension prior to surgery, and one is then faced with two problems: release of the joint and its surrounding structures and exploration more proximally to relieve any anatomical abnormalities that may be present.

I usually approach the joint on the palmar aspect and plan my incisions so that either a Z-plasty or a diamond-shaped full-thickness graft can supply skin cover for the straightened joint after the deeper structures have been relieved.

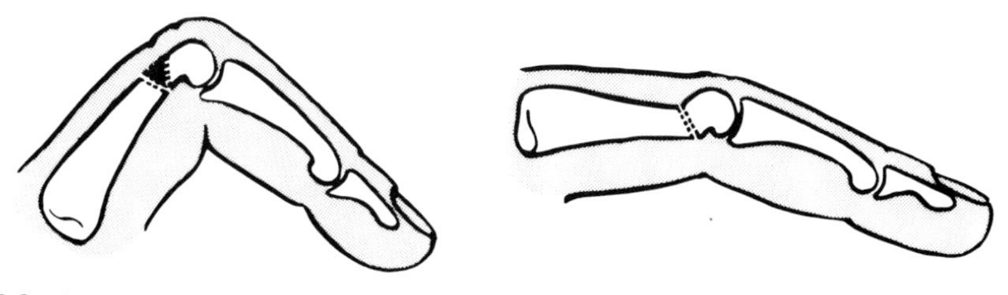

Fig. 9-8. *Camptodactyly—osteotomy of phalanx.* In a fixed deformity, a dorsal angulation osteotomy of the neck of the proximal phalanx will give a postural improvement, but it will reduce the amount of closing of the small finger.

The more proximal exploration is done through a proximal extension of the incision onto the radial side of the finger and up onto the palm. The object of this approach is to identify any abnormality of the intrinsic apparatus or the extrinsic flexor tendons.

Release of the superficialis tendon may be sufficient, but its use as an intrinsic transfer to enhance extension is becoming increasingly popular and makes good sense if the metacarpophalangeal joint has been in hyperextension a long time.

Smith and Kaplan have published an excellent account of their operative procedures for this condition, and in 12 of 14 fingers in which the flexor superficialis tendon was released, the flexion contracture was decreased by at least 33% and there was no loss of excursion or grip strength. There was no difference found in the results whether the tenotomy was performed in the finger, the palm, or the wrist.

No one surgeon will ever accumulate sufficient experience to publish definitive recommendations. At this time my own experience leads me to recommend surgery at a young age, and I would caution against surgery other than release of the superficialis and its possible transfer to the extensor apparatus. Surgery should certainly not be done until after a prolonged period of splinting; and the splinting will certainly have to be continued for many months after surgery. Some experienced hand surgeons recommend 1 to 1½ years of postoperative splinting.

CLINODACTYLY

Bending or curvature of the finger in the radioulnar plane is usually called clinodactyly. It can occur in any finger but is very common in the small finger. In fact, minor degrees of curvature of the small finger in the radial direction are so common as to be considered normal; at least I think they should be since both my small fingers show a 10 degree radial inclination.

Its incidence is hard to define because there is no precise definition of the limits of normal. Probably clinodactyly that is obvious to anyone other than the owner of the affected finger should be considered uncommon. Reports of its incidence must vary with the zeal of the examiner, because its presence in mongoloids has been recorded variously from 35% to 79%, and in normal children its incidence is reported as between 1.0% and 19.5%.

These inclined fingers can be inherited, and the trait appears to depend on an autosomal dominant gene.

The basic cause for the condition is a shift in alignment of the joint surfaces of either interphalangeal joint away from their normal 90 degrees to the long line of the digit (Fig. 9-9). This shift is usually caused by maldevelopment of a phalanx; the middle phalanges are said to be the most commonly involved because they ossify last and there is a greater tendency for shortening in those segments of a ray whose diaphysis ossifies late. Minor developmental abnormalities of the middle phalanx can produce a bent or inclined finger, but the phalanx will not be significantly shortened (Fig. 9-10). This type of clinodactyly is thought to occur with equal frequency in normal and mongoloid children.

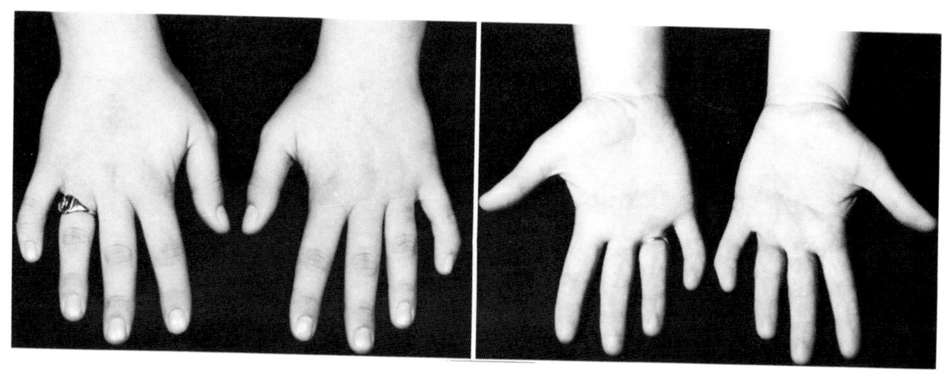

Fig. 9-9. *Clinodactyly—small finger.* The radial inclination of the small fingers is produced by a shift in alignment of the interphalangeal joint surfaces away from their normal 90 degrees to the line of the digit.

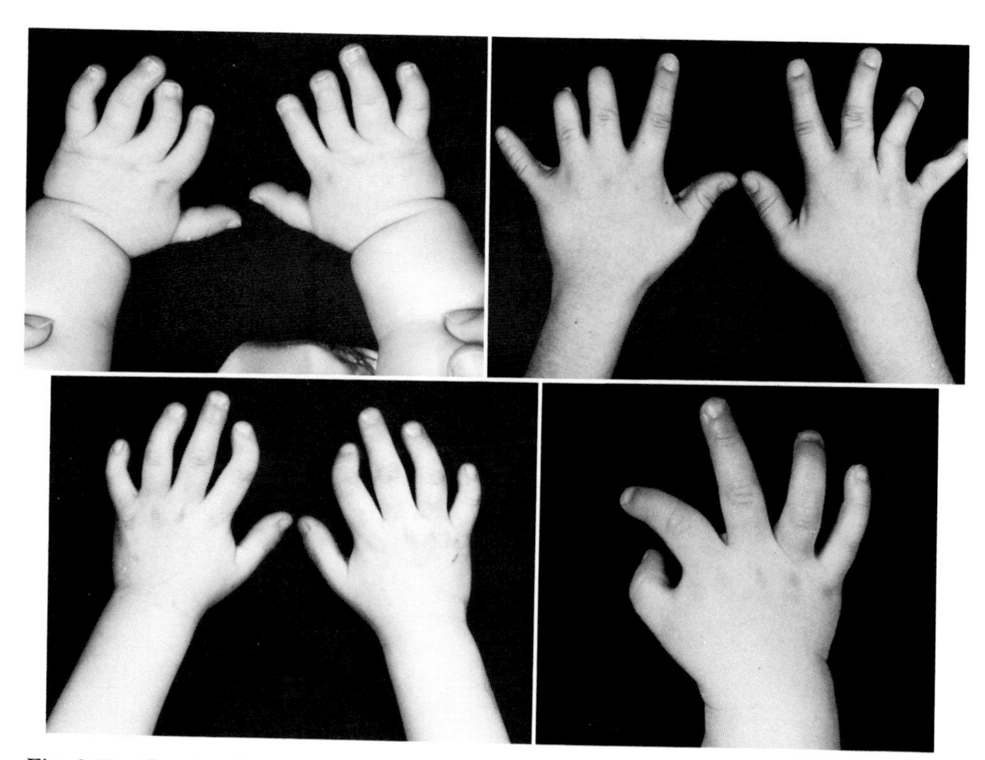

Fig. 9-10. *Clinodactyly—several fingers.* Fingers other than the small fingers can be involved; the deformity is usually produced by developmental abnormalities of the middle phalanx. Examples of multiple digit involvement are shown.

True shortening of the middle phalanx, brachymesophalangia, also occurs and is said to be present in at least a quarter of mongoloid children. Shortening of the middle phalanx only occurs in about 3% of normal children. This shortening of an otherwise normal phalanx must be clearly distinguished from an abnormal delta phalanx in which the triangular shape of the bone can cause a severe deviation of the finger (Fig. 9-11). Marked incurving of the terminal portion of the finger is usually produced by Kirner deformity in which there is a symmetrical bowing of the terminal phalanges (see pages 162 to 163).

Most cases of true clinodactyly are more a cosmetic than functional disability. Parents may insist that corrective procedures be done even in mild cases, but until definite overlapping of the finger occurs during fist making, I do not feel that surgery is justified. I believe that splinting the finger is clumsy and ineffectual and does more good to the parents' cerebrum than the child's finger. Usually one can avoid surgery in fingers in which the middle phalanx is generally well formed. If, however, there is significant angular deformity, then a closing wedge osteotomy taken from the convex, and usually ulnar, border of the finger will give significant correction (Fig. 9-12). The osteotomy site should be held for

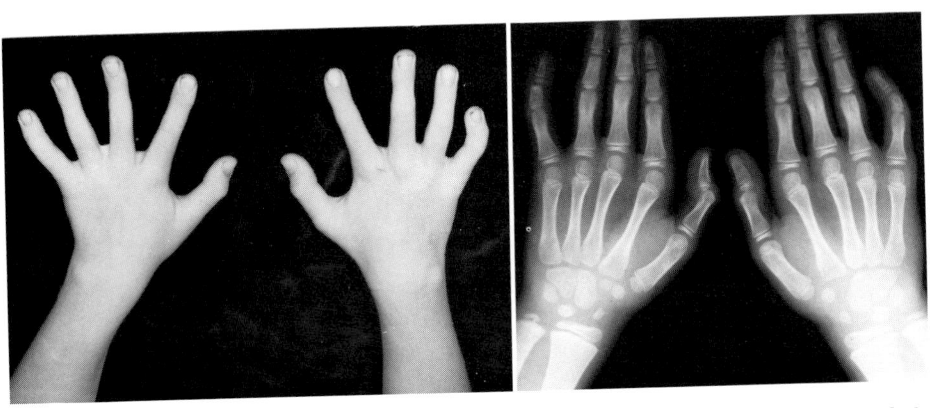

Fig. 9-11. *Clinodactyly—delta phalanx.* In this patient's left small finger there is a slight abnormality at the base of the middle phalanx. On the right side, a classical delta phalanx replaces the middle phalanx of the small finger.

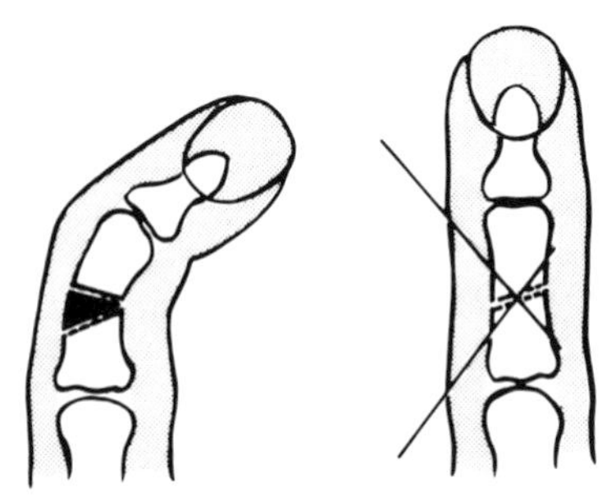

Fig. 9-12. *Clinodactyly—osteotomy.* A lateral closing wedge osteotomy on the convex, and usually ulnar, border of the finger will correct the curvature.

about 1 month with Kirschner wire fixation. The wires should be placed from the ulnar side of the finger in a horizontal plane so they do not penetrate the dorsal extensor mechanism.

Delta phalanx

The Greek letter Δ (delta) is triangular in form, but the abnormal bone that has been given this label is usually trapezoidal in shape.

This abnormality tends to occur in tubular bones having a proximal epiphysis. Their epiphysis is abnormal in the sense that it extends from its normal transverse position along the side of the phalanx in a proximal to distal direction. It is usually C shaped rather than straight and tends to lie on the shorter side of the abnormal bone (Fig. 9-13). This peculiar arrangement of the epiphysis makes longitudinal growth of the digit impossible and progressive angulation inevitable. Ossification of the epiphysis proceeds from proximal to distal, producing an obliquity of the distal articular surface and a retardation of growth, especially on the involved side of the staplelike epiphysis, which result in a triangular deformity of the tubular bone with axial deviation of the segments distal to it.

There is disagreement in the literature about the occurrence of the delta phalanx. Some surgeons believe it is largely associated with polydactyly, others do not. We have recently reviewed 49 patients with 84 delta phalanges. They showed a wide variety of associated anomalies, of which the most common were syndactyly, polydactyly, symphalangia, and triphalangeal thumb. Only six of our patients had no associated abnormalities. Reports in the literature have implied that the deformity is not inherited. There were only two instances of proven familial occurrence, and these were mentioned by Jaeger and Refior. However, we found that 21 of our 49 patients (44%) had a strong family history of inheritance. Some surgeons believe this abnormality occurs exclusively in the

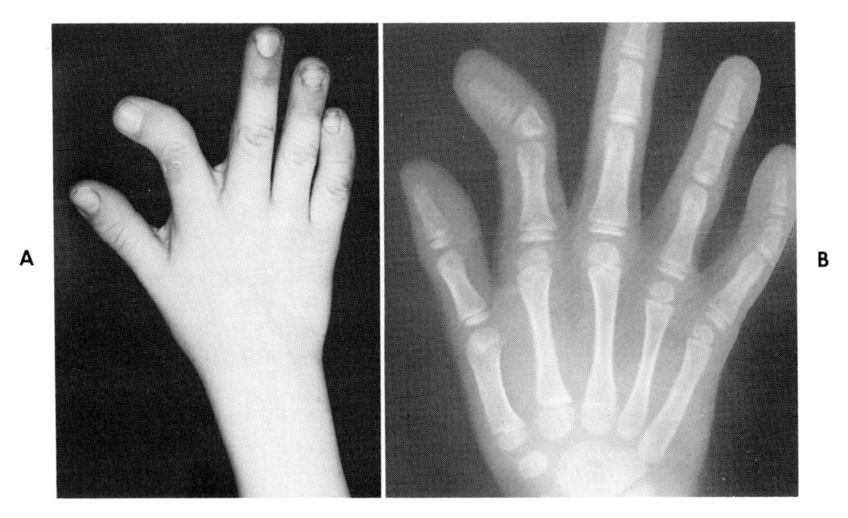

Fig. 9-13. *Delta phalanx.* **A,** The finger is angulated because the proximal epiphysis extends from its normal transverse position along the side of the phalanx. **B,** The epiphysis is C shaped and lies on the shorter side of the abnormal bone.

proximal phalanx; others champion the middle phalanx; I have seen it in both, although never in the same digit. In our series the middle phalanx was involved most often (Fig. 9-14).

Longitudinally bracketed epiphysis

In the early stages of development, before ossification of the epiphysis, it is impossible to define the shape and extent of the abnormal epiphysis. It will be obvious that there is a clinodactyly, which will be found to increase with age. As the patient is followed the abnormal development of the epiphysis can be seen on successive x-ray examinations. A bony bracket will be seen joining two epiphyses situated at opposite ends of the diaphysis. The term longitudinally bracketed diaphysis (LBD) has been coined by Theander and Carstam to describe this abnormality (Fig. 9-15).

Treatment

I believe this description of the abnormality to be correct, and it would seem therefore that early division of the longitudinal bracket, even in the cartilaginous

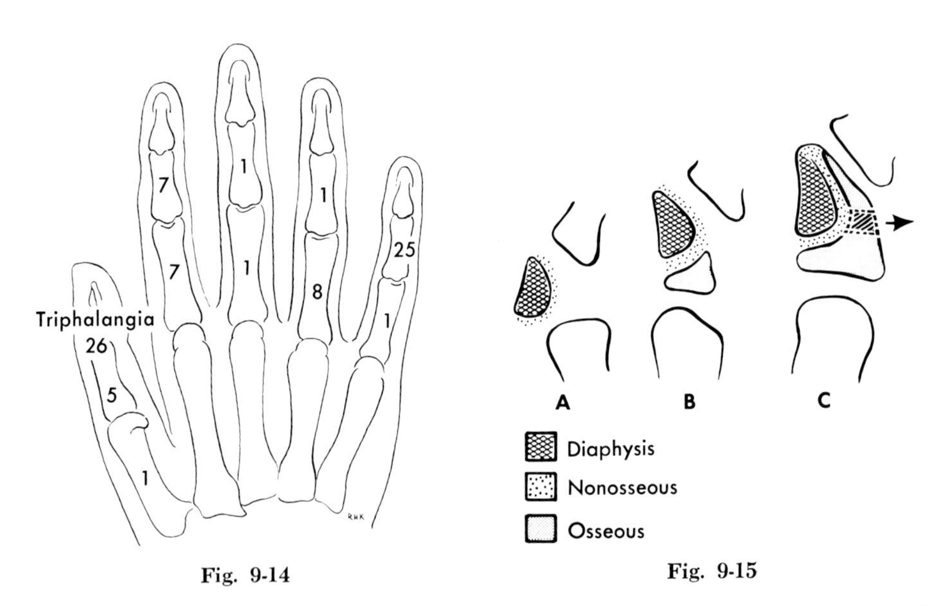

Fig. 9-14 Fig. 9-15

Fig. 9-14. *Delta phalanx incidence.* The distribution of delta phalanges occurring in the hands of 49 patients. The middle phalanx and the small fingers are most commonly involved in otherwise normal digits.

Fig. 9-15. *Longitudinally bracketed epiphysis.* Three stages in the development of a longitudinally bracketed epiphysis. **A,** The bracket is entirely cartilaginous at 7 months of age. **B,** Proximal and distal ossification centers appearing in the bracket at 19 months. **C,** At 8 years, the bony part of the bracket is still separated from the diaphysis by residual nonosseous tissue. The arrow and the crosshatched area show the site and extent of division of the longitudinal bracket. (Redrawn from Carstam, N., and Theander, G.: Surgical treatment of clinodactyly caused by longitudinally bracketed diaphysis, Scand. J. Plast. Reconstr. Surg. 9:199-202, 1975.)

stage, should be curative. I have not tried this, but unfortunately Carstam and Theander report that in the one case in which they used this procedure, the angulation was only reduced from 40 degrees to 30 degrees 1 year later. Osteotomy would therefore seem the operation of choice, and as soon as the shape of the LBD is clearly established, the procedure should be planned.

The longitudinal bracket must be cut and the underlying epiphysis destroyed on the short side of the phalanx. Lengthening can be achieved in two ways: an opening wedge osteotomy can be done and bone packed into the space as described by Richard Smith in Smith and Kaplan (Fig. 9-16), or a wedge osteotomy based on the longer side of the phalanx can create a wedge that is reversed and inserted from the shorter side as practiced by Carstam (Fig. 9-17). I have done very few of these procedures but on the whole have found the reversed wedge osteotomy to be the easier and more satisfactory of the two choices.

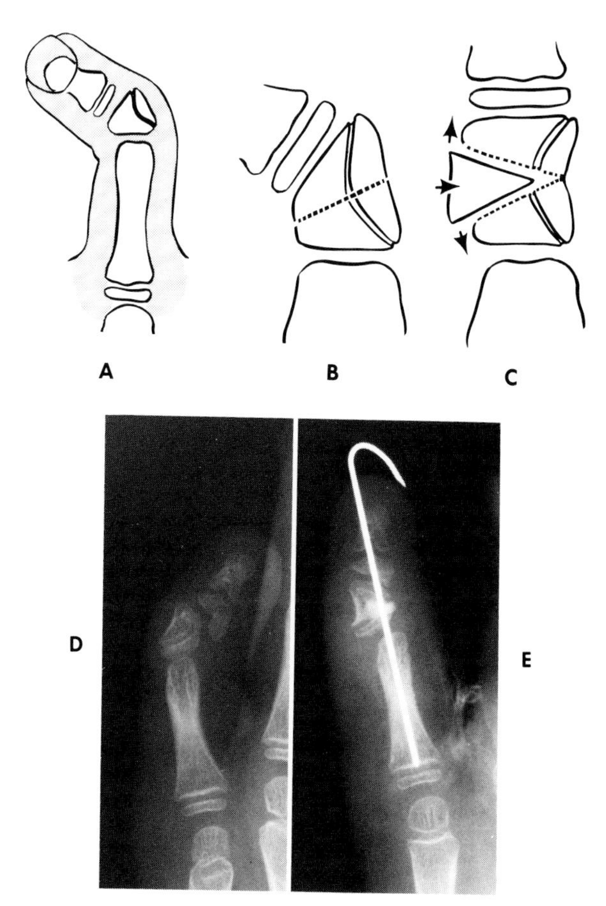

Fig. 9-16. *Delta phalanx lengthening.* **A**, **B**, and **C**, Stages in insertion of a bone graft on the short side of the delta phalanx. **D** and **E**, Before and after the insertion of an iliac crest bone graft. (From Wood, V. E., and Flatt, A. E.: Congenital triangular bones in the hand, J. Hand Surg. **2**[3]:179-193, 1977.)

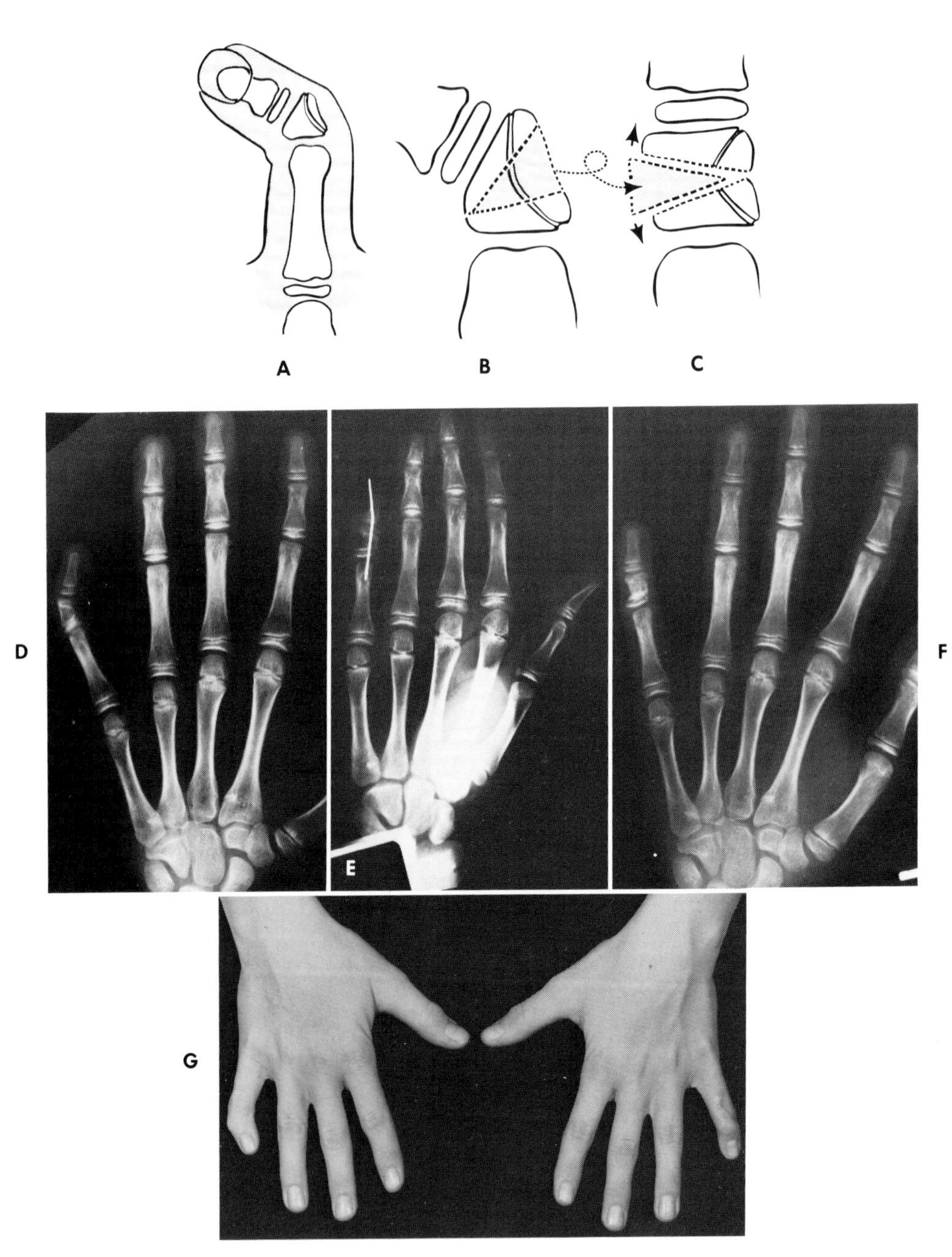

Fig. 9-17. *Delta phalanx lengthening.* **A, B,** and **C,** A wedge osteotomy based on the longer side of the phalanx yields a wedge of bone that, after reversal and insertion on the short side. lengthens the short middle phalanx. **D, E,** and **F,** X-ray films showing a small finger before, during, and after lengthening by wedge bone grafting. **G,** Left small finger corrected; right not yet operated on.

Whichever operation is selected, I believe the key technical point is the method of cutting the bone. The phalanx is small and hard to handle; osteotomes run the risk of jarring the terminal epiphyses and arresting growth in an already short bone. If power tools are used to cut across the bone, the width of the cut, particularly in the wedge osteotomy, creates shortening and a considerable amount of bone dust. Experience has taught me that the best way to cut these small bones is to patiently peck away along the line of the cut with the beak of a small, sharp bone cutter. The bone must not be crushed between the jaws of the cutter; the sharp tips of the ends of the cutter should be used to gradually score a deeper and deeper groove in the cortical bone. Patience and some luck will yield good clear-cut edges without splintering or crushing of the edges. This is particularly important when developing the small wedge for the reverse osteotomy procedure.

For the opening wedge procedure Richard Smith has outlined the successive steps, pointing out that the key feature is to osteotomize the junction between the transverse and longitudinal portions of the epiphysis. Before this the bone must be transfixed with Kirschner wires proximally and distally so that the two small fragments created by the osteotomy can be controlled. After the transverse osteotomy has been done, the distal portion is angled on its Kirschner wire so that the wedge is opened up and the finger lengthened. The longitudinal or extra epiphysis should be totally excised and the remnants packed into the opened wedge of the osteotomy site. Healing usually occurs within a month.

This description has been easy to write, but the operation is hard to do. It is usually made difficult by the seemingly infinite variety of shapes of the longitudinal epiphysis. If, however, the horizontal portion of the epiphysis can be preserved, then excising the remainder and packing it into the opened wedge will usually produce a satisfactory digit. Certainly it will be considerably better than the distorted digit that will result if the condition is left untreated.

In the reversed wedge osteotomy procedure, I approach the phalanx by way of a dorsally curved incision extending from the distal portion of the proximal phalanx over the entire length of the middle phalanx and onto the proximal portion of the distal phalanx. The lateral edges of the extensor mechanism must be mobilized so that both longitudinal borders of the phalanx can be completely seen. The proximal transverse epiphysis and the insertion of the central extensor slip must be protected before the longitudinal bracket is incised at the midpoint of the short longitudinal side. After the extensor mechanism has been lifted away, the V-shaped cuts can be slowly made on the dorsum of the bone, with care being taken to make sure the width of the base of the wedge does not trespass on either transverse epiphysis. The wedge is slowly developed, dissected free of the underlying flexor mechanism, and then reversed and pushed into place (Fig. 9-17). It can usually be held in place by one fine Kirschner wire, which is placed through the center of both distal and proximal epiphyses and finally lodged in the head of the proximal phalanx. I leave the distal end protruding through the skin and withdraw the pin about 4 to 6 weeks later.

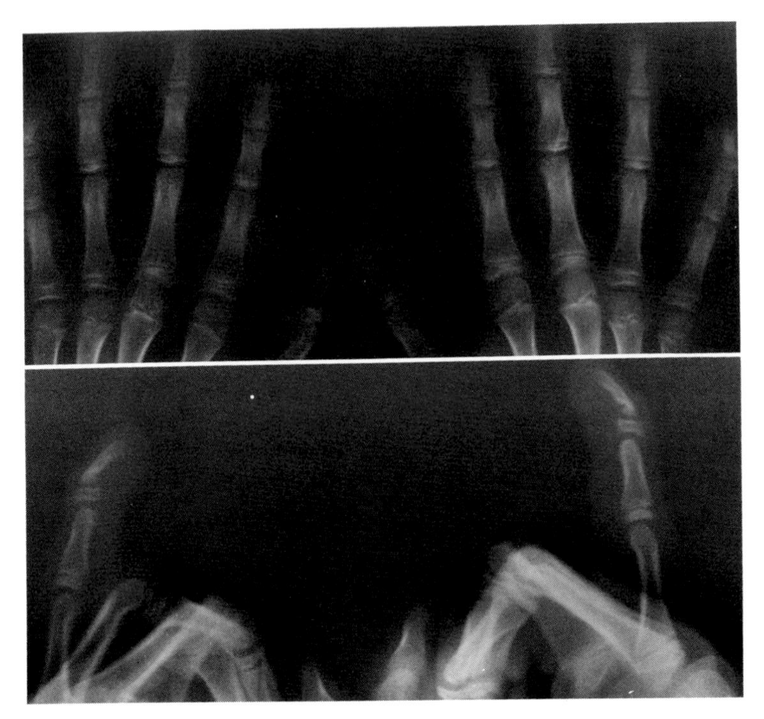

Fig. 9-18. *Kirner deformity.* This inherited trait occurs as a painless swelling and progressive curvature of the terminal phalanx of the small finger. The x-ray films show a palmoradial curvature of the terminal phalanx.

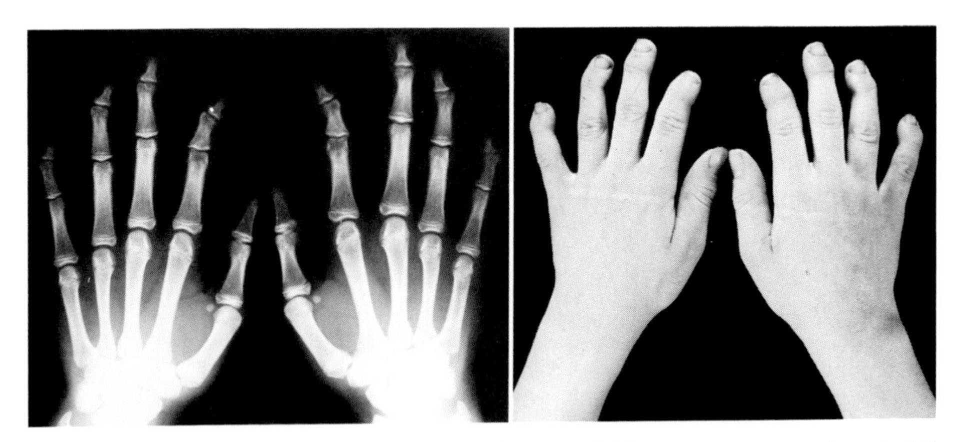

Fig. 9-19. *Frostbitten fingers.* Subclinical frostbite in children can cause a characteristic deviation of the fingers resulting from irregular and premature closure of the epiphyses of the distal phalanges. The middle phalanges may also be affected, as has occurred in both index fingers.

Carstam and Theander report that this wedge osteotomy procedure eliminated or markedly reduced the clinodactyly in all their cases. One patient had been followed for 3 years and showed persistent improvement of the deviation from 60 degrees.

KIRNER DEFORMITY

In 1927 Kirner described a bilateral condition in which the small fingers showed a rather severe palmoradial curvature of the distal phalanx. The condition is probably transmitted as a dominant trait and is not strictly congenital in the sense that it is not usually obvious until about 8 to 12 years of age. It is said that Kirner deformity may be found in the Silver and Cornelia de Lange syndromes although clinodactyly is their more usual association. It is about twice as common in girls as in boys and occurs as a painless swelling and progressive curvature of the terminal phalanx of the small finger.

X-ray examinations show the characteristic bending of the terminal phalanx; there may be a widening of the epiphysis, and its closure may be delayed (Fig. 9-18). The only differential diagnosis that may have to be considered is the deformity of the distal phalanges following cold injury in childhood (Fig. 9-19). In this condition there is premature closure of the distal phalanx epiphysis and all the terminal phalanges are involved rather than just the small finger's.

Treatment is not usually necessary, but if it is, then one or more osteotomies have to be done and the fragments impaled on a Kirschner wire. Carstam and Eiken recommend doing the osteotomy from the palmar aspect and not completing it; a dorsal periosteal hinge should be left to help control the fragments. Correction sometimes cannot be obtained until the obstruction caused by the curved nail has been removed. The skin approach should be through a midlateral incision. Bony healing will occur in 4 to 6 weeks.

CONGENITALLY ABDUCTED SMALL FINGER

Children are occasionally brought for consultation because the small finger persistently lies abducted away from its fellows when the hand is at rest (Fig. 9-20). This deformity can be troublesome and lead to avulsion injuries and sprains of the finger. The usual treatment is calculated neglect or strapping the finger to its more properly aligned neighbor. Neither is curative.

Etiology

Blacker, Lister, and Kleinert have recently pointed out that this deformity can be the result of an anatomical abnormality but have stressed that low ulnar motor loss must be excluded as a cause of the deformity. Their dissections showed that the more ulnar of the two constituent tendons of the extensor digiti minimi usually lies to the ulnar side of the axis of abduction or adduction of the metacarpophalangeal joint of the small finger. In addition, a strong connection was found to pass around the ulnar border of the hand from the more ulnar tendon to the tendon of the abductor digiti minimi. A third important factor was the length

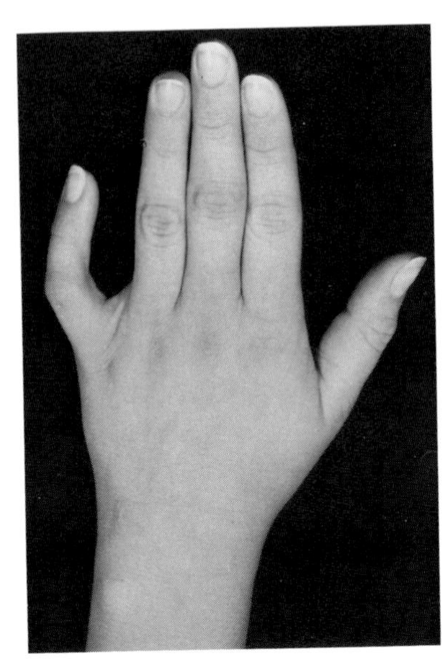

Fig. 9-20. *Congenitally abducted small finger.* Anatomical abnormalities in the arrangement of the two tendons of the extensor digiti minimi can cause this persistent abduction of the small finger.

of the broad junctura that normally passes from the ring to small finger extensors. A combination of these factors sets the stage for persistent abduction of the small finger.

Treatment

Correction is obtained by rebalancing the tendon forces. Two situations can exist at the metacarpophalangeal joint, which need different routes for the transfer of the ulnar component of the extensor digiti minimi. If the metacarpophalangeal joint can be hyperextended, the tendon should be passed to the palm, volar to the transverse metacarpal ligament, and sutured to the flexor sheath. In the absence of hyperextensibility, the tendon is passed deep to the junctura from the ring finger and sutured to the insertion of the radial collateral ligament of the metacarpophalangeal joint of the small finger. Technically neither attachment is easy to do, and I would recommend use of an adequate dorsal incision in the skin of the web. Good exposure is important to allow the transferred tendon to be properly passed through a slit in the flexor tendon sheath or in the radial collateral ligament. The tendon should be brought back and sutured to itself deep in the wound. Strapping the finger to the adjacent ring finger will provide adequate protection for the necessary 3 weeks.

CONGENITAL ULNAR DRIFT

Ulnar deviation of the fingers at birth has been recorded since at least 1897 when Emile Boix used Brissaud's apt descriptive phrase "deviation en coup de

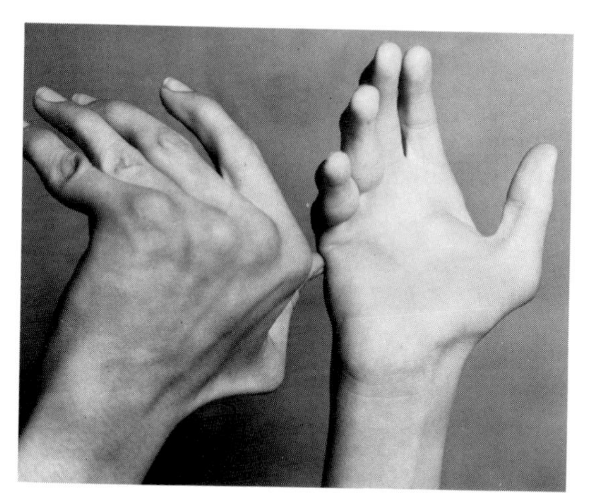

Fig. 9-21. *Congenital ulnar drift.* The characteristic finding in these patients is the constant association of a flexion deformity of the metacarpophalangeal joints with the ulnar drift of the fingers.

vent" (windblown fingers). Isolated case descriptions have appeared in the literature since this time, and by 1976 Powers and Ledbetter reported 44 cases in the literature and added one of their own. Common to all case reports is the existence of flexion of the metacarpophalangeal joints together with deviation of the fingers (Fig. 9-21).

Etiology

Many of these cases are familial and the cause is unexplained (Fig. 9-22). Basically the flexion and ulnar deviation represent an imbalance between extensor and flexor forces acting across the metacarpophalangeal joints. Deformities that occur in the presence of normal extrinsic extensor tendons are hard to explain in the sense that the precipitating factor cannot be identified. In late stages the ulnar dislocation of the extensor tendons is obvious and probably contributes to the deformity.

Treatment

Usually the deformity is bilateral and present at birth. However, advice is usually not sought until the condition becomes progressively more pronounced as the child grows. Whenever a child is seen with the deformity, attempts at splinting are always desirable, because as the hand grows, the surfaces of the developing metacarpophalangeal joints may develop a cartilaginous and bony deformity perpetuating the deviation.

I have seen a number of adults who were born with this condition, but I have never offered them treatment in any form because I am not convinced I can provide them with any significant functional improvement.

Several surgical procedures have been suggested to correct the imbalance. All relocate the extrinsic extensor tendons on the top of the joints and try to

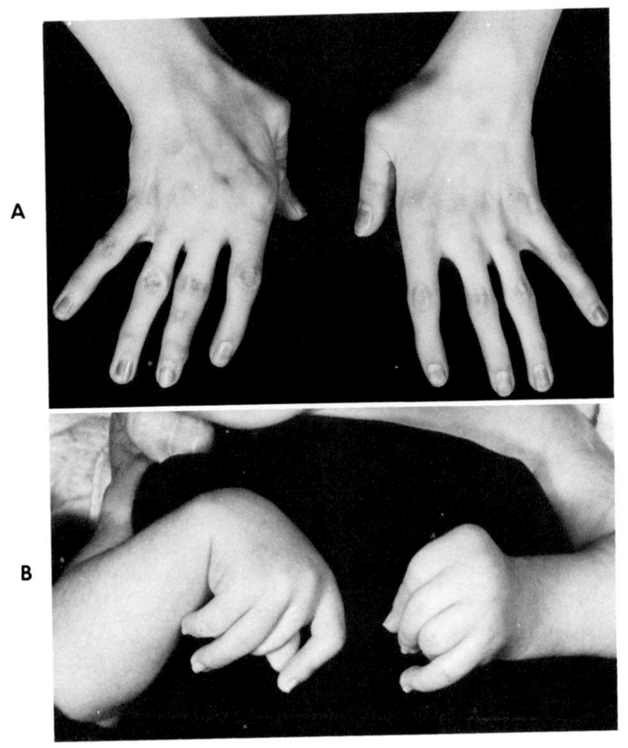

Fig. 9-22. *Congenital ulnar drift.* Many of these deformities are familial, but the etiology is unexplained. In this figure the deformity is worse in the right hands of both mother, **A**, and son, **B.**

develop retaining ligaments from local tissue. Some surgeons advocate additional release of ulnar structures such as the third and fourth dorsal interossei and the abductor digiti minimi. If the ulnar portion of the joint capsule and the ulnar collateral ligament are tight when the joint is held fully corrected, it may be necessary to excise some of the capsule and lengthen the ligament. Kelikian reports that his most enduring results have been obtained following corrective osteotomies of the proximal phalanges or metacarpals.

Ulnar drift, which is associated with a congenital absence or severe weakness of the extrinsic extensors, is caused by the unopposed flexor forces that inevitably pull the fingers both into subluxation and deviation. Children with this deformity are rare, and the few upon whom I have operated have all shown abnormalities of their extensor mechanism.

EXTRINSIC EXTENSOR ABNORMALITIES

Complete absence of all the extrinsic extensor tendons is most uncommon. Tsuge has recorded a collection of five patients in whom the extensor digitorum communis was severely underdeveloped. In his patients, the fingers, which were normally developed, lacked voluntary extension at the metacarpophalangeal joints.

A

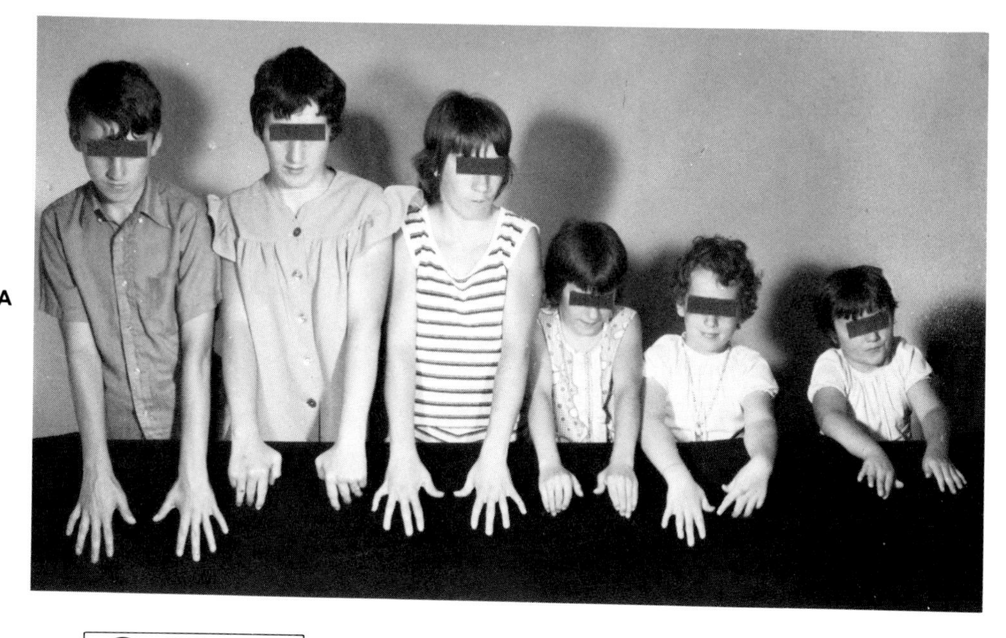

B

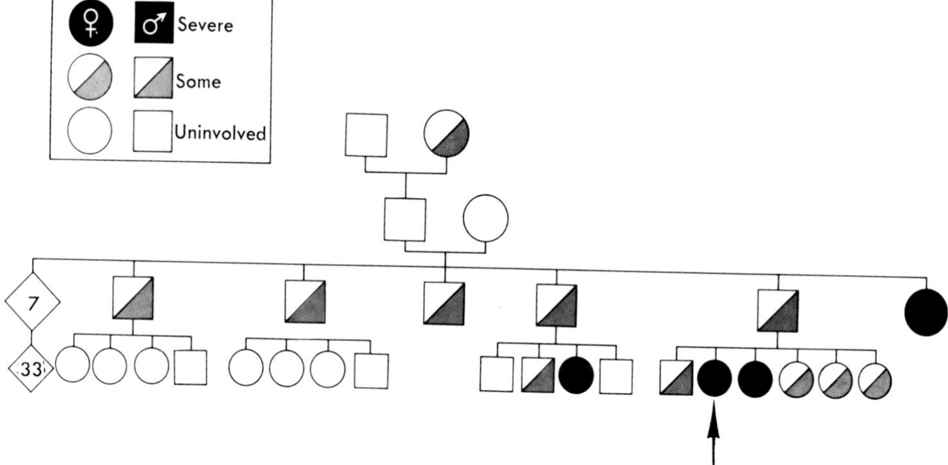

Fig. 9-23. *Extrinsic extensor tendon deficiencies.* **A,** Children of one family showing flexion and ulnar deviation of the metacarpophalangeal joints resulting from absence of the extensor tendons. **B,** The children's deformities began with their father's paternal grandmother and have been transmitted as an autosomal dominant trait over four generations. (From McMurtry, R. Y., and Jochims, J. L.: Congenital deficiency of the extrinsic extensor mechanism of the hand, Clin. Orthop. in press.)

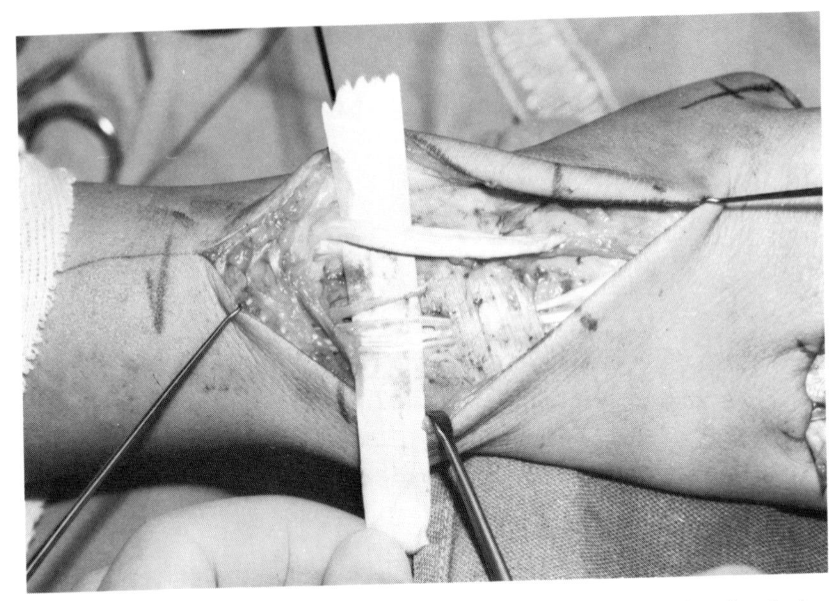

Fig. 9-24. *Attenuated extensor tendons.* Surgical exploration of the hands of the oldest daughter in Fig. 9-23 showed extreme attenuation of the extrinsic extensor tendons. Their size contrasts with the normal extensor carpi radialis longus traversing the upper part of the wound. (From McMurtry, R. Y., and Jochims, J. L.: Congenital deficiency of the extrinsic extensor mechanism of the hand, Clin. Orthop. in press.)

McMurtry and Jochims have recently reported a patient with congenital deficiencies of the extensors to fingers and thumbs. There was a family history over four generations compatible with an autosomal dominant mechanism of inheritance. The deformities began with the patient's father's paternal grandmother; all of the patient's siblings demonstrated a variable degree of involvement (Fig. 9-23). The fingers were held in a flexed and ulnar deviated posture and abnormally small extensor tendons could be palpated in the ulnar valleys of their respective metacarpal heads. Surgical exploration showed threadlike tendons ending in fibrofatty tissue in the forearm (Fig. 9-24).

More limited absence of the extensor mechanism is seen in congenital clasped thumb in which there is hypoplasia or absence of the extensor pollicis brevis and sometimes of the extensor pollicis longus. Absence of the extensor pollicis longus together with the extensor indicis proprius has also been recorded. Lack of these two tendons produces a characteristic posture of the resting hand. In the normal resting hand, muscle tone produces an increasing degree of digital flexion towards the ulnar border of the hand. In these patients the situation is reversed, and there is excessive flexion of the thumb and index fingers (Fig. 9-25). Exploration of these hands shows the usual threadlike attenuated tendons, which when traced proximally end in fibrofatty tissue rather than muscle fibers.

Luckily in patients with either complete or partial absence of the digital extensors, the wrist extensor tendons are normal and can be used as substitute motors. In my patient with complete absence, I used the extensor carpi radialis

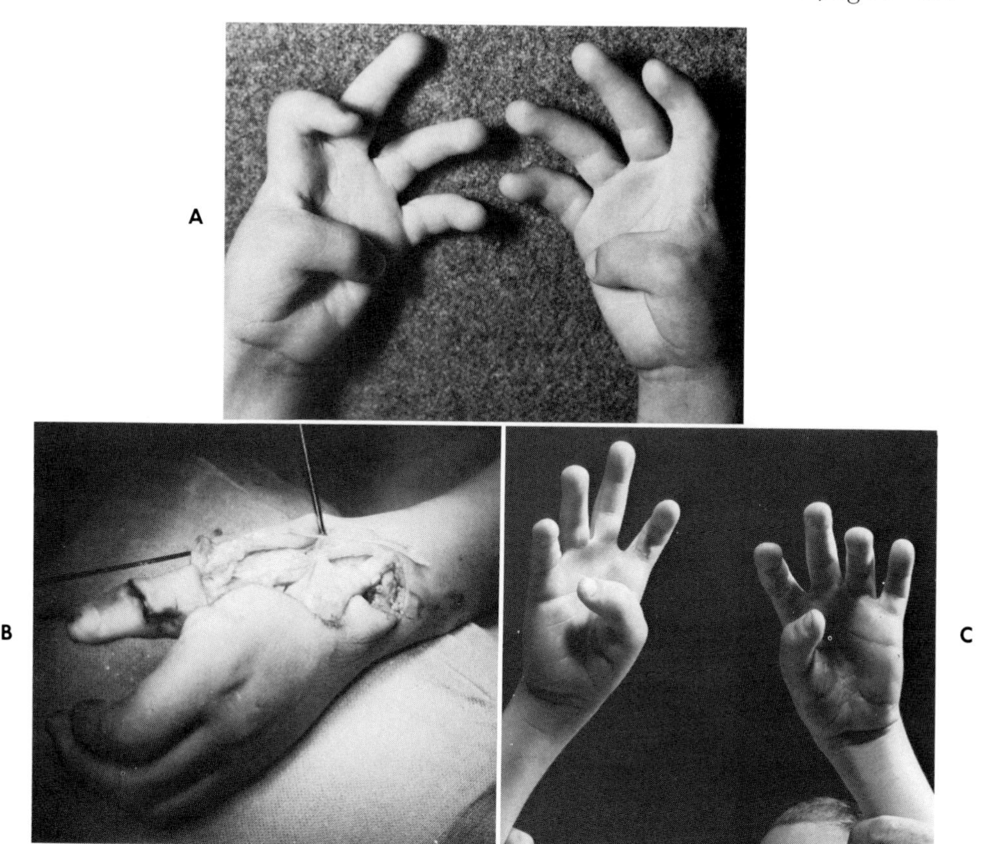

Fig. 9-25. *Absence of radial digital extensor tendons.* **A,** Congenital absence of the extensor pollicis longus and extensor indicis proprius produces a characteristic flexion deformity of the thumb and index finger. **B,** Surgical exploration shows the threadlike affected tendons. The extensor pollicis longus lies along the upper wound edge. **C,** After tendon transfers and supplemental skin grafting to the index finger, extension is improved but far from normal.

longus for the finger extensors by way of a multitailed graft. The long finger superficialis tendon was brought dorsally into the forearm and joined to the extensor pollicis longus.

The literature shows that the transfers that have been most often used in these cases are the extensor carpi radialis longus to the common extensors and the palmaris longus to the extensor pollicis longus after it has been rerouted in a more radial path.

In treating patients with absence of thumb and index extensors, I have always taken the extensor carpi radialis longus and used it as a common motor for the extensors of both the index and thumb.

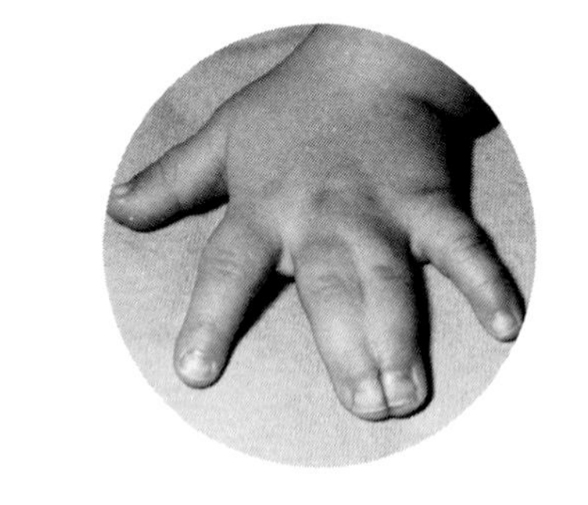

CHAPTER 10

WEBBED FINGERS

Syndactyly is commonly described as a webbing or fusion of two fingers. In fact, it is rarely either of these conditions. The fingers are usually bound closely together so that a true web is rarely present, and actual bony fusion between adjacent fingers is uncommon. Although syndactyly is by far the most common single congenital hand anomaly, it can also be associated with many other important conditions. The external appearance of conjoined fingers may mask such anomalies as polydactyly or brachydactyly, and the condition may be associated with syndactyly of the toes, constriction rings, cleft feet, hemangioma, absence of muscles, spinal deformities, funnel chest, and heart disorders. Syndactyly frequently occurs with other conditions, for example, acrosyndactyly, hypoplasia of the hand, Poland syndrome, symbrachydactyly, and Apert syndrome. The only thing these conditions have in common is a lack of separation between the digits. In some syndromes conjoined fingers are obligatory constituents, but in many others they only participate occasionally. When present with these syndromes, conjoined fingers are erroneously described as syndactylies but are better treated and considered as separate anomalies.

Webbing between digits is not a true entity but is really a result of lack of differentiation between adjacent digits. The separation failure occurs in the mesenchymal blastoma between the sixth and eighth weeks of intrauterine life. This failure can occur spontaneously, but it also is said to show a family history of at least 10%. At the University of Iowa in our own series of cases we have established a family history in nearly 40% of our patients. Dominant inheritance of long-ring finger syndactyly has been shown in several pedigrees. Woolf and Woolf have pointed out that genetic counseling is difficult in these cases since the dominant genes show reduced penetrance and variable expressivity. Thus the abnormality is rarely consistent from one generation to the next, and it is quite likely that the child will be born with a variation of the parent's abnormality. It does appear, however, that paternal genes have a stronger influence than their maternal counterpart.

INCIDENCE

The basic problem in establishing a true incidence of webbing between the various digits is deciding which cases to include under the heading of syndactyly. The incidence of webbing in our patients who have only simple or complex "pure" syndactyly is shown in Fig. 10-1, A. If, however, we include

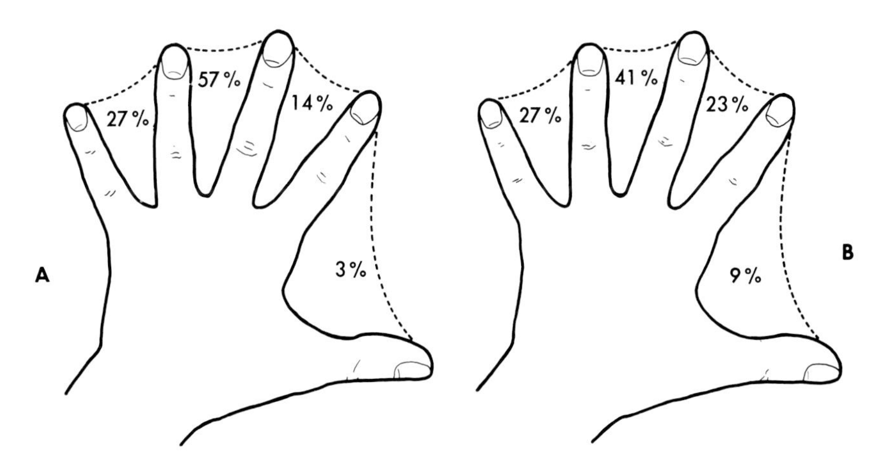

Fig. 10-1. *Site of syndactyly.* **A,** Percentage incidence in Iowa series when only true syndactyly of simple or complex types is considered. **B,** Total count incidence in which associated conditions (all webbed cases) are included. These figures more nearly resemble the usually quoted incidence for syndactyly. (From Flatt, A. E.: Practical factors in the treatment of syndactylism. In Littler, J. W., Cramer, L. M., and Smith, J. W., editors: Symposium on reconstructive hand surgery, vol. 9, St. Louis, 1974, The C. V. Mosby Co.)

all associated conditions except Apert syndrome, the percentage incidence changes to more nearly resemble the usually quoted figures (Fig. 10-1, *B*). Syndactyly is commonly said to occur about once in every 2,000 to 2,500 births. About half the cases are bilaterally symmetrical, and the condition occurs twice as frequently in males.

Although simple webbing in the hand is most common between the long and ring fingers, it is probably less common than a similar condition between the second and third toes. This unimportant anomaly is inherited as a dominant trait, but sporadic cases are more frequent. I was surprised to find that I have incomplete webbing between my second and third toes and have been delighted to find it to be extremely common in friends, acquaintances, and succeeding generations of residents.

CLASSIFICATION

Classification of this deformity is of value only insofar as it aids in treatment. Probably the most useful classification is based on a combination of two components: (1) the degree of webbing and (2) the presence or absence of

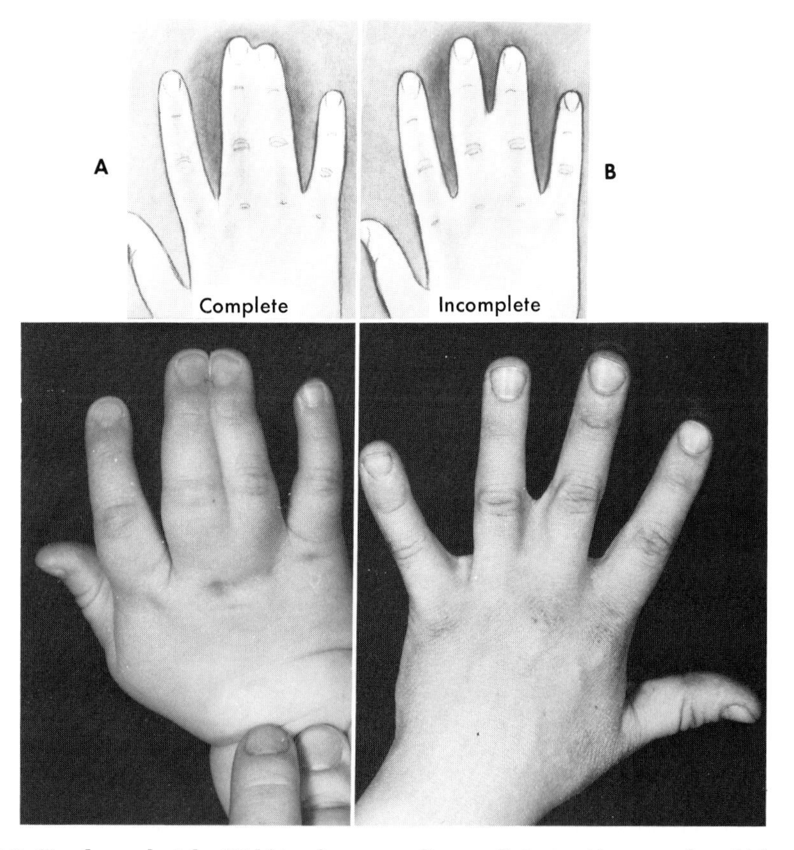

Fig. 10-2. *Simple syndactyly.* Webbing between adjacent digits is either complete (**A**), extending to the tips of the digits, or incomplete (**B**), not reaching the tips.

bony fusion. First the degree of webbing between the fingers is determined. In complete syndactyly the skin web extends to the tips of the involved digits and sometimes produces a common fingernail (Fig. 10-2, *A*). The web can stop at any point between the normal commissure and the end of the fingers, in which case it is classified as incomplete (Fig. 10-2, *B*).

Syndactyly is defined as simple when the web contains only normal soft tissues and as complex when adjacent phalanges are fused or when there is interposition of accessory phalanges. In complex webbing (Fig. 10-3), abnormalities of nerves, vessels, and tendons may also be present.

The distribution and presence of digital nerves is quite unpredictable, but I have the clinical impression that the closer the affected digits are together, the greater is the likelihood of anomalies. The intrinsic muscles may be anomalous or absent, and there is no indication for tendon transfers in an attempt to replace their action.

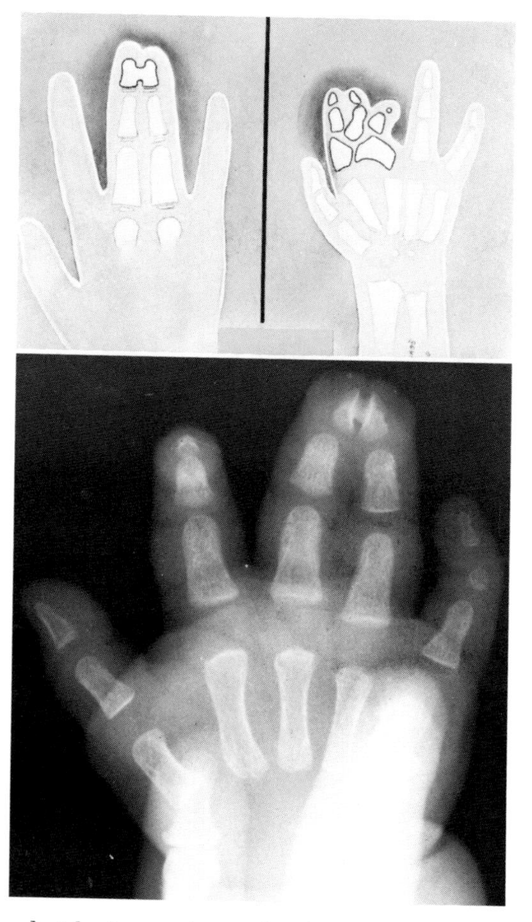

Fig. 10-3. *Complex syndactyly.* In complex syndactyly adjacent digits are fused. There may be interposition of accessory phalanges and even abnormalities of the nerves, vessels, or tendons.

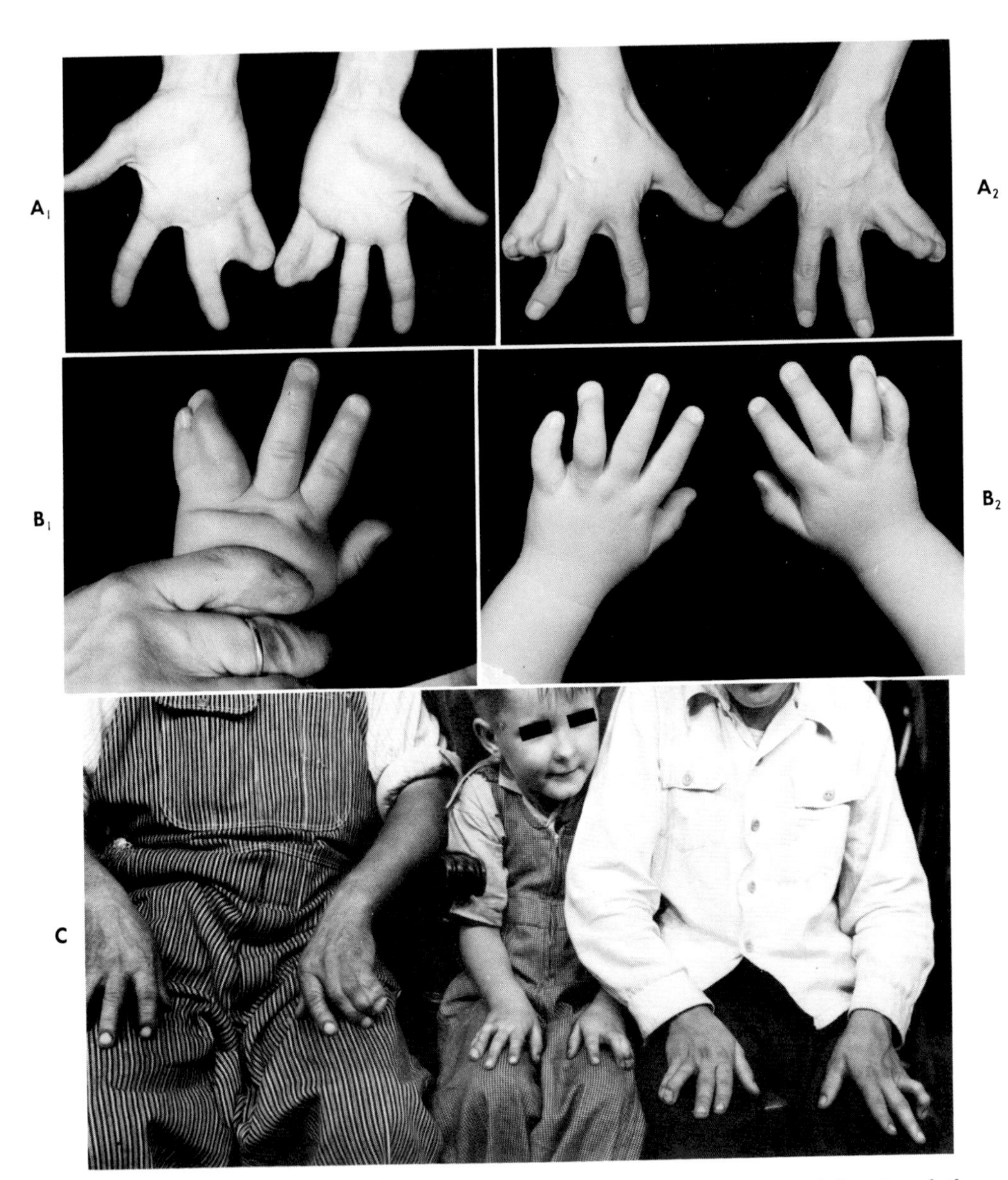

Fig. 10-4. *Syndactyly—ring-small.* **A1** and **2**, Adult hands showing gross deformity of the ring fingers because correction was never performed. **B1** and **2**, A child in whom early bilateral correction has freed the fingers for proper longitudinal growth. **C,** A family with three generations of ring and small finger involvement, none of which was ever surgically corrected. (**B1** and **C** from Flatt, A. E.: Practical factors in the treatment of syndactylism. In Littler, J. W., Cramer, L. M., and Smith, J. W., editors: Symposium on reconstructive hand surgery, vol. 9, St. Louis, 1974, The C. V. Mosby Co.)

PRINCIPLES OF TREATMENT

Webbed fingers are an obvious physical defect, and individuals can and do, even in these times, go through life without having a separation of a syndactyly. For an otherwise normal individual I believe this is unnecessary cruelty. Parents are frequently subjected to social pressures urging early separation, but this also is not necessarily in their child's best interests.

Timing of surgery

The timing of surgery is always a problem. I believe one should ask not how soon the operation can be done but rather how late the functional demands of the hand will allow postponement of surgery. Present day surgical techniques have made obsolete old adages such as, "Wait until the baby fat has gone." or "Do before 7 and revise at 14." It should be explained to the parents that the timing and type of surgery must be related to the particular digits involved and the degree of completeness and complexity of the webbing.

I do not believe it kind or wise to simultaneously operate on both hands of either an adult or a child. It is devastating to be suddenly deprived of all prehensile ability. In the very young it can be justified by the risks of anesthesia, but in general I believe it better to operate on one hand at a time.

For the hand to function and develop as a prehensile organ, the thumb and small finger must be separated from their adjacent fingers early to allow the borders to oppose in grasp. Disturbances of the orderly growth of the long bones of the hand must also be corrected early if normal arch formation is to develop in the hand.

When two digits of unequal length are joined together, the longer digit will inevitably develop a flexion contracture and may also show lateral deviation deformities. Fig. 10-4 shows a variety of patients with bilateral ring-small finger syndactyly, illustrating the deformities that will occur if separation is delayed beyond 1 year of age. Similar uncorrectable deformities arise in triple syndactyly (Fig. 10-5).

Because the thumb develops earlier than the fingers, thumb-index finger webbing is the least common pairing. When it occurs, the deformity of the index finger is profound and the web should certainly be separated by 6 months of age (Fig. 10-6).

It is possible to separate both border digits from their adjacent fingers at the same operation since a single finger will not be denuded on both sides. The combination of ring-small finger and thumb-index finger webbing is rare in pure syndactyly, but it is the rule in acrocephalosyndactyly, and it is our practice to do both dewebbings in such patients at the same time.

There is no great urgency to separate a syndactyly between the long and ring fingers even if the condition is complex. Distal phalanx growth and fusion may continue undisturbed even into adult life (Fig. 10-7). In index-long finger webbing the interphalangeal joints are not on the same levels, and joint contractures will develop if separation is not carried out in the first few years of

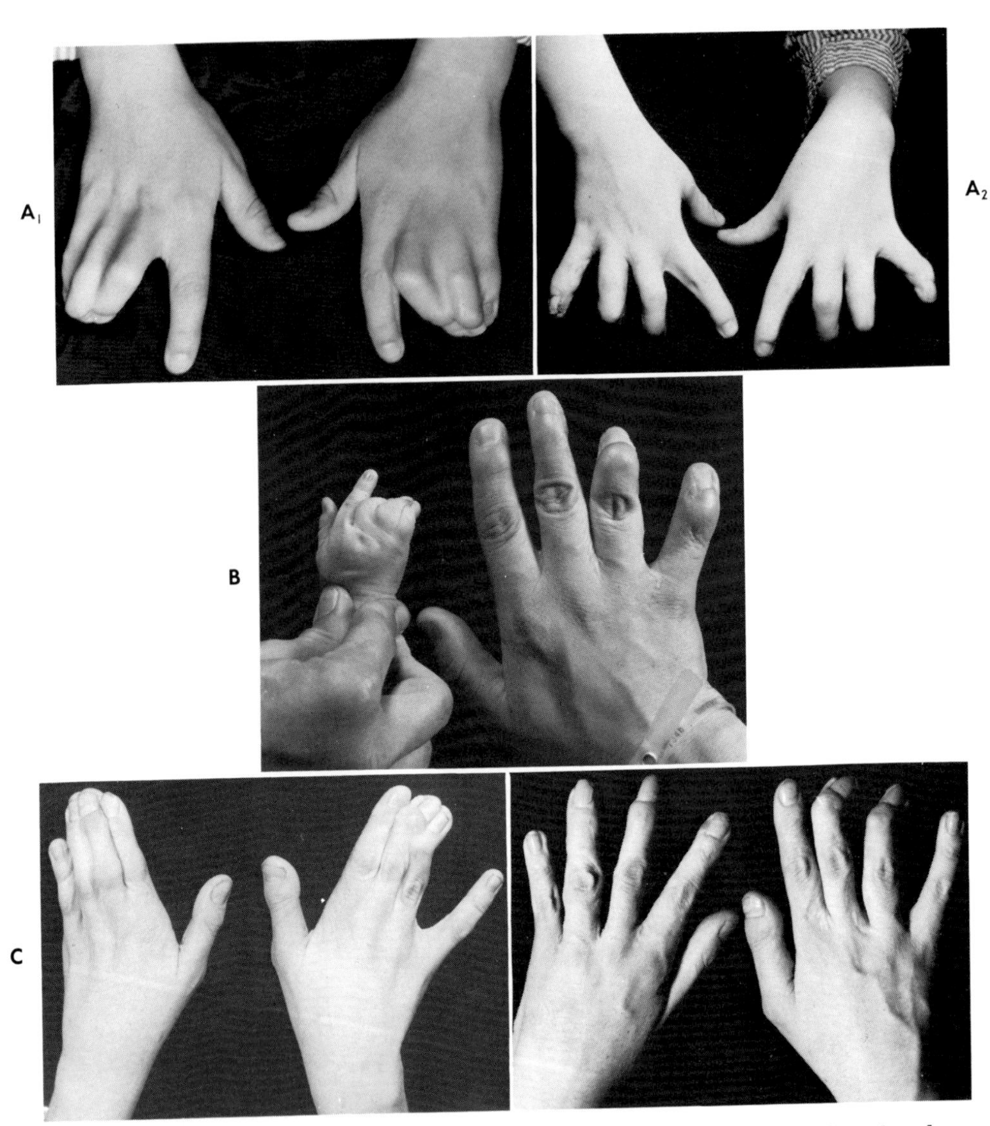

Fig. 10-5. *Triple syndactyly.* When the long, ring, and small fingers are joined, early release is essential if deformity is to be avoided. **A1** and **2**, Release done 25 years ago at the age of 12 years. The long and ring finger deformities have not corrected. **B,** A mother's deformed right hand and her baby's unoperated right hand. **C,** Index, long, and ring finger syndactyly. Release in adult life could not correct the distal interphalangeal joint deformities in the right hand. Because these fingers are nearly equal in length, there is no flexion deformity of the proximal interphalangeal joints.

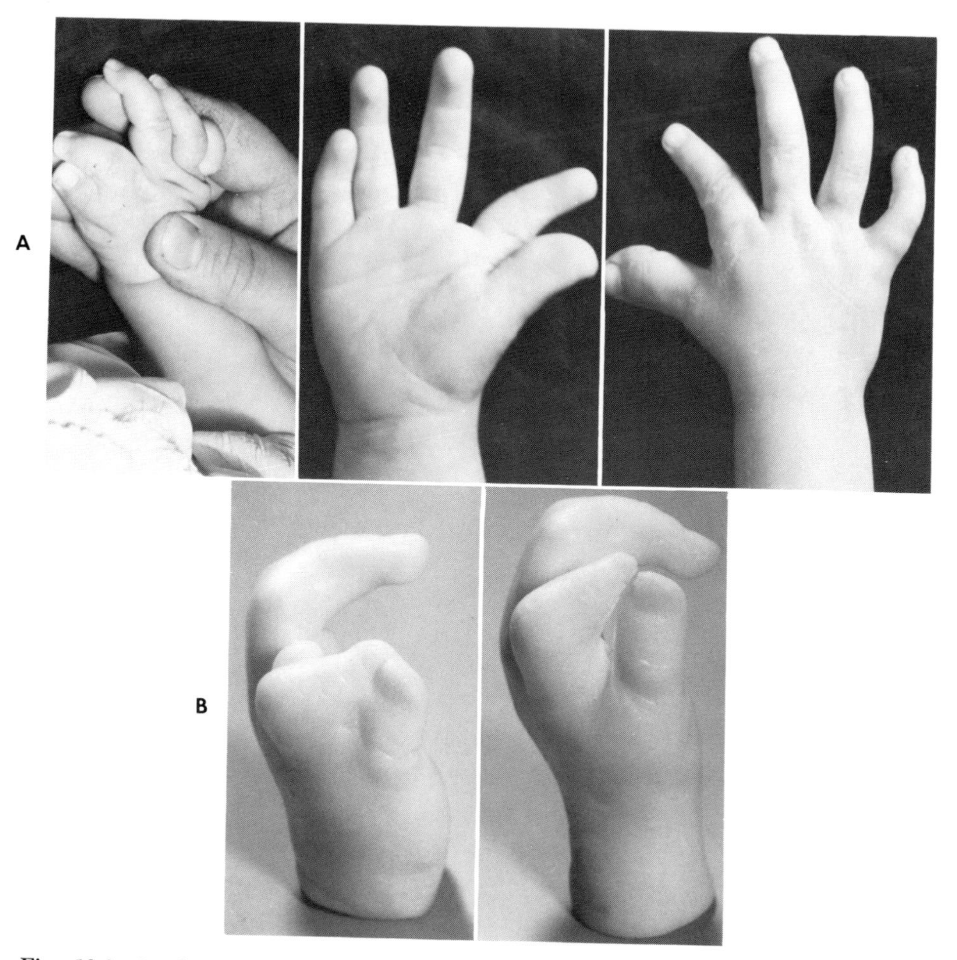

Fig. 10-6. *Syndactyly—thumb-index.* This is the least common webbing, but the flexion deformity of the index finger is usually profound. **A,** Correction in early childhood did not yield a good result. **B,** Casts before and after surgery, which was done at the age of 14 months. The severe flexion deformity of the index finger could not be corrected. (**B** from Flatt, A. E.: Practical factors in the treatment of syndactylism. In Littler, J. W., Cramer, L. M., and Smith, J. W., editors: Symposium on reconstructive hand surgery, vol. 9, St. Louis, 1974, The C. V. Mosby Co.)

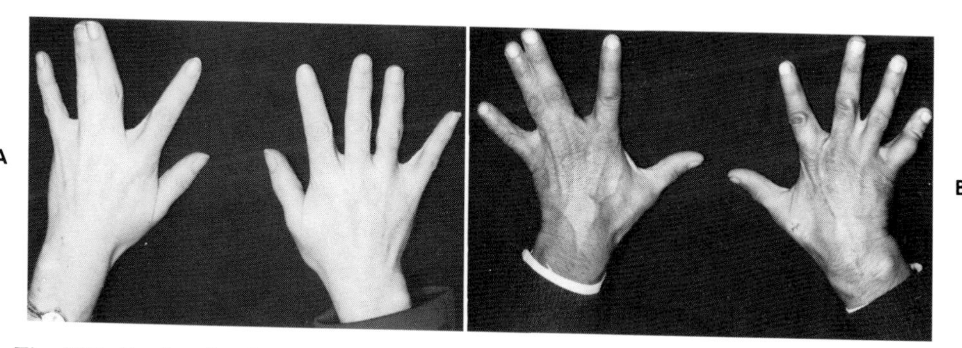

Fig. 10-7. *Syndactyly—long-ring.* In simple webbing of these two fingers there is no urgency in surgical correction because they are of virtually equal length. **A,** This adult came for separation of her left long-ring syndactyly so that she could wear her wedding ring. **B,** This man shows two degrees of incomplete syndactyly; he was not interested in surgical separation.

life. For these three central fingers we usually separate the index-long finger web at about 1 to 2 years of age and the long-ring finger web at about 2 to 4 years of age. In a review of syndactyly repair carried out at the University of Iowa Hospitals between 1946 and 1960, we showed that in those years more postoperative complications and less satisfactory results were obtained in children who were operated on when under the age of 18 months. Surgery should therefore be postponed until after this age if the site and complexity of the condition allow it. During this period, while waiting for growth, the parents can be usefully occupied in stretching the skin of the web. If they are trained to carry out simple massage of the skin between the digits, an effective widening of the skin web can often be obtained. This "extra" skin makes the subsequent surgery considerably easier.

It is therefore prudent to postpone surgery in the early years if possible. I believe that a child is entitled to have all webbing separated before being subjected to peer curiosity in school. I cannot agree with those who feel that surgery can be postponed until the second half of the first decade of life. The growth of the hand is significant in the early years. This rapid increase in skeletal length demands separation of the digits to allow unimpeded growth. It also dictates careful planning of incisions so that scarring after surgery will not lead to contractures.

When all the digits are webbed, the four necessary clefts can be established in only two operations. At the first operation the thumb-index and long-ring separations are carried out. This is followed within 3 to 6 months by separation of the small-ring and index-long fingers at the second operation. This plan avoids the risks inherent in dividing on both sides of one finger at the same time.

OPERATIVE PLAN

General principles

Many techniques of separation have been described since the late 1800s. Since 1956 the principles have become well established and, therefore, the number of operations commonly used has been reduced. It is important to realize that the commissures between the index and long and ring and small fingers are rectangular or square ⊔ shaped, and the central commissure between the long and ring fingers, while frequently a square, may be V shaped. This breadth of the border commissures should always be provided since it is needed for the encircling span of wide grasp.

I have found that the most satisfactory basic operative plan is to establish the depth and width of the commissure by the use of a flap. I use zigzag incisions along the sides of the digits in order to minimize any chance of secondary contracture, and wherever possible, I plan the flaps to completely cover one finger so that a skin graft is needed for the second digit only. If the separation is carried back to the level of the metacarpophalangeal joint, then the likelihood of a revision of the commissure being necessary during the teenage growth spurt is greatly reduced; however, it cannot always be avoided. For this reason

I stress to the parents the need for long-term follow-up, even though all primary surgery has been completed before the child enters school.

Over the years individual surgeons have developed various methods of syndactyly repair, which they have used with satisfactory results. I do not believe there is one single method that will consistently produce superior results. All modern methods have in common the concept that there is a skin deficiency in the proximal part of the wound after separation of the fingers. Some surgeons use split-thickness skin grafts and others prefer full-thickness grafts to supply the extra skin cover. I have used both types and prefer a full-thickness graft taken from the groin. It is important that the location of the donor site is well lateral, in the area of the anterior superior iliac spine. Grafts taken from more medial sites can become embarrassing hair-bearing areas after puberty. I occasionally use the flexor crease of the elbow as a donor site, since a large amount of skin can be readily obtained from there, but follow-up has shown that the scar is frequently unsightly.

Parents often do not understand the need for additional skin. A useful way to demonstrate the necessity for skin grafting is to measure the circumference of the two involved digits in a parent's hand individually and combined (Fig. 10-8). The demonstrable difference between the sum of the two fingers and the combined circumference usually satisfies the parents' questions regarding use of a skin graft.

Treatment of syndactyly is a difficult problem and one that requires good technical ability and judgment. I would therefore recommend to the tyros in

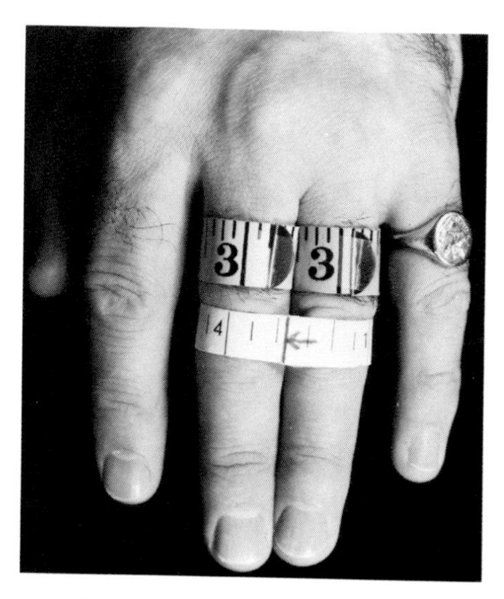

Fig. 10-8. *Skin grafting in syndactyly.* The circumference of my long finger is $3\frac{1}{4}$ inches and of my ring finger, $3\frac{1}{8}$ inches. The total of $6\frac{3}{8}$ inches is considerably greater than the $4\frac{5}{8}$ inch circumference of the two measured together. This difference in measurements can be used to explain to parents the need for skin grafts. (From Flatt, A. E.: Practical factors in the treatment of syndactylism. In Littler, J. W., Cramer, L. M., and Smith, J. W., editors: Symposium on reconstructive hand surgery, vol. 9, St. Louis, 1974, The C. V. Mosby Co.)

this field that they choose one operative technique and become expert at it before they try the many hundreds of modifications and methods that currently appear in the literature.

SIMPLE COMPLETE SYNDACTYLY

The method I find most satisfactory in the treatment of simple complete syndactyly is that published by Bauer, Tondra, and Trusler in 1956. In this method they establish a broad dorsal flap that will pass between the fingers and join a transverse incision in the palm to establish the commissure (Fig. 10-9). I personally prefer this broad dorsal flap since it consists of good, thin, mobile skin that supplies the necessary breadth to the commissure. I have abandoned the use of a palmar flap because of the unyielding nature of the skin.

I do not use the method that employs two triangular flaps, one palmar and one dorsal, because many patients treated by this procedure subsequently develop a narrow V-shaped commissure (Fig. 10-10). Theoretically, the central junction between the two flaps could be stretched forward and would provide a broad commissure, but this has occurred only rarely in the patients I have seen who have been operated on by a variety of surgeons. Dr. Robert Woolf of Salt Lake City has used this method for a number of years and recently called back a number of patients for my review. There is no question that he can routinely achieve excellent results using these twin flaps. I do not know the secret of his success, and the many failures I have seen with this method still influence me to recommend the single broad flap made out of supple dorsal skin.

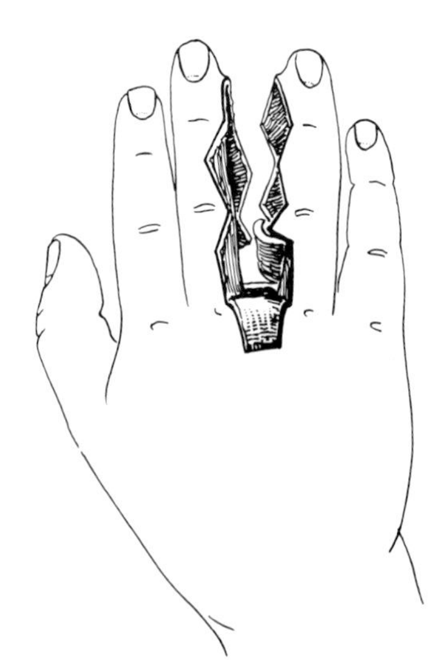

Fig. 10-9. *Syndactyly separation.* The basic flaps raised after the method described by Bauer, Tondra, and Trussler. (From Flatt, A. E.: Practical factors in the treatment of syndactylism. In Littler, J. W., Cramer, L. M., and Smith, J. W., editors: Symposium on reconstructive hand surgery, vol. 9, St. Louis, 1974, The C. V. Mosby Co.)

An added advantage of the broad dorsal flap is that it tends to provide a replica of the normal commissure, which slopes downward toward the palm to give an increased length to the palm compared with the dorsum of the hand.

A second flap is established on the palmar aspect based on whichever digit is selected as the "dominant" digit. In the case of the ring and long finger, I personally would select the ring finger as the dominant one, that is, the one that will carry the flap, because I think wearing a ring over a flap is better than placing the ring over a skin graft. When the ring and small finger or thumb and index finger combinations are involved, I select the border digit as the dominant one. If the index and long fingers are webbed, I usually select the index to carry the palmar flap. Zigzag incisions then are joined and are interdigitated along the sides of the fingers distal to the dorsal and palmar flaps. Inevitably surgeons' methods are altered by others, and not necessarily for the best, but I personally have found that the first dorsal flap, which is designed distal to the distal edge of the dorsal commissure flap, has a somewhat narrow base (Fig. 10-11). I have therefore slightly modified the original plan to slope the end of the dorsal flap so that my first triangular-shaped flap

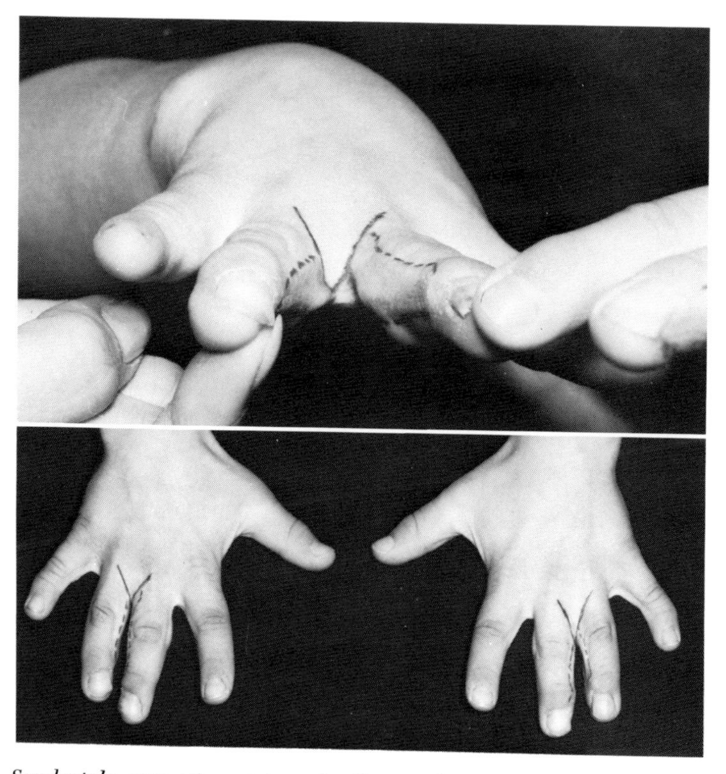

Fig. 10-10. *Syndactyly separation—triangular flaps.* I do not use this method because I have seen too many results like this in which the commissure grows distally with the fingers. Some surgeons get excellent results with this method.

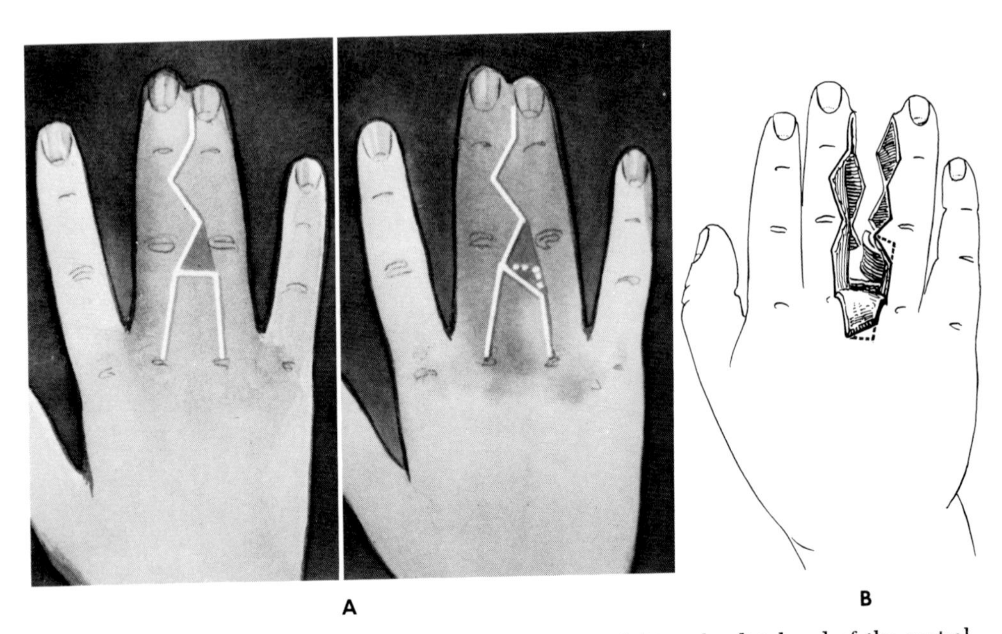

Fig. 10-11. *Syndactyly separation—current method.* By modifying the distal end of the central dorsal flap the first dorsal flap on one finger acquires a broad base. (**B** from Flatt, A. E.: Practical factors in the treatment of syndactylism. In Littler, J. W., Cramer, L. M., and Smith, J. W., editors: Symposium on reconstructive hand surgery, vol. 9, St. Louis, 1974, The C. V. Mosby Co.)

Fig. 10-12. *Syndactyly separation.* A series of operative photographs showing stages in separation of complex syndactyly. **A,** Rectangular palmar flap, *R′*, is based on the ring finger. Its distal edge is sloped to allow the first triangular flap on the long finger to have a broad base. Dotted lines *R* and *L* represent the midlines of their respective fingers. **B,** Dotted lines have been drawn around sides of fingers from apices of palmar triangular flaps. These lines allow proper planning of interdigitating dorsal flaps and slope on the end of the dorsal commissural flap. This flap is made about two thirds the length of the proximal phalanx. **C,** Digital neurovascular bundles are defined, and fat that has been excised is shown on palm. **D,** The dorsal flap has been sewn into the commissure, and the palmar flap based on the ring finger has been brought around the radial side of this finger and sutured to its dorsal skin. **E,** The ring finger has been completely closed without the need for a skin graft. **F,** The long finger has been closed by interdigitating its flaps and with the addition of a square full-thickness graft adjacent to letter *D*. Because there was bony fusion of the distal phalanges, an additional skin graft was used at the fingertip. **G** and **H,** Complete separation and skin closure of the fingers at the end of the operation. (From Flatt, A. E.: Practical factors in the treatment of syndactylism. In Littler, J. W., Cramer, L. M., and Smith, J. W., editors: Symposium on reconstructive hand surgery, vol. 9, St. Louis, 1974, The C. V. Mosby Co.)

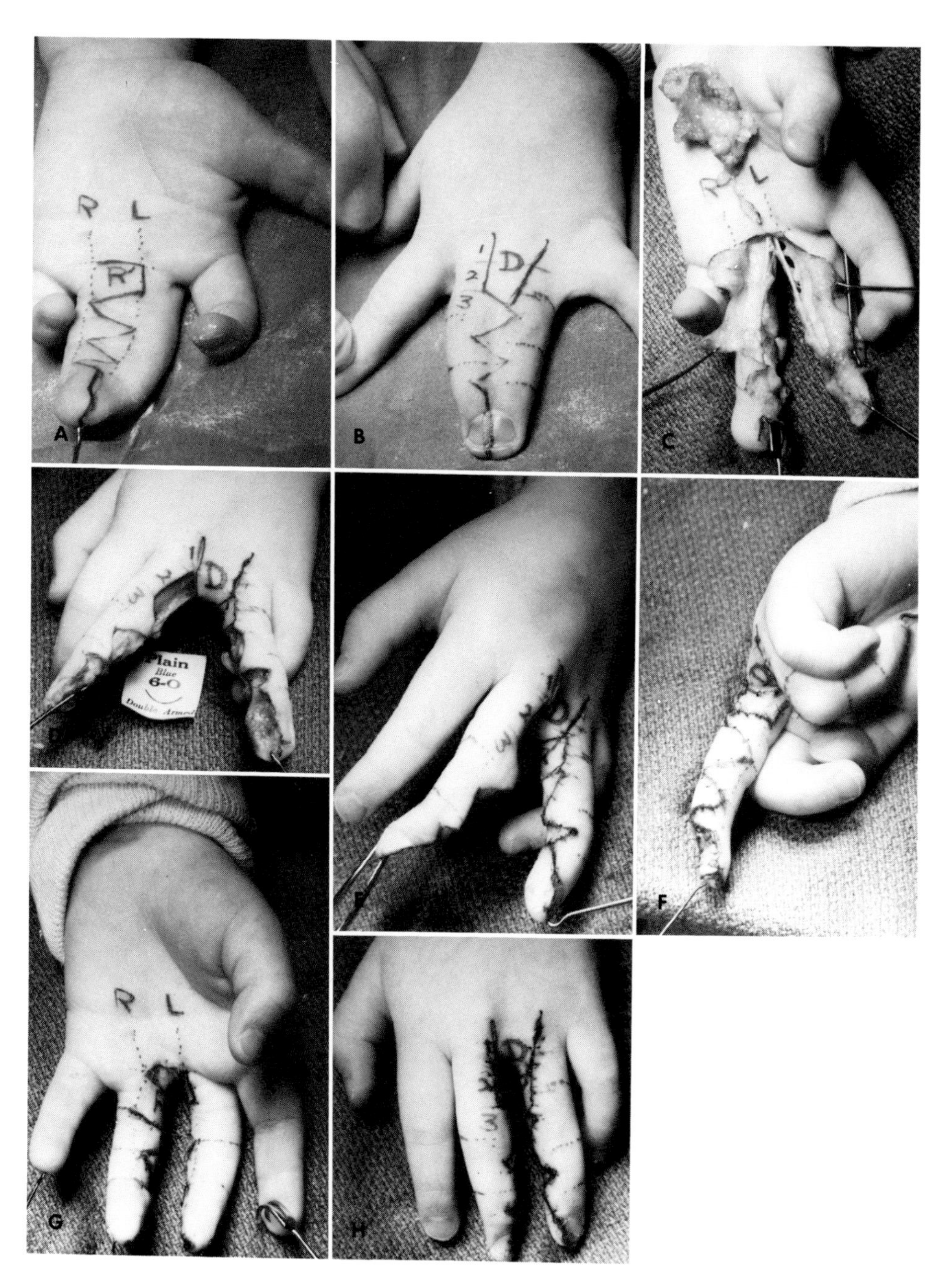

Fig. 10-12. For legend see opposite page.

has a broader base. Because of its elasticity, the dorsal skin accommodates readily to the sloping of its distal edge, and it sews into the palmar edge satisfactorily.

I believe strongly that the key to this whole operation is the judicious defatting of the tissues on the adjacent sides of the fingers and on both palmar and dorsal aspects across to at least the midline of the two involved digits. In small children this fat will be absorbed later, and removing it early allows the flaps to sit in readily without tension. Failure to do this may lead to a considerable amount of edema following surgery and often even the loss or tearing apart of some of the interdigitated flaps.

Actual planning and outlining of the incisions is time-consuming and should be done before the tourniquet is applied. In general I first draw in the palmar flap based on whatever digit I have selected as dominant (flap R′ in Fig. 10-12, *A*). The length of the incision, and therefore the length of the flap, must correspond to the width of the distal edge of the dorsal flap since these two will meet. I find it useful to slope the distal edge of the palmar flap because this will also be an edge of the first of the triangular flaps that will be planned to zigzag distally to the tip of the finger and then transfer onto the dorsum of the hand and work their way back to the distal edge of the dorsal flap. If the dominant flap has been established on the palmar side of the ring finger, then the first triangular flap will be based on the long finger. I find it useful to dot in lines around the sides of the fingers to indicate the tips of the flaps on the palmar aspect so that I may plan on the dorsum the appropriate sites for the interdigitation of the flaps. I then draw the length of the dorsal flap to be about two thirds of the length of the proximal phalanx measured from the line of the metacarpophalangeal joint (Fig. 10-12, *B*).

After the pneumatic tourniquet has been inflated and the flaps have been raised, appropriate defatting should be carried out and a definition of the digital nerves and vessels accomplished (Fig. 10-12, *C*).

There is usually a clear plane of separation between the fingers with very little crossing of fascial fibers and even less of blood vessels. Occasionally there may be a fusion of more distinct fibers, which probably represent Cleland's ligaments. The amount that the dorsal flap may be moved proximally to establish the proximal line of the commissure is dependent on the location of the bifurcation of the digital vessels. If there is a distal Y division of the digital nerves or even a single fused nerve, it can be readily split to fit more proximally (Fig. 10-13). I do not as a general rule divide either vessel between the adjacent fingers to allow a more proximal commissure nor do I divide the transverse intermetacarpal ligament. The dorsal flap is sewn into the palm first, and then the palmar flap is sutured around the side of the dominant finger (Fig. 10-12, *D*). I use No. 6-0 plain catgut throughout so that I do not have to struggle with the child to remove the sutures later. Following this, I continue sewing the interdigitated flaps on the dominant finger and hopefully am able to completely close this finger (Fig. 10-12, *E*). I then close the interdigitated flaps on the nondominant finger and am left with a defect opposite the dominant palmar flap (Fig. 10-12, *F*). A pattern should be cut with tinfoil or some other material exactly reproducing the size of this defect, and then a full-thickness skin graft should be taken from the groin or the elbow in the shape of the pattern. The final closure is obtained by sewing this graft into the defect. Fig. 10-12, *G* and *H* shows the separation of the fingers at the end of the operation.

Postoperative care is simple. I do not impale the fingers on wires or sew them to rigid splints. I use Dacron batting dressings, which supply good compression without shrinking if they become stained with blood. In small, uncooperative children I protect the primary dressings with plaster of Paris and extend this protection proximal to the elbow, which should be flexed to more than 90 degrees *before* the dressings and plaster are applied.

Dressings must be kept dry and are usually left undisturbed for 2 weeks, at which time they are removed. Crustings over the graft and the incisions are common, and there is no urgency in their removal. The catgut sutures disintegrate at different times, and the parents should be taught to remove them when they loosen. If the crusts are attached by a few catgut sutures, these can be cut. Further dressings are applied for about another 7 or 10 days. Following this, there should be complete healing and the child can be allowed freedom of use of the hand.

The full-thickness skin graft frequently takes several weeks to reepithelialize, depending on its thickness. If frank failure of the graft has occurred, I believe the child should be admitted to the hospital, the failure area cleaned up, and a new full-thickness or thick split-thickness graft applied. The area is functionally too important to have to tolerate the scarring of secondary healing.

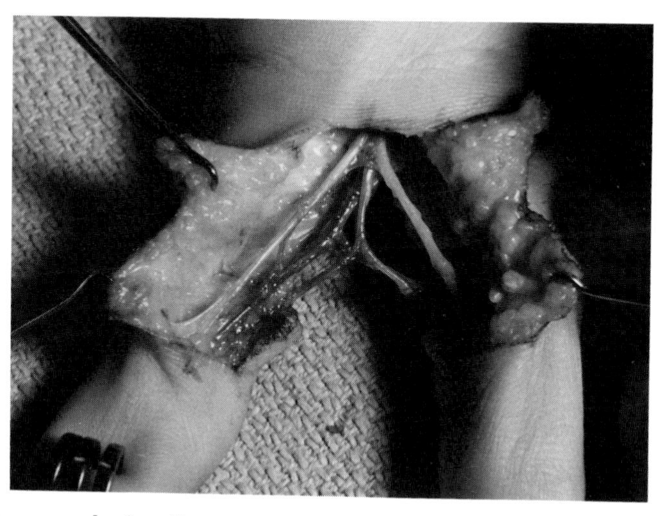

Fig. 10-13. *Neurovascular bundle separation.* The digital nerves can be readily separated in a proximal direction, but the Y division of the common digital artery limits the extent of proximal recession of the central dorsal flap. (From Flatt, A. E.: Practical factors in the treatment of syndactylism. In Littler, J. W., Cramer, L. M., and Smith, J. W., editors: Symposium on reconstructive hand surgery, vol. 9, St. Louis, 1974, The C. V. Mosby Co.)

SIMPLE INCOMPLETE SYNDACTYLY

Finger webs

Deepening the commissure in an incomplete or partial webbing of the fingers can appear to be deceptively easy. These webs frequently appear to have excess skin in their distal portion because the border of the web is not a knife edge but has depth from dorsal to palmar surfaces. But it should be appreciated that the extra skin is not in fact placed at the best site for reconstruction. These webs may have too much skin distally, but they suffer from the same proximal deficiency of skin as does the complete syndactyly. Extra skin still is needed to provide width to the top and sides of the commissure, and it is difficult to plan procedures that move the distal skin in a proximal direction.

When the distal edge of the web extends up to or beyond the level of the proximal interphalangeal joint, one has to use a shortened version of the standard syndactyly separation operation. The extra depth of skin at the distal edge may prove an embarrassment, and I often find it necessary to excise this skin. Unfortunately, the excess skin is usually not large enough nor of the correct shape to be used as the skin graft on the side of the commissure.

Operative procedures

Z-plasty operation. Traditionally Z-plasty has been used to deepen the shallow web. The two flaps do provide a widening of the span of the web but tend to sit back proximally in a narrow V-shaped apex. For this reason I prefer to use the four-flap Z-plasty frequently used in deepening the thumb web.

Transposing these flaps provides a very satisfactory, well rounded commissure in cases in which there is mild distal extension of the web. Technically it can be hard to do because the flaps are often very small and suturing may be difficult in the narrow space between the two fingers.

Butterfly flap. When the edge of the web is too distal but does not approach the proximal interphalangeal joint, I have found that excellent coverage is provided by the interdigital butterfly flap devised by Shaw.

This well thought out operation provides a broad dorsal flap that carries a considerable amount of skin on its distal portion to provide width and laxity for the commissure. The concept behind this operation is the use of two opposing Z-plasties in contrast to two in-line Z-plasties (Fig. 10-14). The use of a pair of opposing Z-plasties creates a common dorsal pedicle from the two dorsal halves of each Z-plasty, thus providing a broad dorsal flap with an elongated distal edge that can be moved palmarward and proximally to create a good commissure.

Two variations are possible. In one an inverted, wide angle V is created on the palmar skin by the edges of the palmar halves of each Z-plasty (Figs. 10-15 and 10-16). In the other, which can be used when the depth of the web is large, the inverted V is planned as an inverted Y. This alteration causes a blunting of the tips of the palmar halves of each Z-plasty, thereby providing a greater

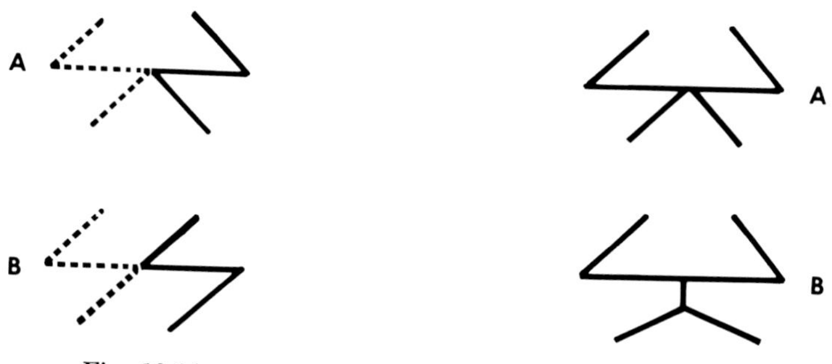

Fig. 10-14

Fig. 10-15

Fig. 10-14. *Double opposing Z-plasty.* **A,** A pair of opposing Z-plasties creates a common dorsal pedicle from the two dorsal halves of each Z-plasty. **B,** Two in-line Z-plasties do not yield this broad dorsal flap.

Fig. 10-15. *Butterfly flap.* This operation uses the principle of the double opposing Z-plasties to create a dorsal flap in incomplete syndactyly. **A,** The standard method. **B,** When the depth of the web is large the inverted V can be drawn as an inverted Y yielding broader lateral flaps than in **A.**

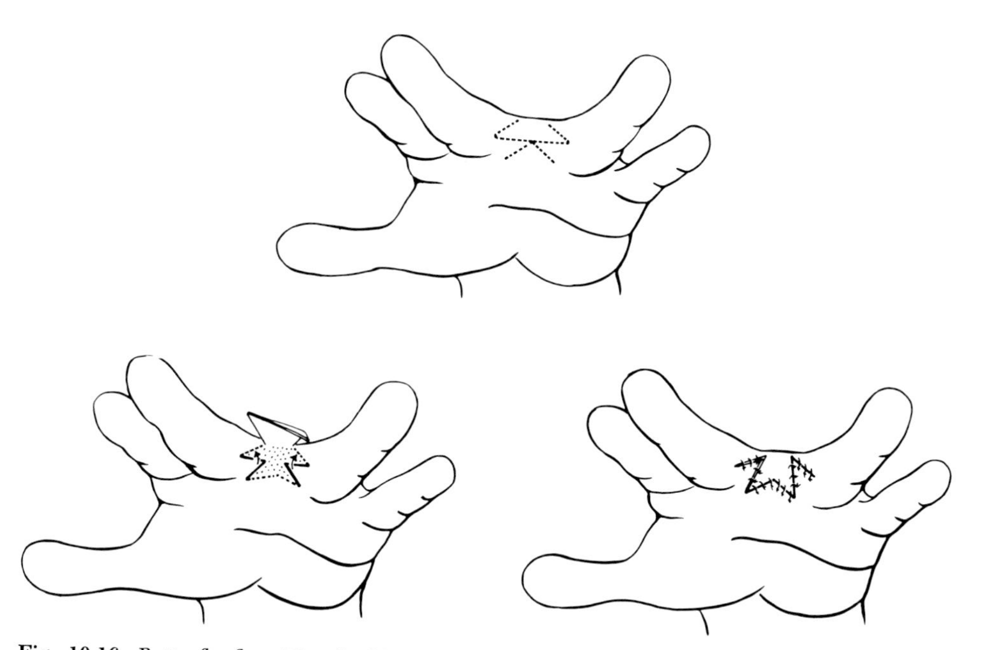

Fig. 10-16. *Butterfly flap.* The double opposing Z-plasties are drawn on a hand to illustrate the creation of a dorsal flap and the dorsal and proximal migration of the two triangular flaps.

amount of skin that can be transposed onto the sides of the fingers (Fig. 10-17). When planning the flaps, care must be taken not to make the distal tips of the dorsal flap too narrow. If this is done, the tips are extremely hard to suture and sometimes do not survive. The butterfly flap procedure deepens the web while relieving tension in both a transverse and vertical direction.

Dorsal-palmar flap combinations. When the webbing reaches to or near the proximal interphalangeal joint the basic scheme of using a broad dorsal flap

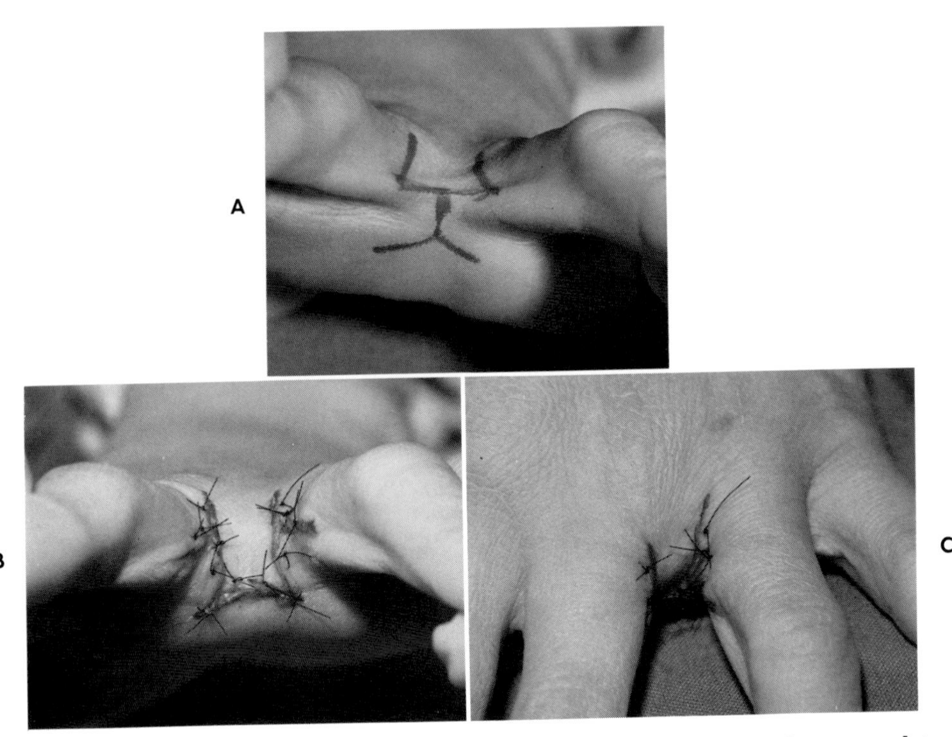

Fig. 10-17. *Butterfly flap.* **A,** The inverted Y was used in this hand. **B,** The flaps sutured in place. **C,** Satisfactory recession of the web space.

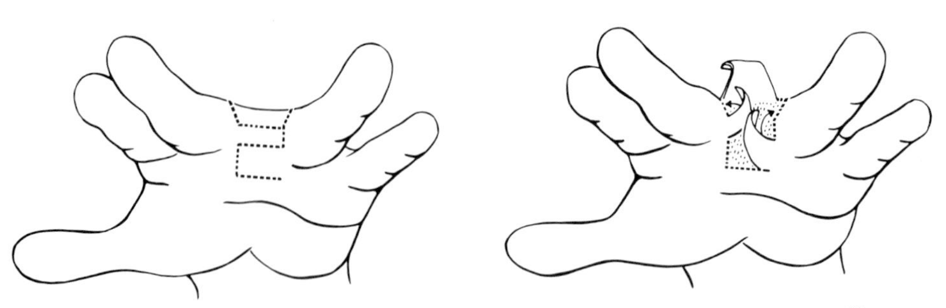

Fig. 10-18. *Dorsal-palmar flap combinations.* Occasionally the depth of the web is sufficient to yield an additional lateral flap on the nondominant finger and thereby reduce the size of the skin graft needed.

and a proximal palmar flap based on the dominant finger can be used, but it is the more distal portion that is difficult to close satisfactorily. One is often tempted to try to obtain total flap coverage, but this usually causes constrictions in the fingers, particularly on the proximal portion of the nondominant finger. Experience has taught me that better results are obtained by following the basic plan of using a supplemental skin graft even though excess distal skin may have to be excised.

Occasionally the dorsopalmar depth of the distal web edge is sufficiently great that one can make a rectangular flap, based distally, on the side of the nondominant finger, which will fold proximally and supply good lateral skin cover for the nondominant finger (Fig. 10-18). However, this is usually not adequate for total coverage of the nondominant finger, and it is much better to use a skin graft than to drag the skin edges together under tension. The key to the use of this lateral flap is the length of the dorsal flap. If the dorsal flap can be made a proper length by using only true dorsal skin without encroaching on the more vertical end surface of the web, then a lateral flap can be designed. If there is the least doubt about dorsal flap length, its lateral edges should be extended distally over the dorsal edge of the web to include the dorsopalmar vertical skin and the distal edge of the flap should be drawn at the junction with the palmar skin.

COMPLEX SYNDACTYLY

In complex cases, the syndactyly is often a minor problem in relation to the restoration of function to the whole hand. Many cases are complicated by bizarre skeletal deformities and by varying degrees of hypoplasia. The overriding principle is to establish as normal a skeleton as possible at a very early age. If this is done, then formation of digits and definitive work on the webs can be postponed until later.

When the complexity is a simple union between nails and distal phalanges

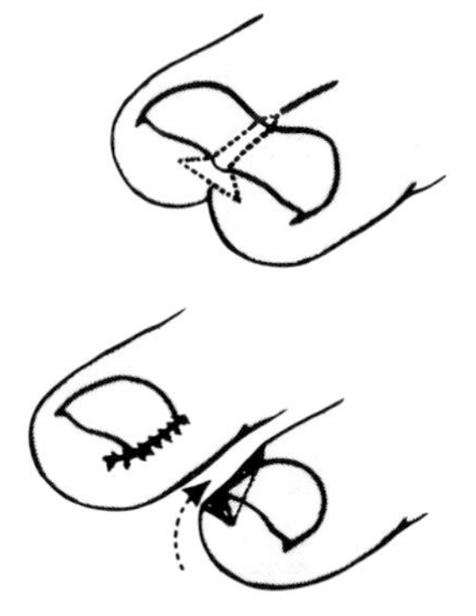

Fig. 10-19. *Complex syndactyly—common nail.* A central strip of nail wide enough to reduce each nail to a normal width should be removed. An additional wedge excision of finger pulp will allow good closure of the skin against the cut edge of the nail.

of fingers nearly equal in length, separation to the level of the distal interphalangeal joint and skin grafting of the defect will prevent the development of flexion contracture and allow full web correction at a more appropriate age. A common nail for the two fingers should not be simply cut in half; it will be too broad. The cut edges should be removed together with a strip of nail bed wide enough to narrow the nail to match the other digits. It is often useful to take a longitudinal wedge of finger pulp as well (Fig. 10-19). By this means the fingertip is narrowed, a spatulate end is avoided, and the terminal skin can sometimes be brought up adjacent to the cut nail edge.

If border digits are fused at their tips, separation to the level of the distal interphalangeal joint is usually not sufficient. The flexion contracture of the more proximal joints is usually so severe that complete separation is needed as soon as it is feasible. Even if separation is done early, secondary procedures to overcome the flexion contractures are still frequently necessary. The timing and extent of these secondary operations are often extremely difficult to decide because of the risk of epiphyseal growth arrest.

When the syndactyly conceals a polydactyly, the interdigital extra phalanges have to be removed. The only advantage of this particular problem is that the broad width between the digits supplies a considerable amount of extra skin that allows easy interdigitation of the skin flaps after the two digits have been separated (Fig. 10-20).

It is always tempting to try to maintain four fingers in a hand, and some complex cases of syndactyly may appear to present a relatively simple technical problem.

At operation, however, additional complexities can be found that may lead to bad results later (Fig. 10-21). When severely complicated syndactyly is present, I believe it ridiculous to attempt to obtain four normal looking and moving fingers; I go straight to a plan that will allow formation of three good digits. Fig. 10-22 illustrates a complex polysyndactyly on the radial side of the hand involving the index and long fingers. It was elected to trim the double proximal phalanx of the index finger and to totally remove the extremely disturbed skeleton of the long finger. The early end result of this showed that the epiphysis of the index proximal phalanx was growing satisfactorily, and long-term follow-up showed that growth continued and a satisfactory three-fingered hand was obtained.

• • •

Syndactyly is a frequent feature in skeletal dysplasias and various malformation syndromes. In fact, Poznanski published a list of 28 syndromes associated with hand syndactyly. It is said to be commonly associated with the following seven conditions: Aarskog, acropectorovertebral, Apert, brachydactyly B, Carpenter and Poland syndromes and oculodentodigital dysplasia. These syndromes in themselves are not all equally common, and I have chosen to discuss the technical problems of only a few of them.

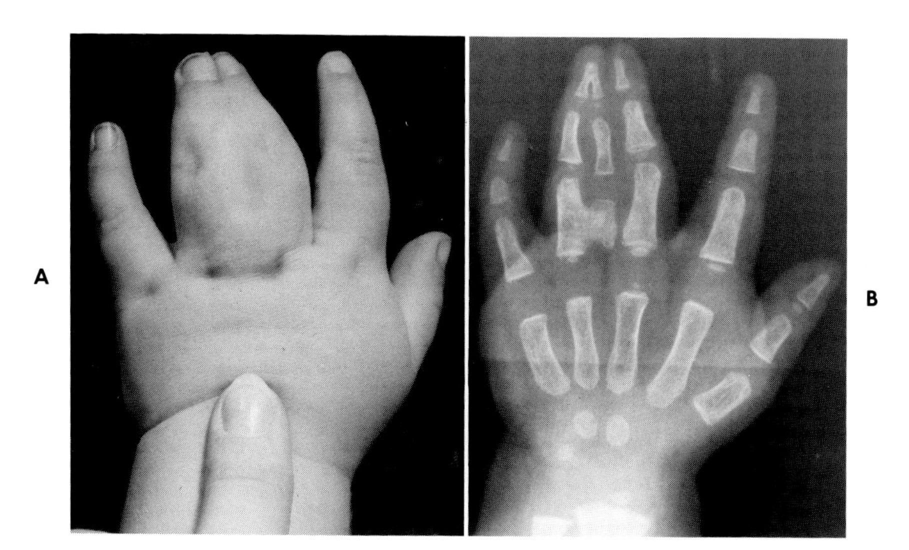

Fig. 10-20. *Concealed polydactyly.* **A,** In this patient a Type II polydactyly is hidden in the web between two apparently normal digits. **B,** The polydactyly that could be missed if an x-ray film is not made. (From Wood, V. E.: The treatment of central polydactyly, Clin. Orthop. **74**:196-205, 1971.)

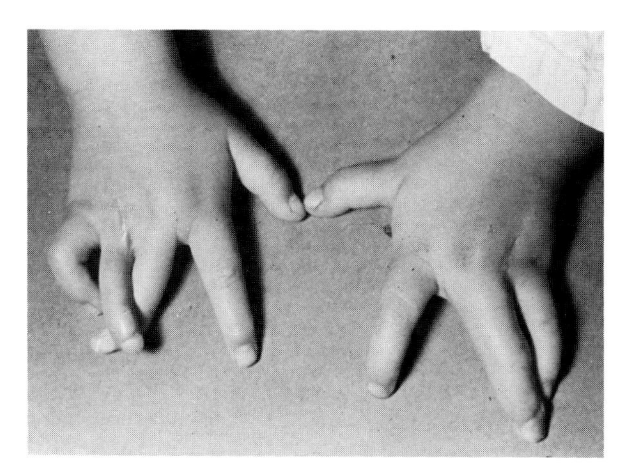

Fig. 10-21. *Complex syndactyly—bad result.* The left hand has been provided with a functional three-fingered hand. Attempts to preserve four fingers in the right hand led to functional and cosmetic disaster. (From Wood, V. E.: The treatment of central polydactyly, Clin. Orthop. **74**:196-205, 1971.)

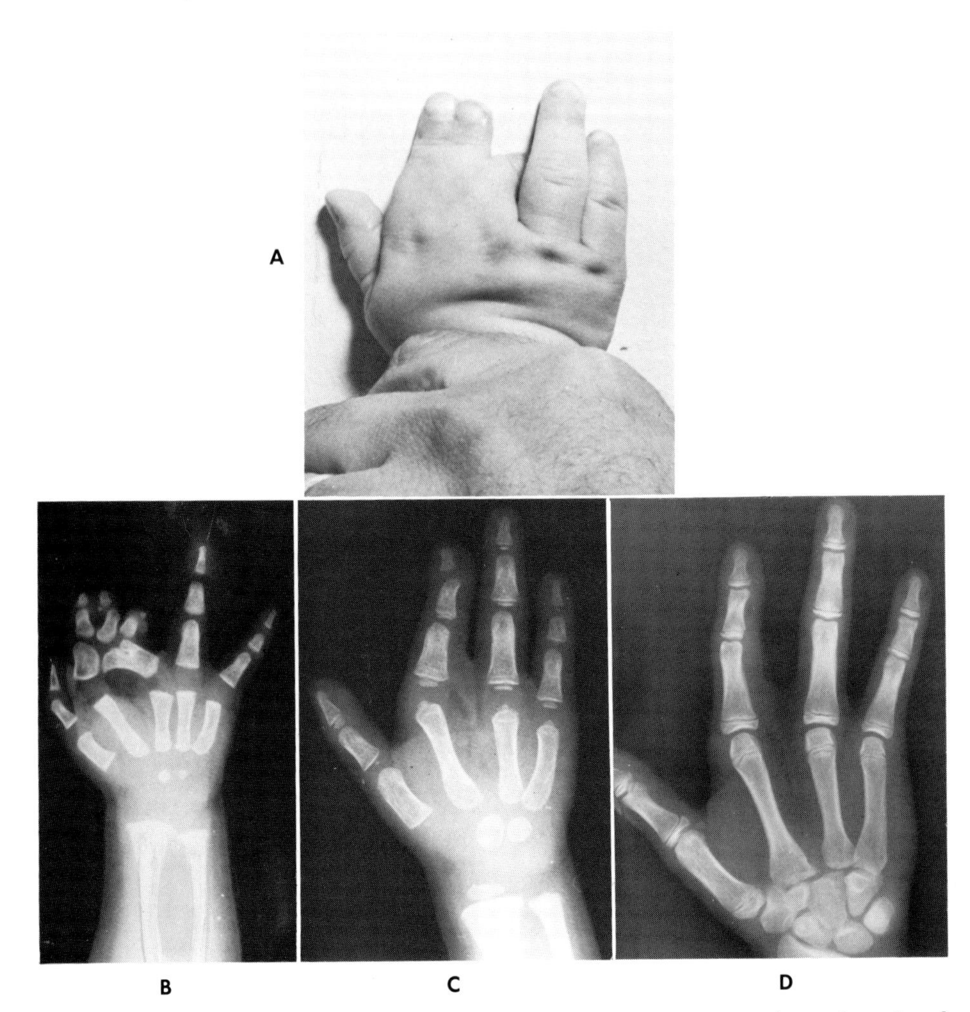

Fig. 10-22. *Complex syndactyly—three-fingered hand.* **A** and **B**, Complex polysyndactyly involving index and long fingers. **C**, The double proximal phalanx of the index finger was divided longitudinally and the deformed long finger removed. Remolding of index proximal phalanx is proceeding satisfactorily 2 years after the operation. **D**, Nine years after the operation the phalanx has remolded completely and epiphyseal growth has continued normally. A good functioning three-fingered hand has been produced. (From Flatt, A. E.: Practical factors in the treatment of syndactylism. In Littler, J. W., Cramer, L. M., and Smith, J. W., editors: Symposium on reconstructive hand surgery, vol. 9, St. Louis, 1974, The C. V. Mosby Co.)

ACROSYNDACTYLY

In acrosyndactyly there is a fusion between the more distal portions of the digits with the space between the digits varying from broad down to pinpoint in size, but there is always a communication between the dorsal and palmar aspects of the conjoined digits (Fig. 10-23). This condition, which has been variously called terminal fenestrated, exogenous, or amniogenous syndactyly, is not hereditary and occurs spontaneously.

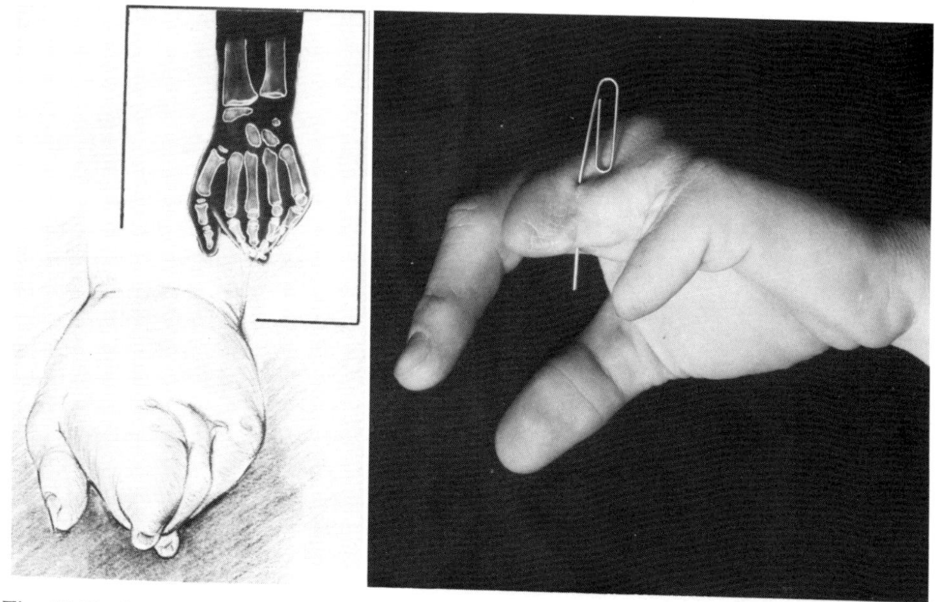

Fig. 10-23. *Acrosyndactyly.* A condition in which there is fusion of the terminal portion of two or more digits with proximal epithelial-lined clefts or sinuses between the digits. (From Walsh, R. J.: Acrosyndactyly: a study of twenty-seven patients, Clin. Orthop. **71**:99-111, 1970.)

It is probably caused by an intrauterine insult that occurs after the fingers have fully separated. The dissections of Losch and Duncker have shown that healing after the insult produces scar tissue that draws together the affected fingers and is subsequently covered by an ingrowth of ectoderm. Acrosyndactyly is usually associated with shortening of the affected fingers, and Patterson, who has studied this condition extensively, has established the criteria that these cases exhibit no real proximal webbing, no bony fusion between the fingers, a high degree of association with constriction rings, and various abnormalities of the affected fingers apart from the soft tissue fusion.

When constriction rings are present, Patterson postulates that the intrauterine insult is an ulceration occurring at the rings with subsequent fusion. He has also pointed out that there may be a more fundamental developmental failure because of the frequent presence of associated but unrelated congenital abnormalities such as clefts of the lip and palate, clubfeet and cardiac abnormalities.

Separation of the less complicated forms of acrosyndactyly is simple, and the cosmetic and functional results are usually satisfactory. When, however, acrosyndactyly is associated with congenital constriction rings, the technical problem can be severe. The basic principle of establishing a proper longitudinal skeletal pattern must be combined with an early decision as to whether to salvage three or four fingers.

Walsh, in 1970, reviewed 18 University of Iowa patients who had 24 involved hands with a total of 68 involved digits. Every one of these digits showed underlying distortion or absence of bony structure although no symphalangia was found. He established a useful clinical distinction of mild, moderate, or severe cases based on the skeletal abnormalities (Fig. 10-24).

Mild cases are those in which three phalanges and two interphalangeal joints are present in the affected digits. Moderate cases are those with two phalanges and only one interphalangeal joint, while the severe cases usually show only stubby fingers containing one phalanx and no interphalangeal joints. According to these criteria there were 4 digits with mild acrosyndactyly, 44 with moderate, and 20 with severe. This low proportion of mild involvement has continued to be present in our later patients, who now total 66.

There is no standard pattern in the degree of adherence of the digits nor is there a standard site for the epithelial lined cleft or sinus that is always present between adjacent digits. Occasionally the fingers are joined by a narrow skin bridge. Frequently this connection is snipped or tied off in the newborn nursery (Fig. 10-25).

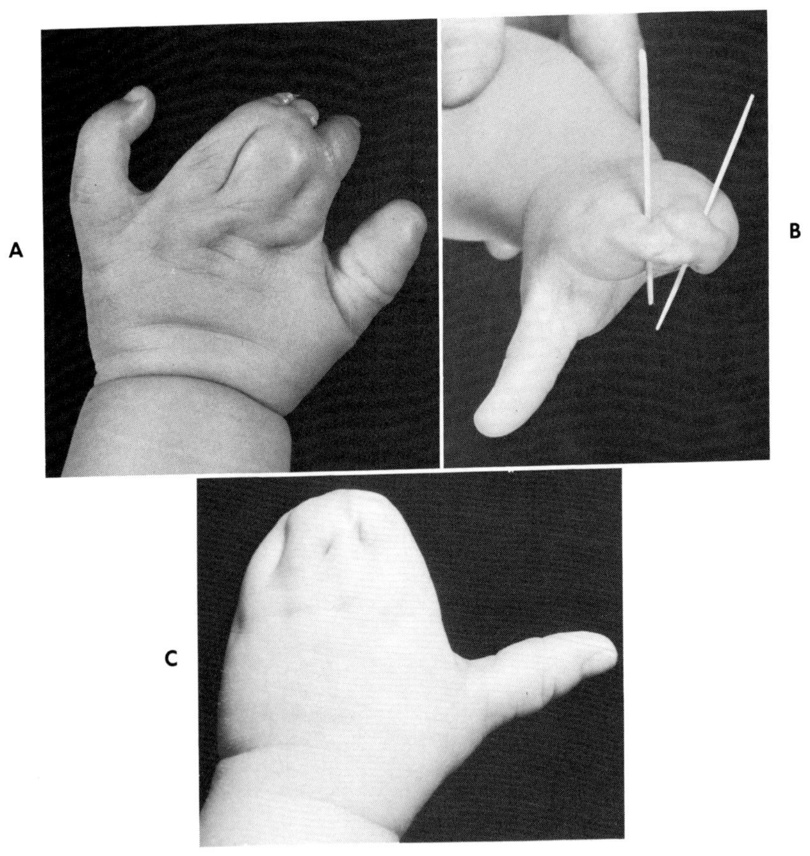

Fig. 10-24. *Acrosyndactyly.* **A,** Mild. **B,** Moderate. **C,** Severe.

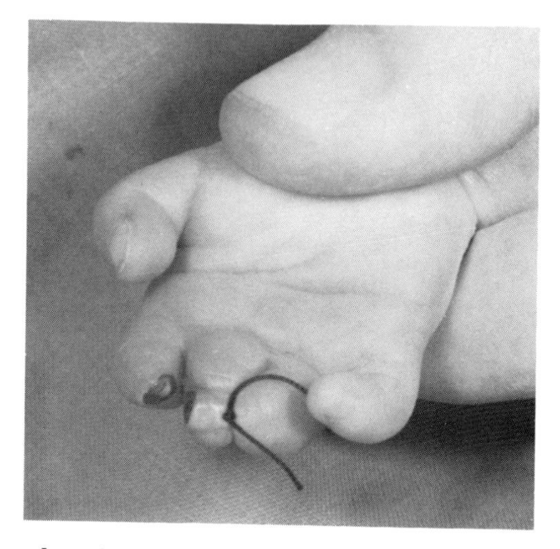

Fig. 10-25. *Acrosyndactyly.* When the fingers are joined by a narrow skin bridge, this connection can be cut or tied off with a black silk suture.

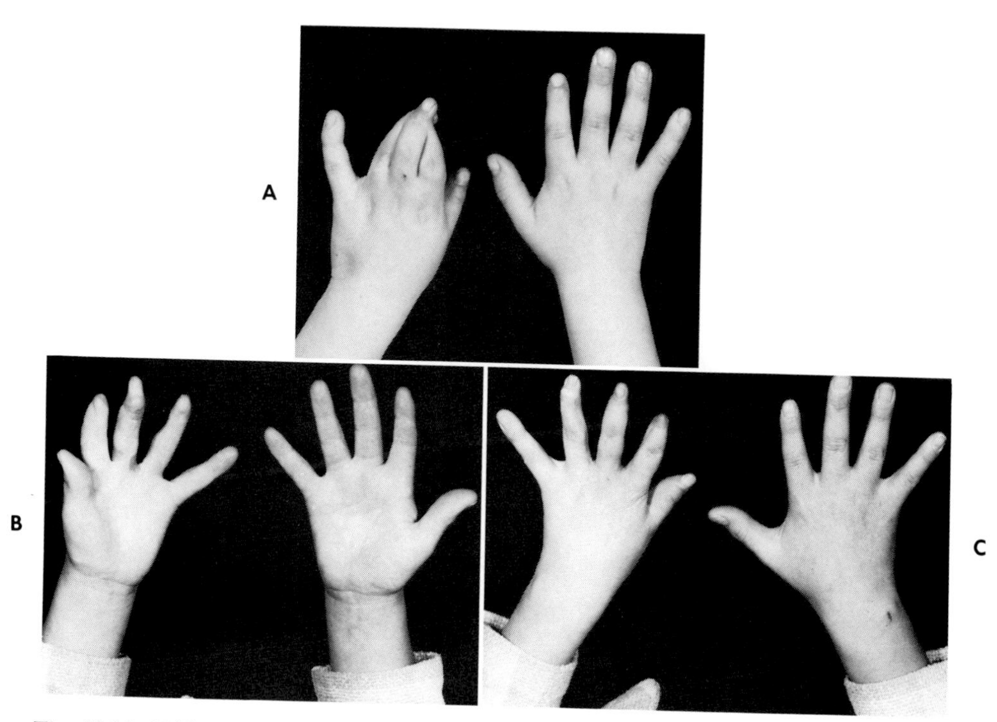

Fig. 10-26. *Mild acrosyndactyly.* **A,** Before surgery. **B** and **C,** After surgical release. Separation should be done at an early age. Split-thickness skin grafts give satisfactory cover for the raw areas of the finger tips.

Operative plan

General principles. The extent of the fusion is important in determining whether the tips of the digits are to be released at a preliminary operation and the commissures reconstructed later or whether tip release and commissure construction can be planned as one operation.

The other important factor in operative planning is the site and extent of the clefting or sinus between the digits. Whatever the size of the epithelial lined space, it will invariably be more distal than the optimal site for commissure reconstruction. Sinuses will therefore have to be excised, but longitudinal clefts can often be incorporated in the operative plan. These sinuses tend to run a sloping course from dorsal to palmar surface, and more often than not, one or even both openings will occur in the middle of a potential skin flap. Skin coverage of the separated fingers is obtained by local flaps, but planning these flaps around the sinuses can be a most frustrating experience. When constriction rings are present, local flaps can often be constructed out of the fatty skin protuberances. A surprising amount of coverage can often be obtained after these protuberances have been defatted and rotated into place. Despite the use of these local flaps, skin grafts frequently have to be used, and to a greater extent than in a simple syndactyly. A significant finding in our long-term follow-up study has been the limitation of finger flexion. Only poor or fair motion is obtained in most of the proximal interphalangeal joints of fingers with moderate involvement. A similar limitation of motion was found in the distal interphalangeal joints of patients with mild degrees of acrosyndactyly.

Mild cases. The basic aim in surgical correction of these hands is to free the digits early enough to allow the fingers to grow parallel and without articular deformities, particularly at the metacarpophalangeal joints. Frequently these cases have relatively long clefts rather than narrow cylindrical sinuses between the fingers, and skin grafts are needed on the more distal portions of the fingers.

Usually I prefer to separate these fingers around 6 months of age or as soon as possible if the patient is seen at a later age. Split-thickness skin grafts provide perfectly satisfactory skin cover at this primary operation. The technical problems involved in using multiple, small, full-thickness skin grafts are great, and the ultimate result is not significantly better than that obtained with split-thickness skin grafts (Fig. 10-26).

Subsequently proximal recession of the web spaces using the standard syndactyly plan will provide a good, functioning hand.

Moderate cases. In most of these patients' hands only two phalanges and one interphalangeal joint can be demonstrated on x-ray examination. Fusion of these shortened digits towards their tips will adduct the border fingers at a considerable angle and cause early joint distortion unless the digits are released (Fig. 10-27).

Frequently there appears to be a jumbled collection of fingertips distal to the site of fusion, since it is rare for fusion to occur at the most distal portion

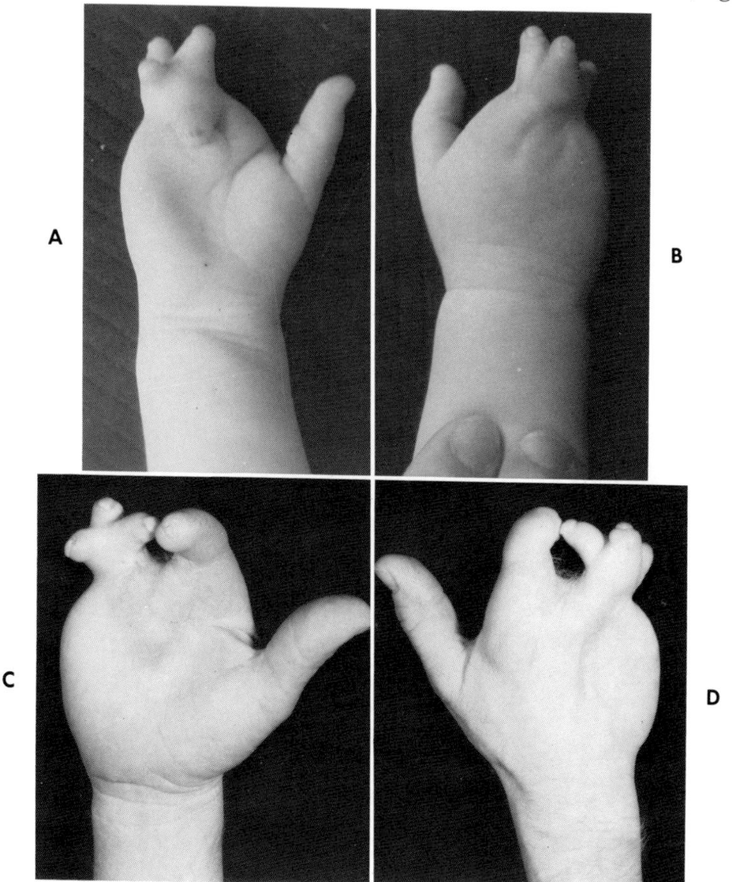

Fig. 10-27. *Moderate acrosyndactyly.* **A** and **B,** This patient was first seen as a child in 1932 when the index finger was separated from the others. **C** and **D,** Thirty-seven years later he returned, but no further surgery was advised.

of the fingers (Fig. 10-28). These fingertips may not carry nails and often cannot be directly matched to an appropriate finger; however, every attempt should be made to do this rather than to excise these tips because they can contain phalangeal buds that may be associated with articular spaces not visible on the x-ray film.

The principles of surgical correction are the same as for mild cases—early liberation of the trapped fingers followed by deepening of the commissures (Fig. 10-29).

Severe cases. The variety of deformities possible in these cases makes it impossible to discuss their treatment except in generalities. Usually they are associated with varying degrees of constriction rings, and bulbous fatty protrusions occur distal to the rings. Often the fingers are affected to different degrees, and the varying lengths of adherent digits make the planning of skin coverage for the separated digits a formidable task.

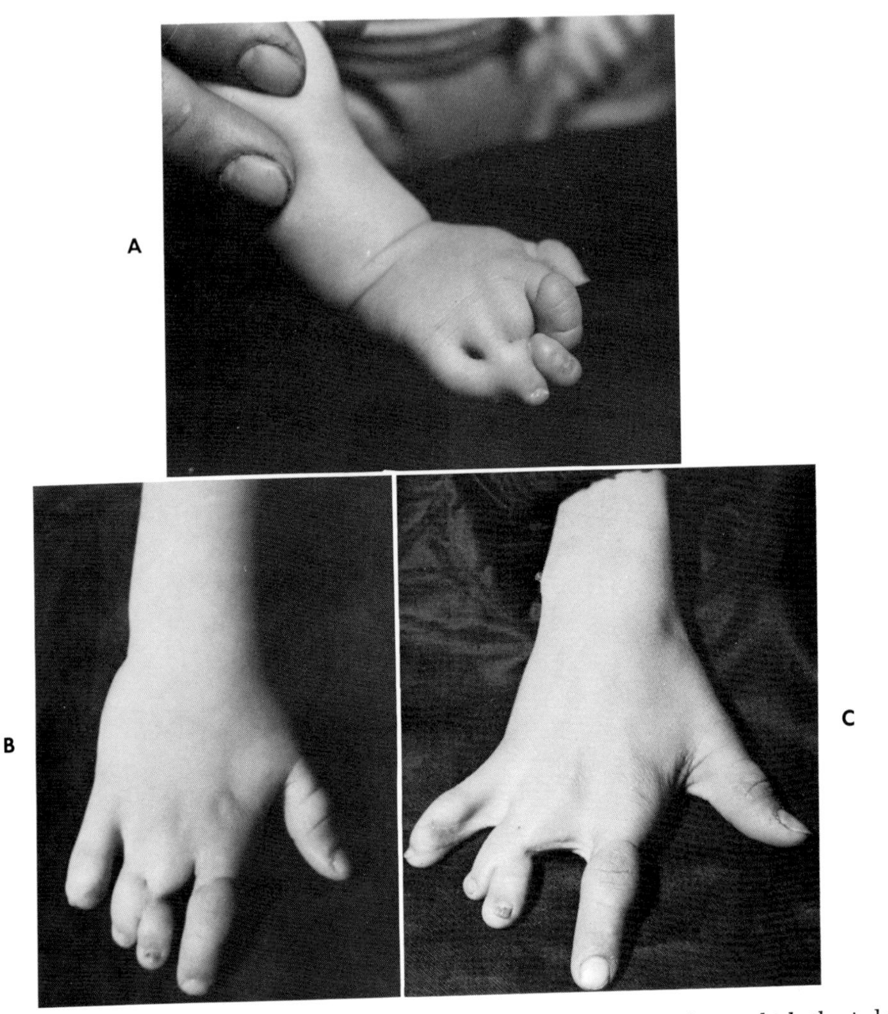

Fig. 10-28. *Moderate acrosyndactyly.* **A,** A boy with ulnar acrosyndactyly in which the index was not involved. **B,** Surgery released the other border finger—the small. **C,** In adult life he has a good functional hand.

Since the fingers are always short and are frequently only stubs, it is important to separate them as far proximal as is possible. Usually the separation can be combined with the deepening of the commissures, which should be carried well proximal, at least to the level of the metacarpal heads (Fig. 10-30).

Early release of these stubby fingers is mandatory since it will allow the maximum use during the growth years. These little hands frequently present a much greater technical challenge than the actual separation of the fingers. When a ballooned and bulbous fingertip has to be amputated, the skin should be used as a graft after it has been defatted and flattened. Revision and further deepening of the commissures in teenage is often necessary. The problems caused by the constriction rings are fully dealt with in Chapter 11, page 213.

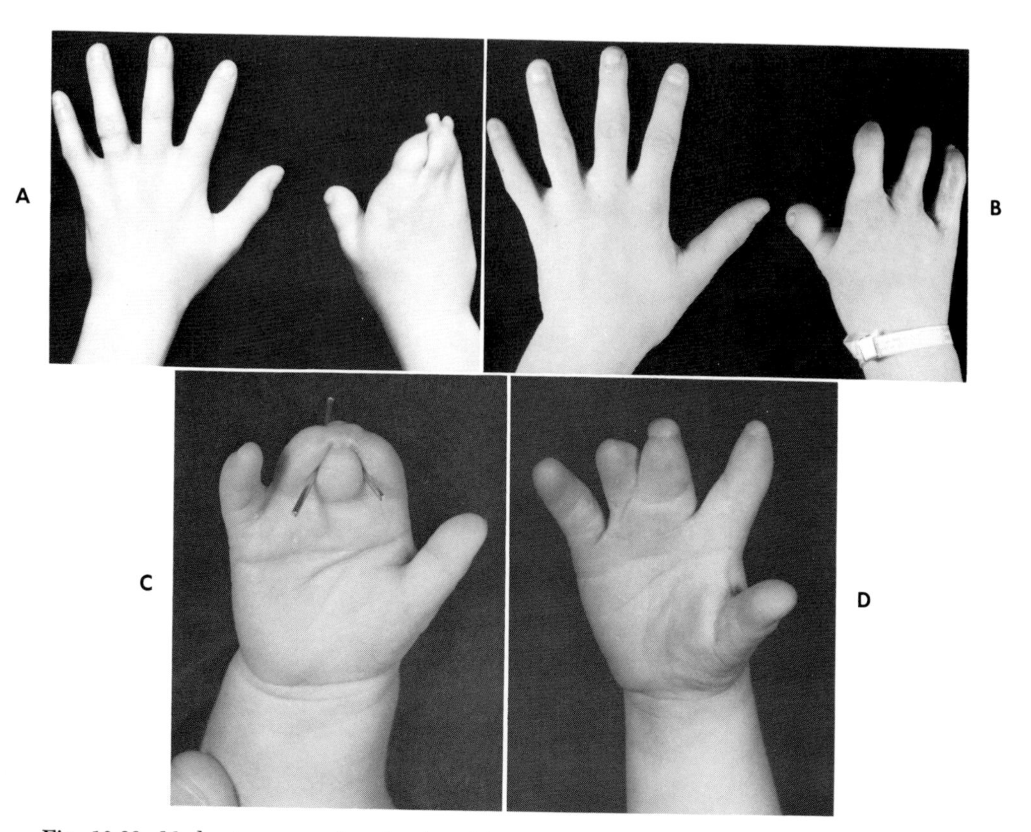

Fig. 10-29. *Moderate acrosyndactyly.* **A** and **B,** A patient in whom a three-fingered hand was created. **C** and **D,** A similar type of deformity in which it was possible to develop four fingers.

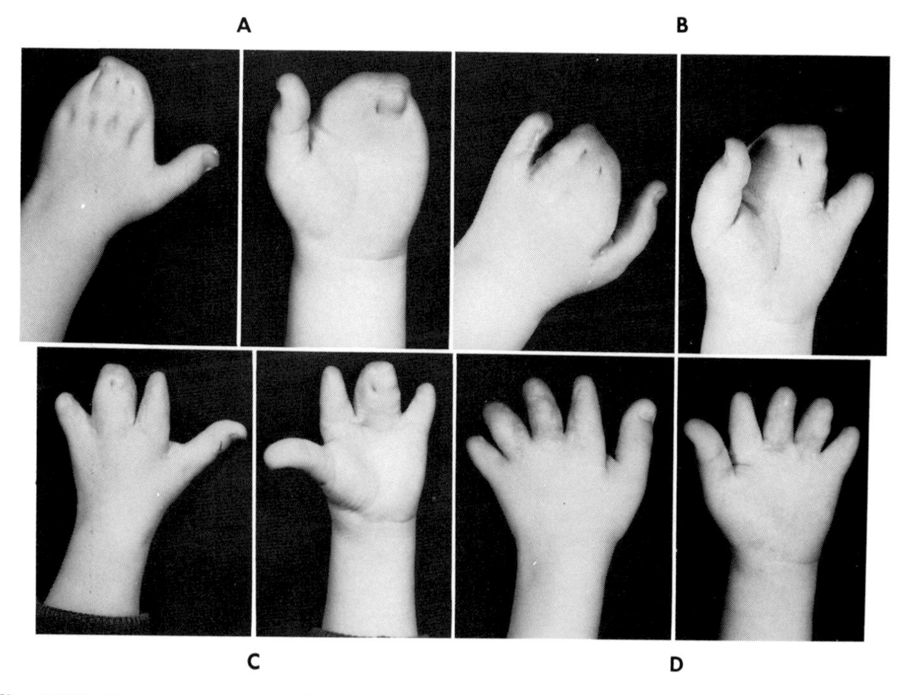

Fig. 10-30. *Severe acrosyndactyly.* A severe involvement of the hand in which by staged surgery four fingers were obtained. **A,** Before surgery. **B,** Border digit freed. **C,** Index finger separated. **D,** At completion of surgery.

SYMBRACHYDACTYLY

The infinite variety of these conditions makes it extremely difficult to describe the therapy for this group of deformities, but the principles are, I believe, well established. The skeleton must be separated early to allow longitudinal growth. Frequently this dictates that a three-fingered instead of a four-fingered hand be obtained. It matters little if in the early months of life a false or elective syndactyly is established by skin grafting if in turn this allows longitudinal definition of the skeleton. Later separation can produce hands that function well but are frequently unattractive (Fig. 10-31).

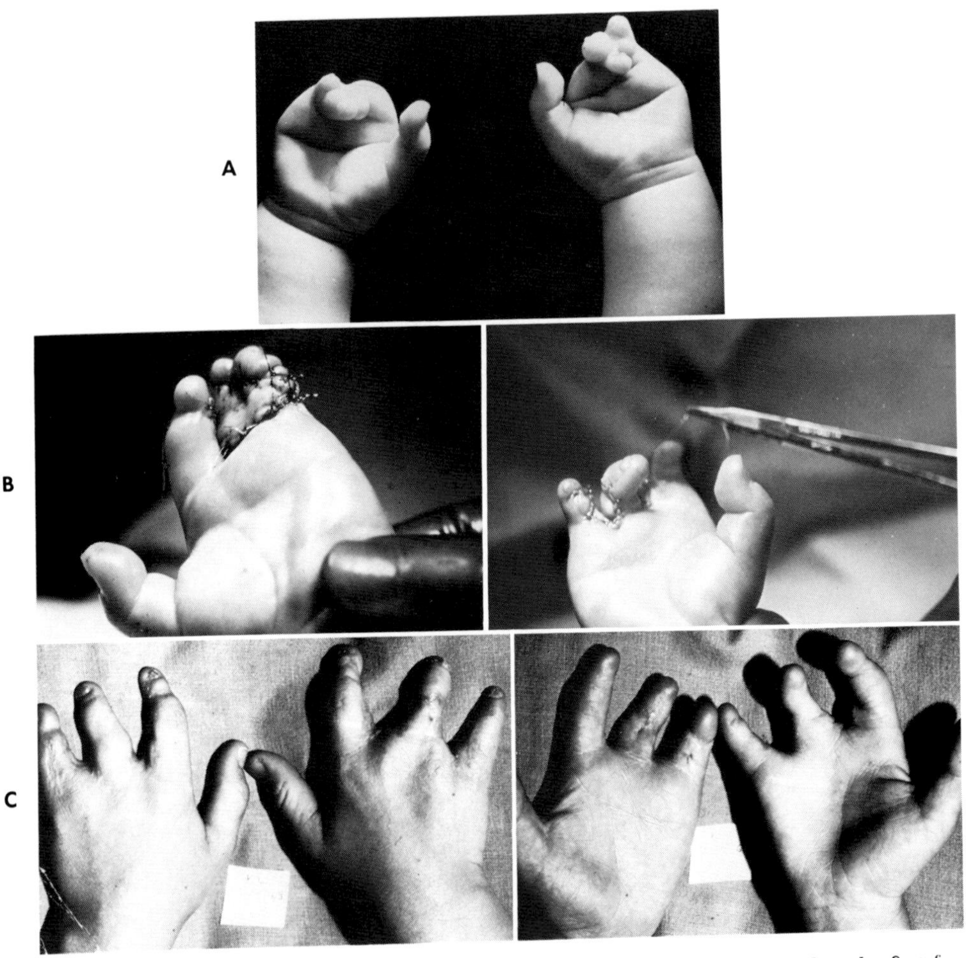

Fig. 10-31. *Symbrachydactyly.* **A,** Coalesced fingers such as these must be freed in the first few months of life. **B,** At primary surgery, fingers must be separated sufficiently to allow normal configuration of hand skeleton and to be properly covered with skin. It is often advantageous to create a syndactyly at this time, as was done in the left hand. **C,** The fingers were separated later, and a reasonable pair of three-fingered hands was salvaged from the original jumble. (Courtesy Dr. D. C. Riordan, New Orleans, La. From Flatt, A. E.: Practical factors in the treatment of syndactylism. In Littler, J. W., Cramer, L. M., and Smith, J. W., editors: Symposium on reconstructive hand surgery, vol. 9, St. Louis, 1974, The C. V. Mosby Co.)

APERT SYNDROME

Acrocephalosyndactyly was first properly described as a syndrome by Apert in 1906. Since that time a variety of closely related genetic disturbances characterized by premature fusion of the cranial sutures and varying degrees of syndactyly of one or more of the hands and feet have been described (Fig. 10-32).

All individuals, whether Oriental, black, or white, show similar deformities. Mental retardation occurs but is certainly not inevitable. Many patients with this condition can be found in mental institutions, some because their mental status justifies it and others because their parents have abandoned them as a result of their repulsive appearance.

Two main clinical categories have been distinguished: the true or typical Apert syndrome and an atypical group. The hand in true Apert syndrome shows severe interdigital bony union, typically with a mid-digital bony mass and a

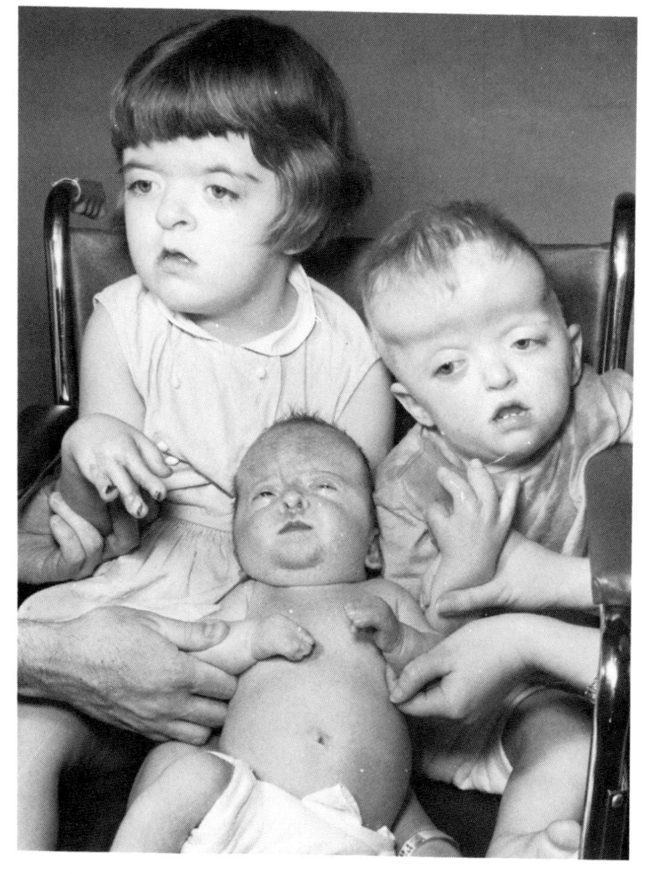

Fig. 10-32. *Apert syndrome.* This syndrome is characterized by premature fusion of the cranial sutures and varying degrees of syndactyly of the hands and feet. The hands of these children are at varying stages of surgical correction. (From Hoover, G. H., Flatt, A. E., and Weiss, M. W.: The hand and Apert's syndrome, J. Bone Joint Surg. [Am.] **52**:878-895, 1970.)

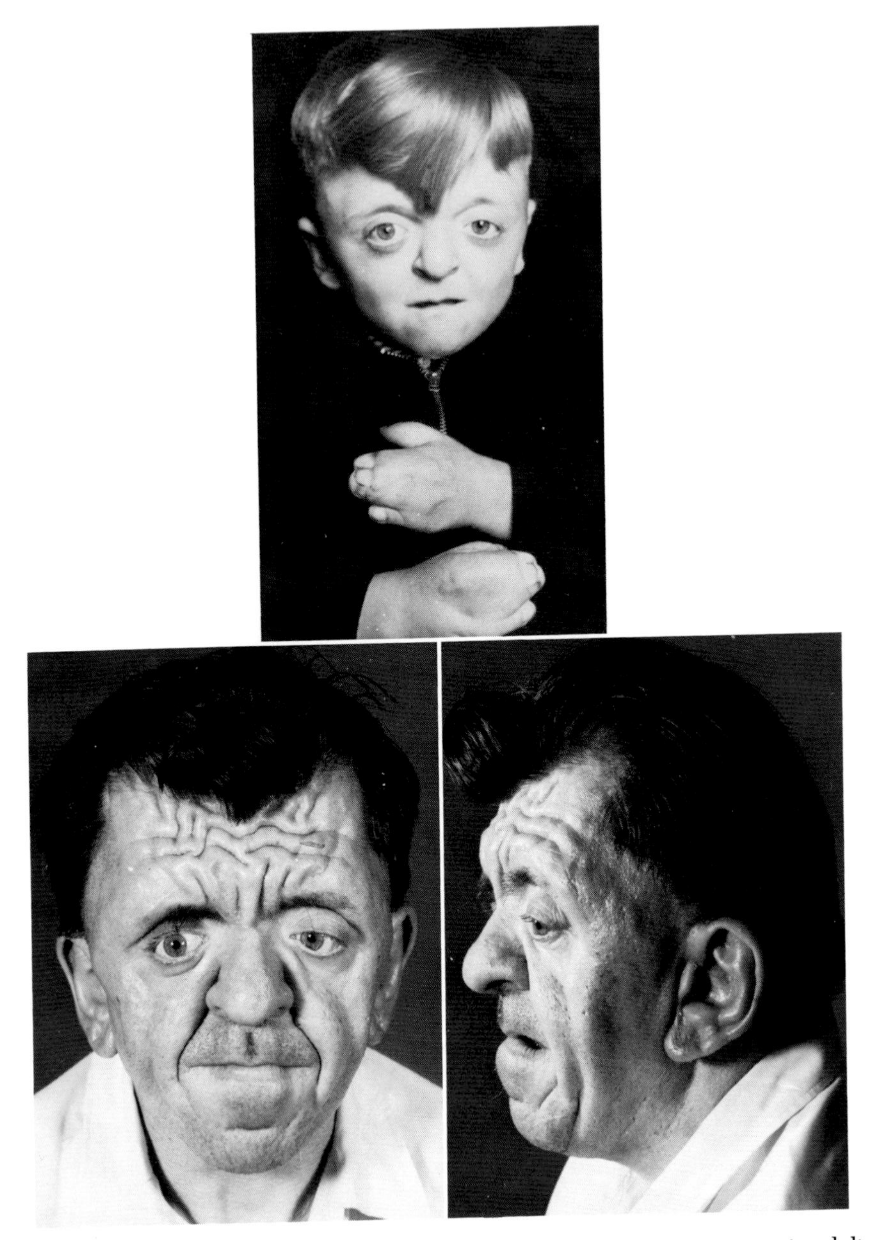

Fig. 10-33. *The head in Apert syndrome.* The fully developed deformity is seen in adult life. The apex of the skull is at the bregma, and the occipital area is flattened. The eyes are bulging and wide set. (From Hoover, G. H., Flatt, A. E., and Weiss, M. W.: The hand and Apert's syndrome, J. Bone Joint Surg. [Am.] **52**:878-895, 1970.)

single nail common to the index, long, and ring fingers. This severe mitten involvement of the hand is not seen in the atypical cases, although varying degrees of hand abnormality can be found. McKusick has subdivided these cases into various types such as Apert-Crouzon or Vogt, Saethre-Chotzen, Waardenburg, and Pfeiffer. As more patients are recognized with these conditions, the variety of abnormalities and their combinations increases, and these distinctions are becoming somewhat blurred.

A similar but distinct pair of syndromes can be recognized since they are characterized by acrocephalopolysyndactyly. In both Noack and Carpenter syndromes the cranial and facial features are similar to Apert but the hands differ because of the presence of polydactyly.

Apert syndrome is rare, but some reports suggest an incidence as high as 1 in 200,000. The syndrome can probably be attributed to a single gene in heterozygous form, and the sporadic cases that appear are occasioned by mutations in the germ cells of one or the other parent. Once the condition has appeared, it shows a strong dominance. A married woman with Apert syndrome is known to have produced three children who all have the typical syndrome. Two other instances have been reported in which mothers with acrocephalosyndactyly have produced children with deformities identical to their own.

Physical signs

In Apert syndrome the forehead is high and broad (Fig. 10-33), the apex of the skull is at or near the bregma, and the occipital area is flattened. The eyes are bulging and wide set with the lateral canthus lower than the medial. There is a short broad nose and a prominent lower jaw. The maxilla is sunken and the upper teeth are crowded, causing them to appear in a double row in some patients. The anterior palate is high and arched, and occasionally it is cleft posteriorly. Recent advances in facial reconstructive surgery have produced operative plans that, although formidable, can create dramatic improvement in the facial appearance.

The most consistent finding in the hands is syndactyly of all the fingers (Fig. 10-34). The fingers are usually stiff in adult life, but in early childhood may retain some interphalangeal joint motion. The palmar aspect of the hand is usually spoon shaped because the fingers are held tightly together at their distal ends. Since independent motion of the fingers is not possible, the hand can only be used for gross movements in a paddle-like fashion. The thumb is very short and is always deviated radially at the metacarpophalangeal joint; it may be included in the webbing with the fingers. There is frequent shortening of the arm and forearm with limitation of elbow motion and shoulder abduction. The lower extremities show foot deformities and usually complete syndactyly of all toes. Abnormalities of the gut have been described in autopsy findings.

We at the University of Iowa have a total of 25 patients in our study group and have reported the detailed clinical findings in 20 cases of typical Apert syndrome. With few exceptions the hand deformities in these patients were

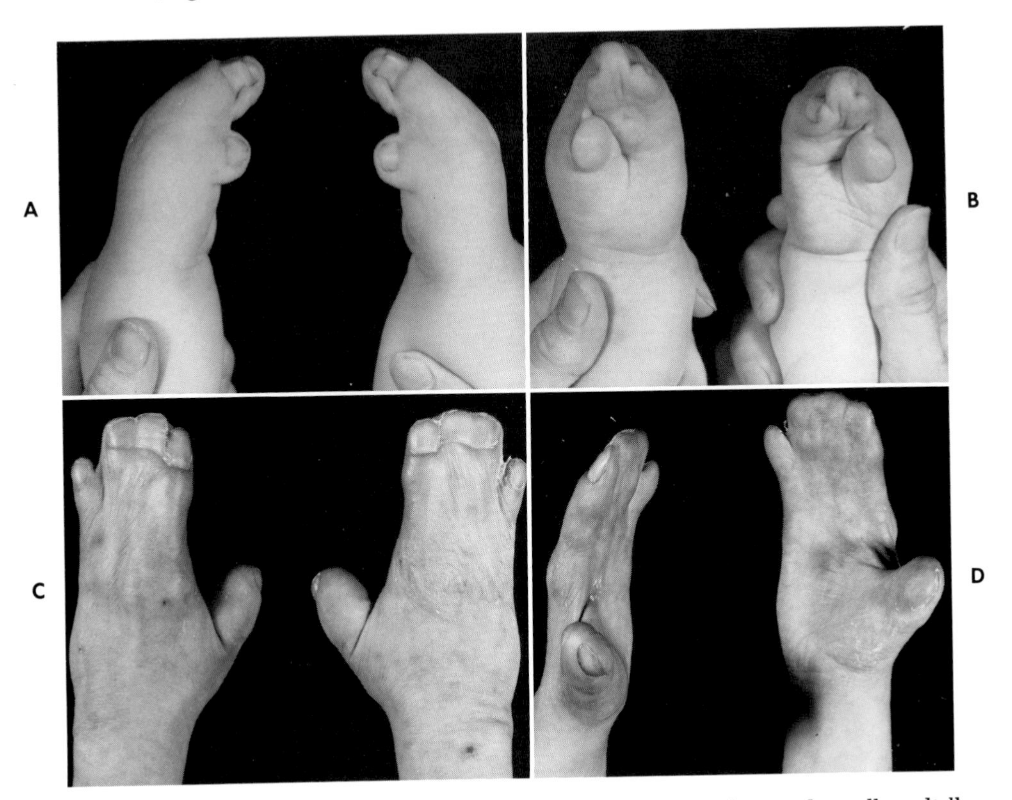

Fig. 10-34. *The hand in Apert syndrome.* **A** and **B,** The thumbs are short and usually radially deviated, and the palm is spoon shaped because the fingers are fused at their tips. **C** and **D,** In adults the hand is often flat and the interphalangeal joints are completely stiff. (From Hoover, G. H., Flatt, A. E., and Weiss, M. W.: The hand and Apert's syndrome, J. Bone Joint Surg. [Am.] **52**:878-895, 1970.)

symmetrical. All the hands had a complex syndactyly of the index, long, ring, and small fingers. In all of the adult patients the interphalangeal joints of the fingers were stiff. All of the fingers in children over 4 years of age showed no active movement at the interphalangeal joints. In the children 3 years old and younger, either small finger and one index finger retained some motion. Most of the fingers were much shorter than normal. The palmar aspect of the hand was occasionally flat but usually spoon shaped because the fingers were held tightly together at their distal ends.

The index, long, and ring fingers always demonstrated complete syndactyly and frequently had a common nail. The small finger generally showed complete soft tissue syndactyly with the ring finger and an independent nail. Occasionally the syndactyly was present only to the level of the distal interphalangeal joint. Syndactyly between the thumb and index fingers occured in over 30% of our patients.

Disturbances of the skeleton are profound and become more apparent as the child grows. Bony fusions in the hand become evident with increasing age. It is probable that there is complete cartilaginous continuity across the digital

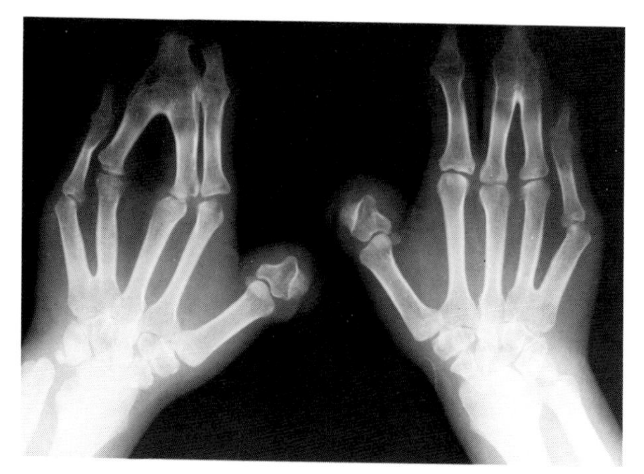

Fig. 10-35. *X-ray film of the hand in Apert syndrome.* Characteristic findings are fusion of the bases of the fourth and fifth metacarpals, obliteration of the interphalangeal joints, complex syndactyly, and short thumbs. (From Hoover, G. H., Flatt, A. E., and Weiss, M. W.: The hand and Apert's syndrome, J. Bone Joint Surg. [Am.] **52**:878-895, 1970.)

joints, and the ossification occurring with growth allows symphalangia and synostosis to become evident on x-ray films (Fig. 10-35).

Hand function

As originally described by Apert, hand function prior to surgery typically consists of using one hand against the other in spoonlike fashion.

Our patients who have only their thumbs partially free have a prehensile pattern consisting of the ulnar aspect of the thumb (usually the interphalangeal or the metacarpophalangeal joint surface) opposing the shaded area of the proximal radial border of the index finger as depicted in Fig. 10-36. All prehensile tasks are accomplished by using this lateral type of pinch and by using one hand against the other. Only one patient could not, by some ingenious means, perform the preoperative tasks expected of anyone of her intelligence and level of motor development. She had complete syndactyly of the thumb and all fingers, and her function consisted entirely of holding objects between her two hands and of batting at toys.

The operations I have used improved the versatility of prehensile patterns that these patients could use rather than added to the number of tasks they could perform. The three-fingered hands seem to have a greater versatility of prehensile patterns as compared to the four-fingered hands. This is probably because of the ability to utilize more of the movement at the metacarpophalangeal joints.

All our patients with three-fingered hands can oppose the thumb to at least the base of each finger, and two patients can oppose the thumb to the tips of all fingers. Tasks that had been accomplished by using one hand against the other prior to surgery are now accomplished utilizing more of the prehensile

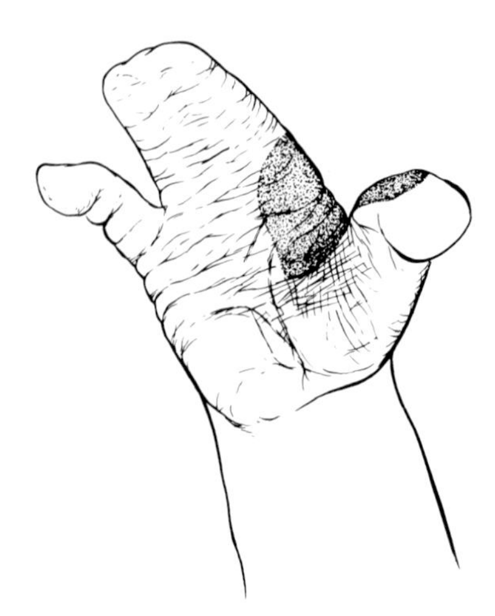

Fig. 10-36. *Prehension in Apert syndrome.* Grasp and any prehensile activity are difficult for these patients. The shaded areas are the surfaces against which objects are held. (From Hoover, G. H., Flatt, A. E., and Weiss, M. W.: The hand and Apert's syndrome, J. Bone Joint Surg. [Am.] **52**:878-895, 1970.)

patterns that a normal hand uses to perform the same task. Use of the body as a stabilizing surface for objects is eliminated to a great extent. Radio or television dials can be turned without being approached with the hand or body in the awkward position previously needed. Buttoning and bow-tying require fewer movements, and awkward hand positioning is eliminated. The grosser movements needed in self-care skills do not seem to change in pattern after surgical treatment.

Operative treatment

Although many of the patients with Apert syndrome are mentally retarded to a varying degree, most of them are able to, or can be taught to, care for themselves. A hand with individual fingers, even though they may be stiff, is more useful and more cosmetically acceptable than a fused, spoon-shaped, relatively immobile hand. Mental retardation is no reason to withhold surgical treatment of the hand in Apert syndrome.

The principle in planning reconstruction should be to provide the digits with skeletal elements as normal as possible covered by supple skin with normal sensation. In hands with complex syndactyly, particularly in which bone fusion between phalanges is present, early operation is necessary. The operation should be planned to define the digits and allow unimpeded skeletal growth. If surgery is postponed, incongruities of length and growth between fingers usually produces additional deformities in an already malformed hand.

My experience with this type of deformity leads me to believe that the most satisfactory functional and cosmetically acceptable hands can be provided for these individuals by the following care plan.

In the previously unoperated hands of a child below the age of 2 years,

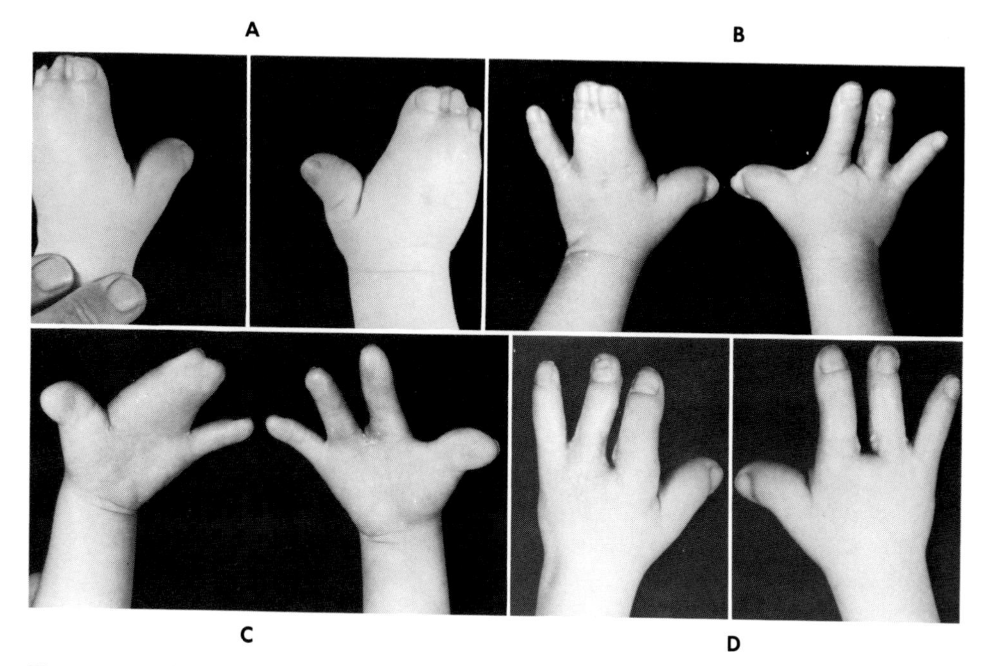

Fig. 10-37. *Three-fingered hand in Apert syndrome.* The most satisfactory functional and cosmetic hands can be provided for these individuals by amputating the long finger and producing a three-fingered hand. **A,** Before surgery. **B** and **C,** Completion of right hand surgery and release of border digit on the left hand. **D,** Final result. (From Hoover, G. H., Flatt, A. E., and Weiss, M. W.: The hand and Apert's syndrome, J. Bone Joint Surg. [Am.] **52:**878-895, 1970.)

both hands are operated on at the same time. In the older patient, who must feed himself and care for his body functions, one hand at a time is operated on. In general the surgical approach to these hands is as follows:

1. The border digits, the thumb and small finger, are released. This should be performed before the patient is 1 year of age if the patient is referred soon enough. A full syndactyly release of the thumb-index and ring-small finger webbing is done according to the modified principles of Bauer, Tondra, and Trusler. If the thumb is not included in the syndactyly a Z-plasty deepening of the thumb-index web space is sufficient. A full-thickness skin graft from the antecubital area or the groin will be needed to complete the syndactyly releases.

2. After 6 to 9 months the hand is again operated on. This allows time for adequate revascularization and softening of the tissues of the fingers previously operated on. Release of the index, long, and ring fingers is done and amputation of the long finger at the metacarpophalangeal joint performed. This provides good skin coverage, with proper sensibility, for the remaining index and ring fingers (Fig. 10-37).

I believe it advisable to leave the third metacarpal in place and not to narrow the hand by removing it and shifting the index ray ulnarward. Our long-term follow-up studies show that more deformity develops in hands so treated than in those in which the metacarpal was left in place.

Under usual circumstances, in patients below the age of 2 years, only two surgical procedures will be needed to produce bilateral, adequately functioning three-fingered hands. In older patients in whom one hand at a time is operated on, a minimum of four operations will be needed.

Some deformities, such as radially deviated thumb, clinodactyly produced by epiphyseal malalignment, or delta phalanx and finger stiffness induced by symphalangia, may require surgical correction at a later date. In addition, it is sometimes necessary to deepen a web space, particularly after the adolescent growth spurt.

POLAND SYNDROME

Characteristically this syndrome includes a congenital dysplasia of the whole of the upper limb, absence of the sternocostal portion of the pectoralis major muscle, and varying degrees of syndactyly. The fingers are frequently shortened because of lack of growth, or even absence, of the middle phalanges.

The term Poland's syndactyly was first used to describe this group of congenital anomalies by Clarkson in 1962. He reported that in 1841 Alfred Poland first described the association of congenital thoracic anomalies with ipsilateral syndactyly. He dissected a male cadaver at Guy's Hospital which lacked the sternocostal head of the pectoralis major and of the pectoralis minor, while the serratus anterior, the external oblique, and the muscles of the left arm were hypoplastic.

> In the left hand the middle phalanges were absent in all the fingers except the middle finger, where a ring of bone ¼ inch in length supplied its place. The web between the fingers extended to the first phalangeal articulation so that only one phalange remained free on the distal extremity of each finger.*

The thoracic anomalies—absent pectoral muscles, hypoplastic or absent breast and nipple, hypoplasia of a rib, pectus excavatum, pectus carinatum, elevated scapula, and scoliosis—have been reported in the absence of hand anomalies, and these deformities occur with greater frequency than does the full-blown complex. Because of the persistent findings of anomalies other than syndactyly with this condition, I prefer the term Poland syndrome, and in 1976 we reported in detail on 43 consecutive cases in our study of congenital anomalies. In the same year Sugiura reported his experience with 45 cases observed over a period of 20 years. He also prefers the use of the term syndrome, and I agree with his description of the four essential features: (1) unilateral shortening of the index, long, and ring fingers; (2) syndactyly of the affected digits; (3) hypoplasia of the hand; and (4) absence of the sternocostal portion of the ipsilateral pectoralis major muscle.

The cause of Poland syndrome is unknown, but it is generally thought to

*Poland, A.: Deficiency of the pectoralis muscle, Guys Hosp. Rep. 6:191, 1841.

occur sporadically. Although hereditary traits have been demonstrated for some anomalies of the hand such as polydactyly only one report of Poland syndrome described familial incidence, and we were unable to establish a positive family history in any of our 43 patients.

Physical signs

The involved hand is always hypoplastic, the forearm is usually, and the arm is infrequently (Fig. 10-38). The extent of the involvement of the upper extremity tends to be proportional to the extent of the syndactyly and the hypoplasia of the hand. The syndactyly and the involvement of the bones vary in type and severity. The syndactyly may be complete or incomplete and tends to be simple, involving soft tissues only; aplasia of the nail often occurs. Usually all of the fingers are involved, making a mitten hand, and the first web space is usually shallow. Although all digits may be hypoplastic, the thumb often appears disproportionately small and is usually malrotated. It tends to lie in supination in the same plane as the fingers. The characteristic bone anomaly is hypoplasia of the middle phalanges. It is often so severe that there is effectively only one interphalangeal joint. Shortening of the middle phalanges is not necessarily confined to the digits included in the syndactyly; it is more common on the ulnar side of the hand than on the radial.

The thoracic deformity is variable, but in all patients with the syndrome the costal head of the pectoralis major is absent and in many the pectoralis minor is absent. Pectoral aplasia is the most frequent skeletal muscle deficiency in the general population, but only 13.5% of patients with this defect have Poland syndrome. Absence of the pectoral muscles seems to cause little func-

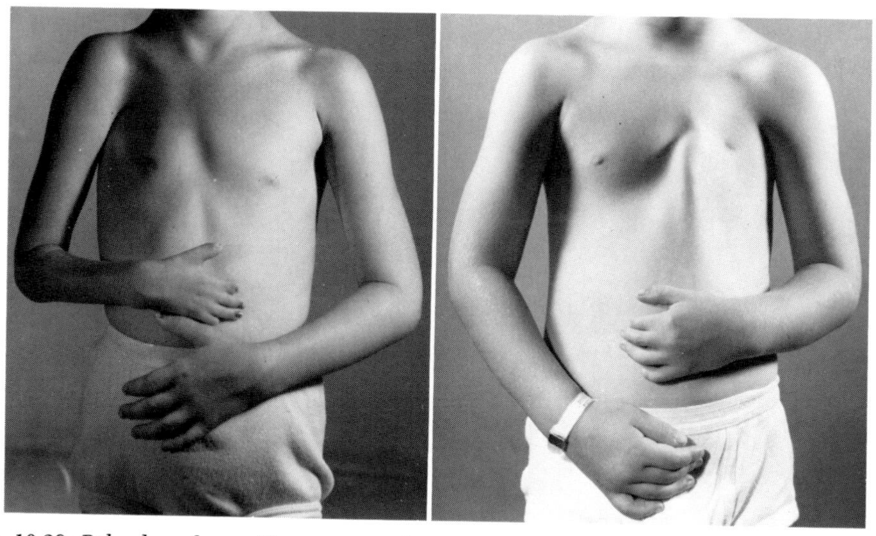

Fig. 10-38. *Poland syndrome.* Two patients showing the typical involvement in which the hand is always hypoplastic, the forearm usually is, and the arm infrequently is. The thoracic deformity varies, but the costal head of pectoralis major is always missing.

tional impairment despite Poland's original clinical description that implied an inability to internally rotate the arm.

> George Elt, a convict, aged 27, respecting whom no history could be obtained; except that it was remarked that he could never draw his left arm across his chest; and that when asked to give his left hand, in order that his pulse might be felt by one standing on the right side, he invariably turned round to do so.*

Brown and McDowell mention the case of a professional fencer whose pectoralis major was missing on the dominant side, and Christopher reported on a left-handed pitcher who had absence of the left pectoralis major muscle.

In our 43 patients the pectoralis minor was absent with certainty in 15 and present in 4. In one patient the deltoid, latissimus dorsi, and pectoral muscles were absent. In most patients the nipple was hypoplastic and elevated, but it always was present. The pubertal females tended to have hypoplasia of the breast (Fig. 10-39). The axillary web was normal in 20 patients, and in 20 others the records did not contain specific notations about it. It was so contracted in 3 patients that surgical correction was required. Three patients had x-ray evidence

*Poland, A.: Deficiency of the pectoralis muscle, Guys Hosp. Rep. **6**:191, 1841.

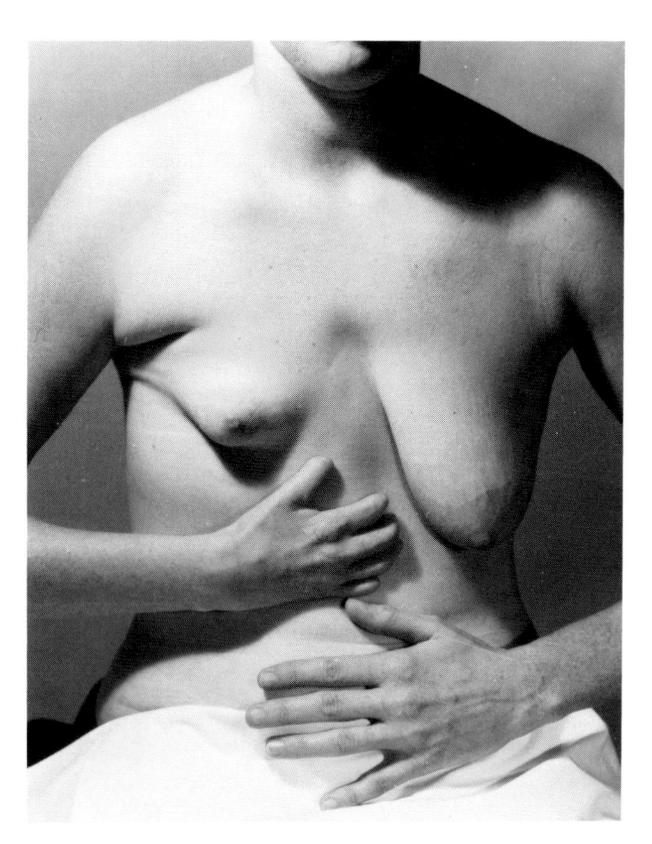

Fig. 10-39. ***The breast in Poland syndrome.*** In the female patient the breast on the affected side is usually severely underdeveloped.

of rib deficiencies, usually the anterior halves of the second through fourth ribs on the affected side; 4 had associated herniation of the lung; and 2 had elevated scapulae. No cervical or thoracic spinal anomalies were seen. One patient had a right thoracic idiopathic scoliosis as well as cervical ribs. Two patients had dextrocardia.

Operative treatment

It is important that the parents realize early in the treatment plan that two features of this syndrome will be untreatable and will remain with their child throughout his or her life. The affected limb will always be smaller than a normal limb, and the fingers will usually be stiffer and smaller than normal. When these limitations have been accepted, a realistic treatment plan can be made.

I think that the surgical reconstruction procedures should be completed by the time the child is of school age. The optimum age for starting treatment depends on the type of deformity. For example, border digits need early separation, while separation of the index and long fingers may be delayed. Generally, separation should be initiated by the first year of life, before abnormal functional patterns have developed and before deformity has progressed.

The webbed fingers can be separated by the standard syndactyly method and the thumb-index web space can be satisfactorily separated by Z-plasty or the four-flap modification of the Z-plasty when the web is only moderately shallow (Chapter 5, page 64). For more complete first-web syndactyly, in which the thumb lies adducted and supinated in the same plane as the fingers, a single or multiple, large, dorsal rotation-flap technique (Chapter 5, page 73) gives a deep, durable web space. The parents must be informed that follow-up will be necessary until the child attains skeletal maturity since re-deepening of webs is often required after the pubertal growth spurt.

The sequence of surgery in these mitten hands depends on the severity of the anomalies. If all the fingers are to be saved, I start by separating the thumb from the index finger and the long from the ring finger followed 6 months later by separation of the index from the long finger and the ring from the small finger. All the webs can thus be deepened in two operations, and simultaneous separation of three adjacent digits, which is never advisable, can be avoided. If construction of a three-fingered hand is anticipated, initial separation of the border digits is best. The three central digits can then function as a paddle until the surgeon has to select the digit for amputation. The selection depends not only on the deformity but also on the need for an adequate first web space to allow unhampered opposition and abduction of the thumb. A period of 6 months should be allowed between consecutive operations to allow scar tissue to mature and to provide time for digits deficient congenitally or from prior surgery to become revascularized.

Fig. 10-40 illustrates the sequential treatment of a typical 6-year-old boy with a left-sided Poland syndrome.

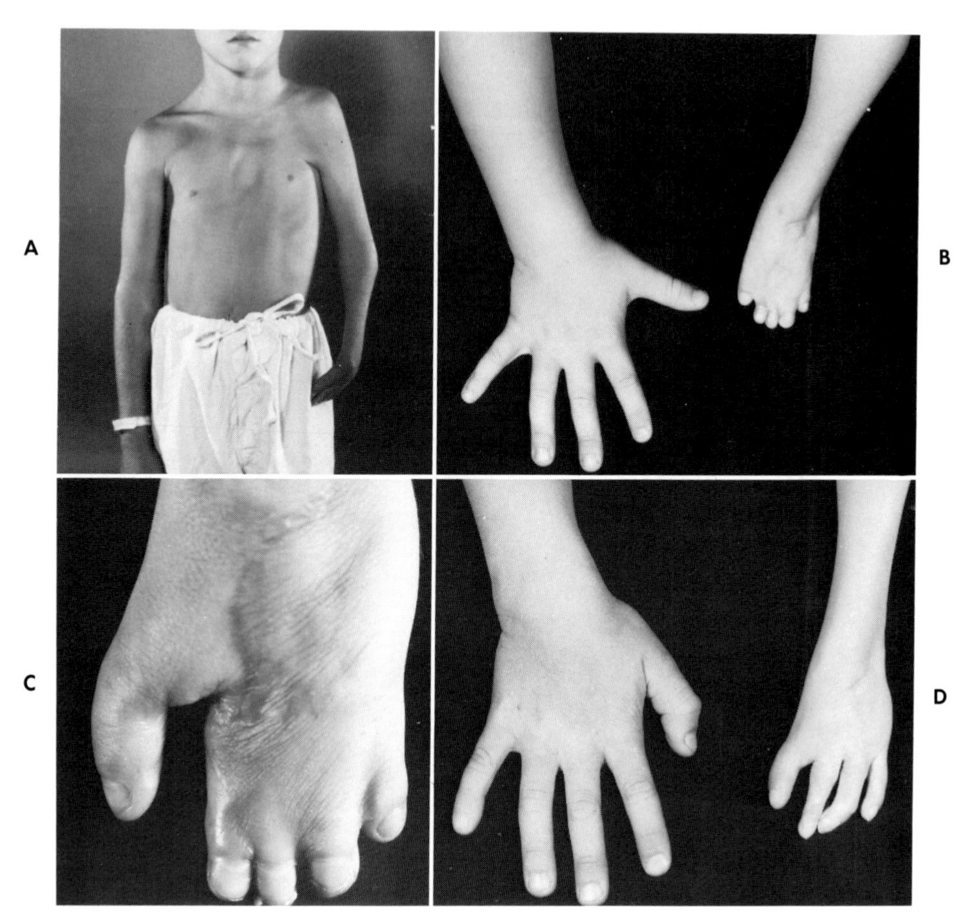

Fig. 10-40. *Poland syndrome—operative treatment.* **A** and **B,** Before surgery, showing the hypoplasia. **C,** Multiple dorsal rotation flaps have been used to release the thumb. **D,** Final result after development of a three-fingered hand. (NOTE: This is the same patient illustrated in Fig. 5-15.) (From Ireland, D. C., Takayama, N., and Flatt, A. E.: Poland's syndrome: a review of forty-three cases, J. Bone Joint Surg. [Am.] **58:**52-58, 1976.)

We have reviewed many of our patients a number of years after completion of surgery and feel that although the end result is a small, somewhat stiff hand, it is usually of functional value. The parents are uniformly pleased with the cosmetic improvement in the hand, and the children appreciate the upgrading of their affected extremity from being merely an assisting hand to being one with useful, independent hand function.

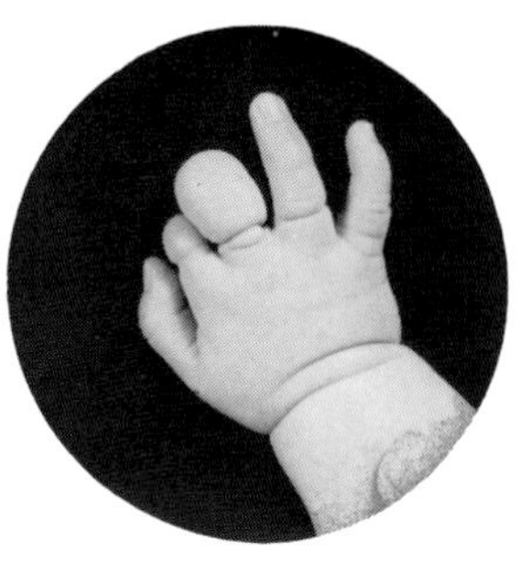

CHAPTER 11

CONSTRICTION RING SYNDROME

The fascination of constriction ring syndrome lies with its cause rather than with its treatment, which is not usually difficult. There seems to be general agreement that the use of the word syndrome is appropriate because of the frequent, but not constant, association of this disorder with either acrosyndactyly, transverse arrest of the digits, or both of these conditions. There is no consensus as to why these conditions occur alone or in association with each other.

A review of the literature is entertaining but unenlightening. Hippocrates first suggested that amniotic bands might press on a limb and produce deformities or amputations. Since then the etiological battle has swayed back and forth between intrinsic and extrinsic causes, with iatrogenic lesions of chickens and rats being used as evidence. In 1964 one proponent of the mechanical etiological group felt so strongly that he wrote "I have been maintaining for many years in many publications that if the Baconian principle of assembling every possible item of information about a question is followed, it becomes obvious that mechanical factors play a very important part in the pathology. . . ." The writings of Patterson come nearest to the Baconian method of setting out all the factors, and he does not agree that mechanical factors play a very important part. He stresses the close histological similarity between a normal skin crease and a constriction ring. The normal skin crease is caused by selective lack of development in the mesodermal mass that makes up the subcutaneous tissues. Aberrations of this process could result in abnormally placed creases, and more severe failures would produce a typical ring, whereas gross mesodermal failure would lead to a deep ring.

Others believe that the condition is produced by prenatal environmental factors, of which premature rupture of the amniotic sac is probably the precipitating cause. The sudden reduction in fluid volume can induce excessive uterine muscle contracture and consequent production of amniotic bands and even penetration of the amnion by the more prominent distal portions of the limb. In this regard it is interesting to note that in Table 11-1 the short border digits show the least incidence of ring formation. I have nothing constructive to contribute to this debate, but I have found it interesting to assemble the undisputed clinical facts that are associated with this condition.

CLINICAL FEATURES

Characteristically these cases occur at random, and there is no evidence of an inherited defect. Patterson states that the incidence is about one in 15,000 births. There is no pattern to the deformity, and the lesions are asymmetrical but tend to occur toward the periphery of the limb.

A significant and frequent anomaly is the transverse arrest or absence of digits. This shortening of the digits is probably the result of a progressive increase in the severity of the ring constriction eventually leading to intrauterine necrosis and autoamputation (Fig. 11-1).

The circumferential extent and depth of the grooves vary, and consequently the amount of lymphedema distal to the constriction also varies. The grooves usually lie at 90 degrees to the long axis of the digit, and their depth ranges from a faint groove to a deep gutter. The depth is usually not equal around the whole circumference; the dorsum of the digit is usually more deeply grooved. Distal to the constriction the swelling is diffuse throughout the subcutaneous tissues and usually does not pit on pressure.

Involvement site

At the University of Iowa, we have records of 48 patients who had constriction rings of their upper extremities with the great majority concentrated in the hand. There were only 5 rings at the wrist, 7 in the forearm, and 4 around the arm

Table 11-1. Frequency of digital involvement

Digits	*Kino*		*Iowa*	
Thumb	7	3.8%	31	11.1%
Index	47	25.7%	60	21.5%
Long	57	31.1%	71	25.4%
Ring	46	25.2%	78	28.0%
Small	26	14.2%	39	14.0%

Data from Kino, Yoshitake: Clinical and experimental studies of the congenital constriction band syndrome, with an emphasis on its etiology, J. Bone Joint Surg. (Am.) **57**:636-643, 1975; and Flatt, A. E.: Private study.

(Figs. 11-2 and 11-3). One of the rings around the arm was so tight that it led to severe edema, and a midhumeral amputation had to be done. One hand had to be amputated because of a severe wrist constriction that created a so-called balloon hand (Fig. 11-4). One other constriction amputation had to be done on the right long finger of a ten-year-old boy because of excessive edema. From the 48 patients, we have studied a selected group of 45 patients with upper and lower limb involvement. Twenty-four of these patients showed involvement of the upper extremity alone, with the arm being involved in three, the forearm in four, and the wrist in another three. Fourteen patients showed involvement of the

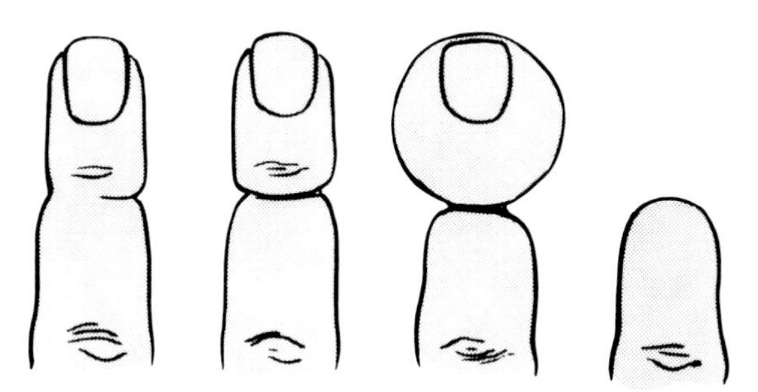

Fig. 11-1. *Extent of constriction rings.* The circumferential extent and depth of these rings varies. Progressively increasing severity of the constriction can lead to ballooning of the finger tip and eventually to autoamputation. (Redrawn from Fischl, R. A.: Ring constriction syndrome, Transactions of the International Society of Plastic and Reconstructive Surgeons, 5th Congress, Australia, 1971, Butterworth Pty. Ltd., pp. 657-670.)

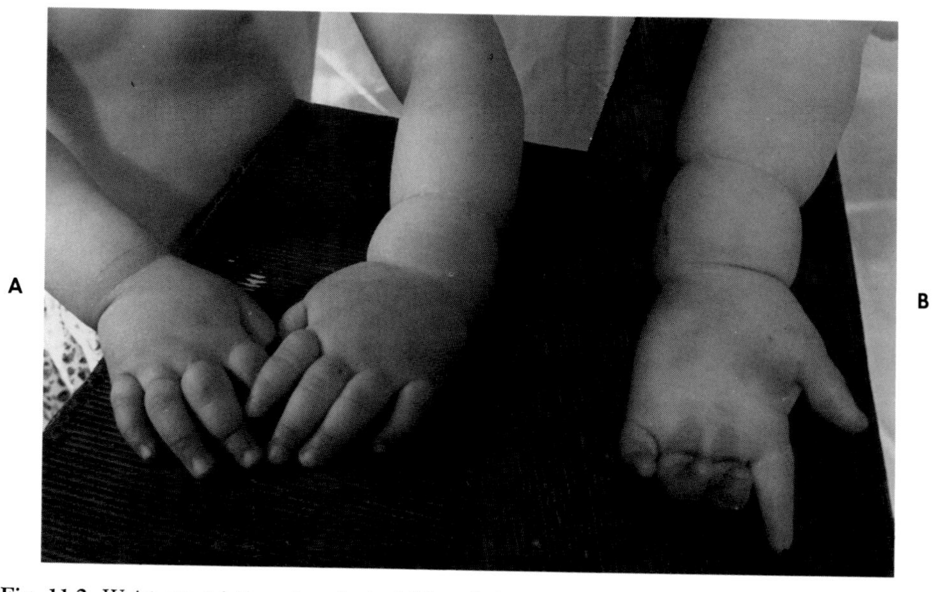

Fig. 11-2. *Wrist constriction ring.* **A,** A child with forearm and wrist rings. **B,** The forearm ring has been excised and the wound resutured with no improvement.

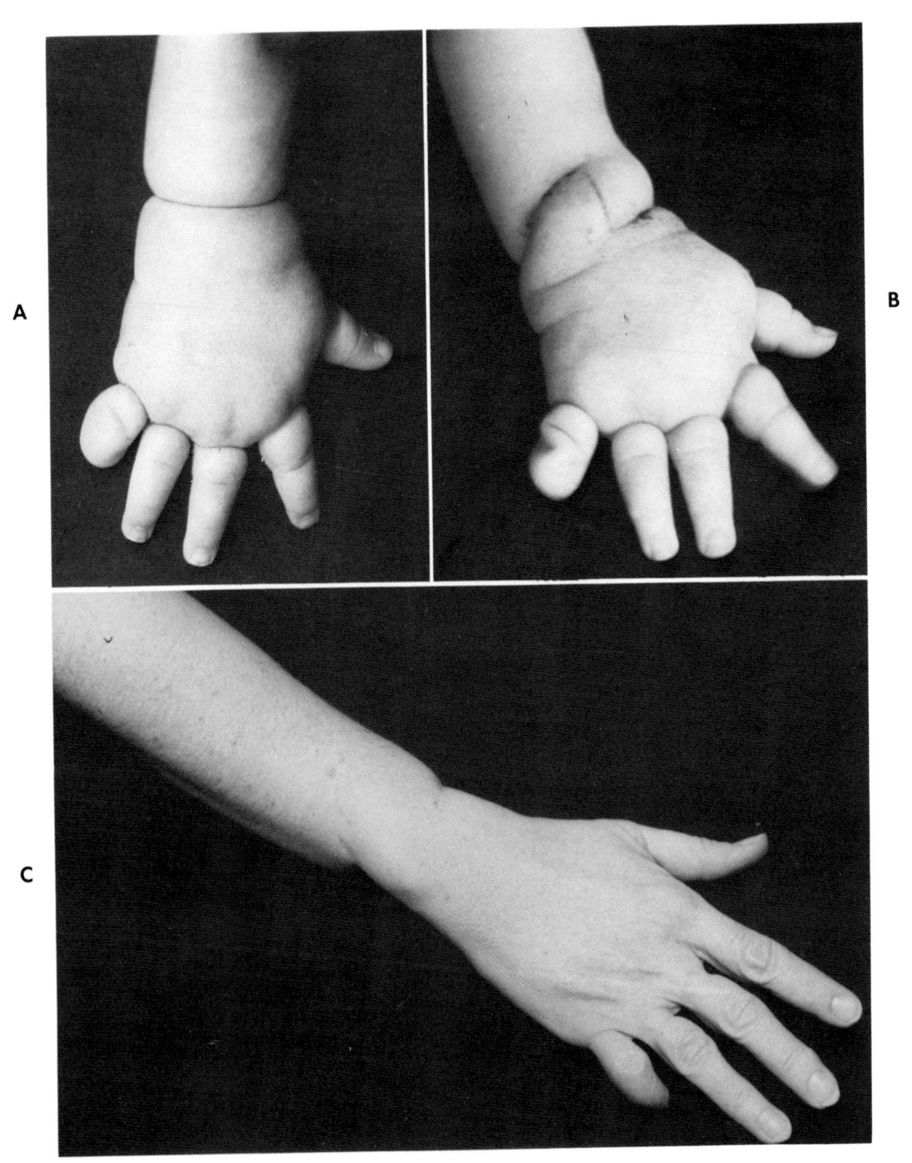

Fig. 11-3. *Wrist constriction ring.* **A** and **B,** A deep wrist ring treated by the Z-flap excision method. **C,** Forty years later the result is still satisfactory, and the edematous fingers have returned to normal without surgery.

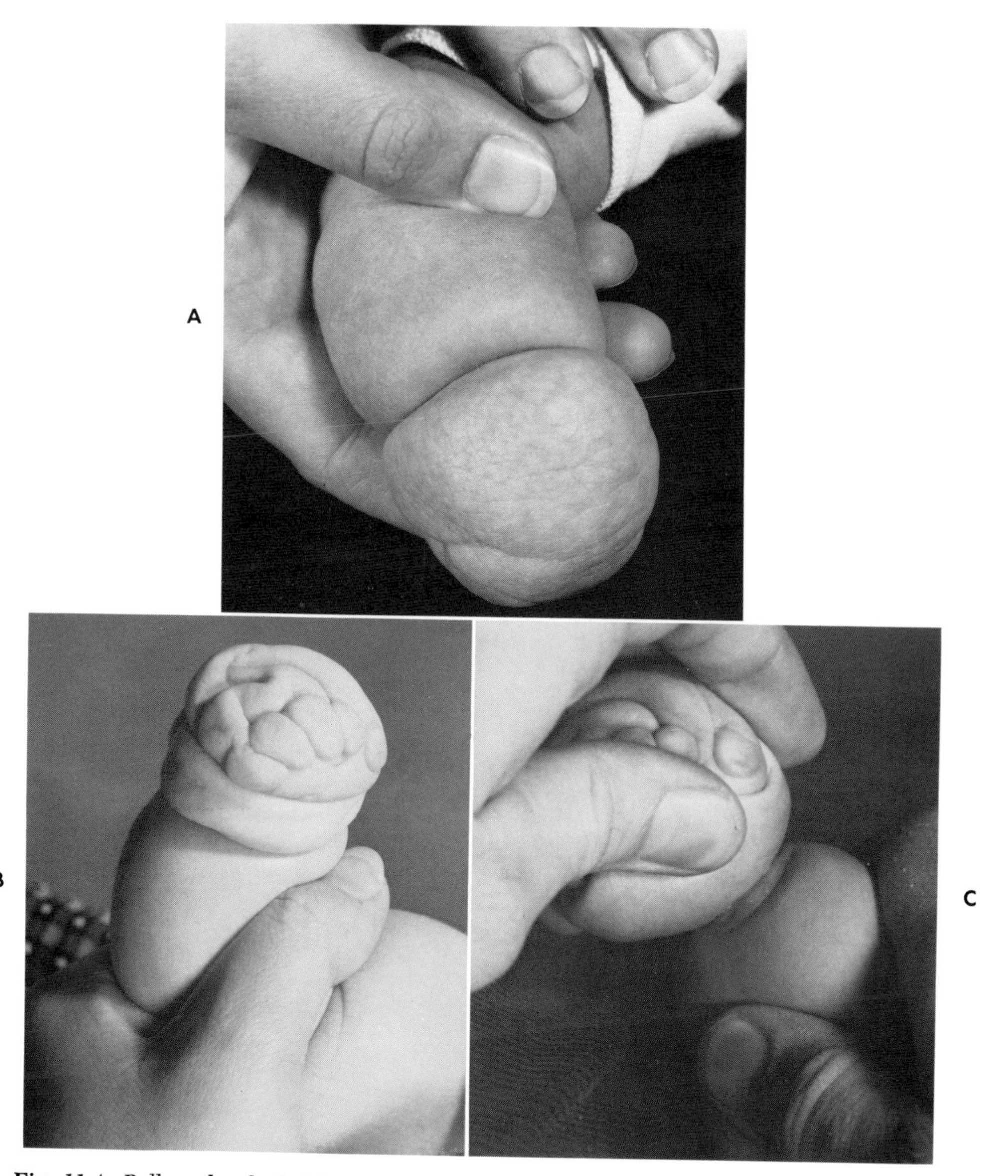

Fig. 11-4. *Balloon hand.* **A,** The dorsum of the hand. **B,** The fingers were almost unrecognizable and did not function. **C,** The ring had contracted down to the forearm bones. Two Z-plasty operations were done at the parents' request, but amputation was eventually necessary.

hand alone. Ninety-three percent of the 45 patients showed involvement distal to the wrist or ankle or both (Fig. 11-5).

Frequency of involvement

Kino has reported the frequency of digital involvement in a series of 59 patients, and we have established the frequency of digital involvement in 82 of our patients with upper limb involvement (Table 11-1). There is general agreement between these two series, although the Iowa group shows a greater involvement of the thumb. It may be of significance that the longer digits are affected most frequently.

Concomitant hand anomalies

Constriction rings are commonly associated with digital shortening, but they are also associated with a variety of other anomalies. Table 11-2 shows the concurrent anomalies within the hands in 39 cases in the Iowa series.

Associated anomalies

Associated anomalies are also common; clubfeet and cleft lip and palate are the most frequent. Various series quote a range of from 40% up to 56% in-

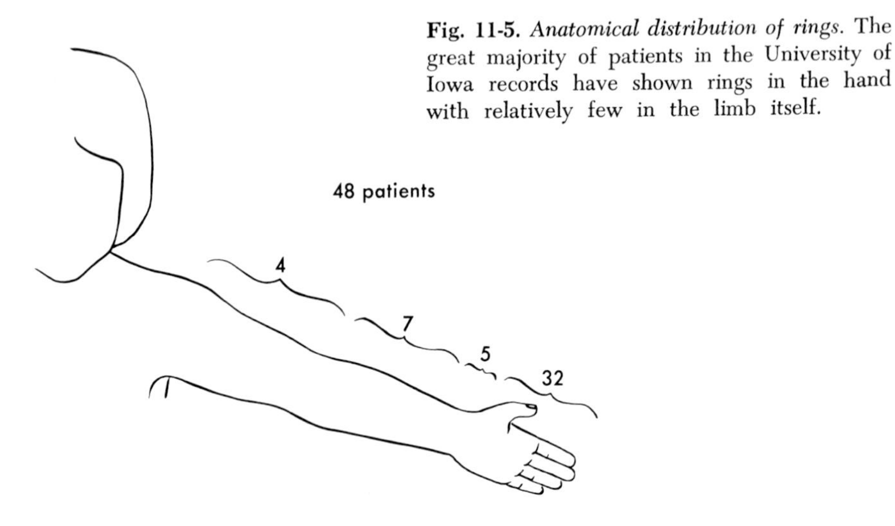

Fig. 11-5. *Anatomical distribution of rings.* The great majority of patients in the University of Iowa records have shown rings in the hand with relatively few in the limb itself.

Table 11-2. Concomitant hand anomalies

Condition	*Number*
Syndactyly	26
Acrosyndactyly	13
Hypoplastic phalanges	10
Brachydactyly	7
Symphalangia	3
Symbrachydactyly	1
Camptodactyly	1

cidence of associated anomalies. In our own series, 80% had concomitant hand deformities; 40% showed a combination of foot and hand anomalies, the most frequent combination being syndactyly of the hand and clubfoot (24%). Eleven percent of our series had oral cavity anomalies. It is usually said that there is no associated malformation of internal organs, but in our series one patient had a patent ductus arteriosus.

Abnormalities of the nails

Anomalies of the nails have been largely ignored in the literature, but in our experience, 40 patients with constrictions distal to the wrist or ankle had nail deformities. Those with proximal rings did not show any abnormality of their nails (Table 11-3).

Vascular and neurological involvement

A temperature gradient across the constriction band has been described by Ramakrishinan and Nayak. Barenberg and Greenberg reported decreased sensibility distal to a band in one patient. We paid particular attention to these factors in our 45 patients and found significant involvement of both elements.

Our examination of the circulatory system was retrospective in the sense that all but one patient were seen postoperatively. Comparison of the findings above and below the band showed that 16% of the patients had edema and decreased capillary refill distal to proximally located rings. Eleven percent of the patients

Table 11-3. Nail deformity

	Cases	*Percentage*
Location of rings in patients with nail deformity		
Distal (beyond wrist or ankle)	40	100
Proximal (above wrist or ankle)	0	—
Type of deformity		
Hypoplastic and slow growing	32	80
Absent nail	8	20
Total number of patients with nail involvement	40	80*

*Of 48 patients studied.

Table 11-4. Neurological involvement

Parameter	*Ring location (cases)*	
	Proximal	*Distal*
Motor deficit	0	0
Sensory deficit		
Pin prick	9	3
Light touch	9	3
Two-point discrimination	0	2
Pallesthesia	2	0
Temperature discrimination	4	0
Proprioception	0	0

with proximal rings showed peripheral cyanosis after prolonged exertion, dependency, or exposure to cold temperatures. Neurological deficits were also quite common and showed varying degrees of diminution of sensibility (Table 11-4).

All patients with neurological deficits across proximally located rings demonstrated significant temperature gradients at the same site (Table 11-5).

Table 11-5. Skin temperature gradients

Location of rings	*Cases*	*Gradients*
Proximal	20 (44.45%)	2.4° C (range 0.8° to 4.1° C)
Distal	2	1.3° C

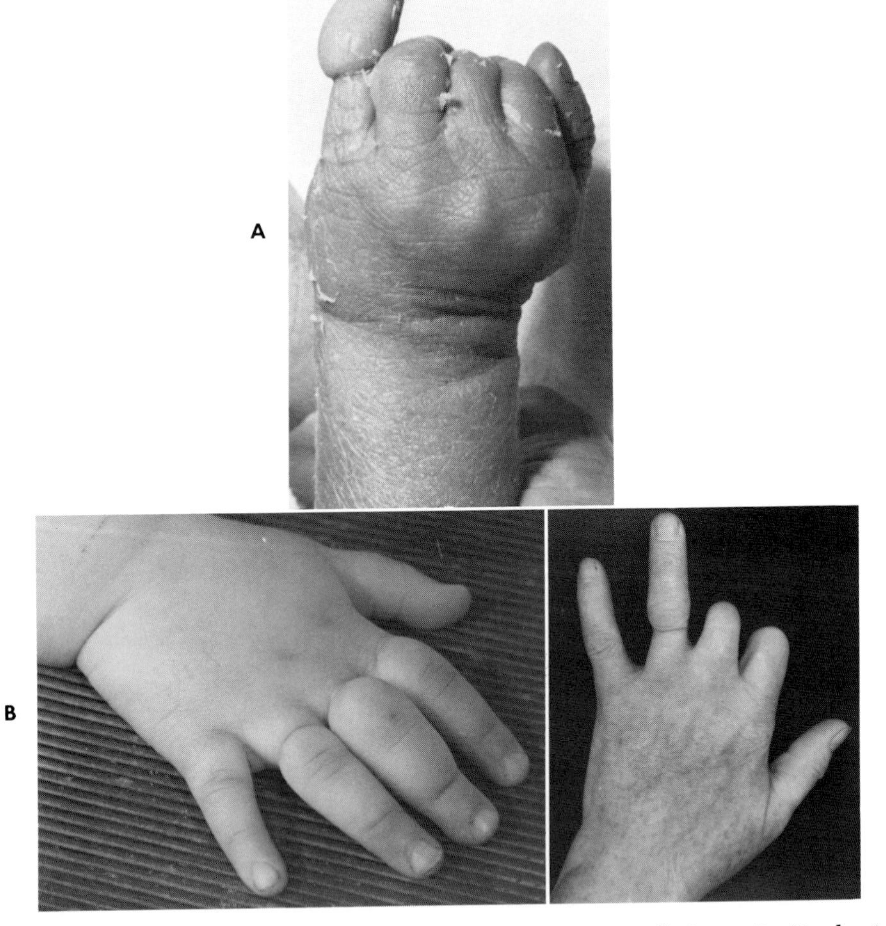

Fig. 11-6. *Simple constriction rings.* **A,** A simple ring on the small finger. **B,** Simple rings involving the index, long, and ring fingers. **C,** A simple ring on the ring finger and autoamputation of the index and long fingers.

CLASSIFICATION

Patterson has devised a useful grouping for these cases. The first three groups have different treatment requirements.

1. Simple constriction rings (Fig. 11-6)
2. Constriction rings accompanied by deformity of the distal part, with or without lymphedema (Fig. 11-7)
3. Constriction rings accompanied by fusion of distal parts ranging from mild to gross acrosyndactyly (Fig. 11-8)
4. Intrauterine amputations

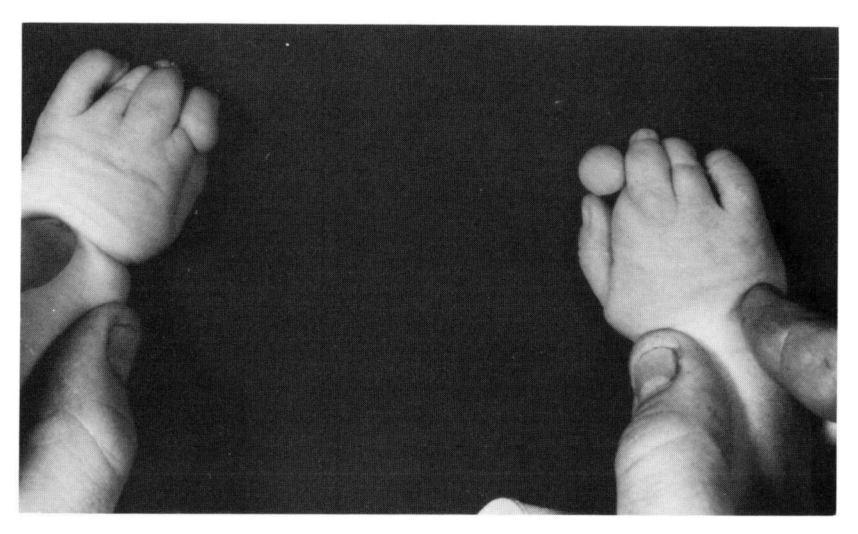

Fig. 11-7. *Ring with lymphedema.* The left index finger shows moderate swelling on its radial side. The right index finger is so badly constricted that it has become spherical distal to the ring.

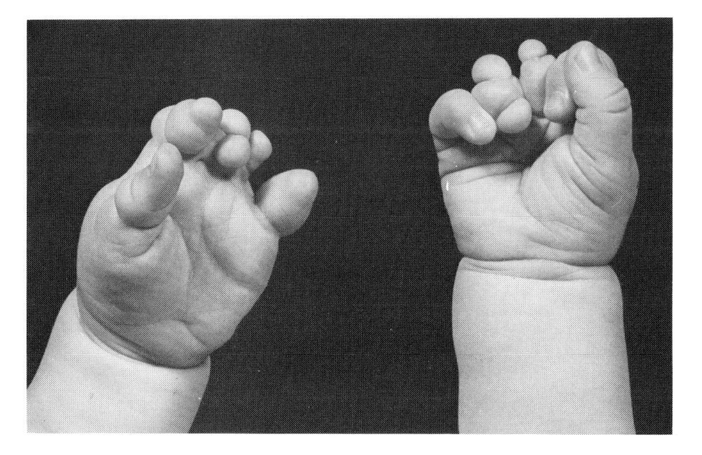

Fig. 11-8. *Acrosyndactyly and constriction rings.* Acrosyndactyly is frequently associated with constriction rings. In the left hand all fingers are affected. Epithelial pits can be seen near the edge of the shadow of what is probably the tip of the long finger. In the right hand the index and small fingers are not affected.

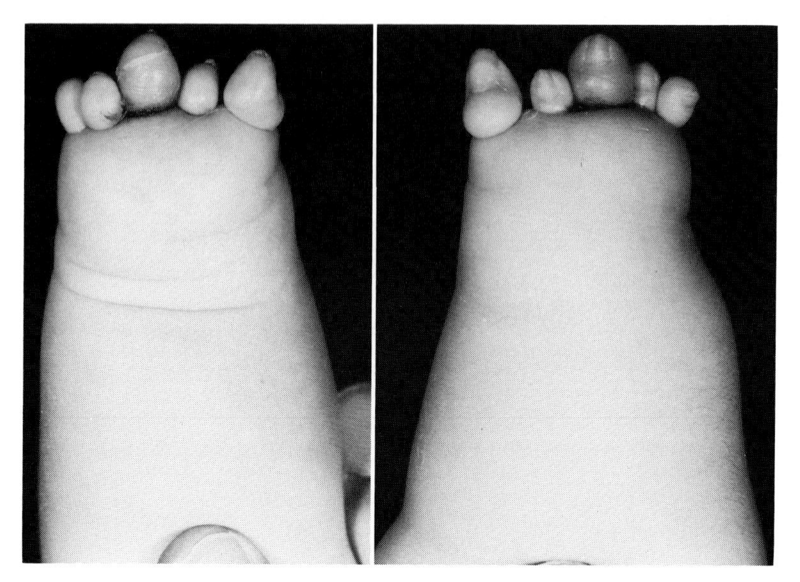

Fig. 11-9. *Constriction rings and granulation.* All digits are affected in this rudimentary hand, but the middle digit is badly congested because of its tight granulating ring.

These deformities may be present in any of a variety of combinations in one hand and show completely different groupings in the other hand.

It is important to appreciate that these rings are not necessarily static in their effect. Frequently there is granulation tissue in their depth; subsequent healing may cause further constriction and consequent distal vascular embarrassment (Fig. 11-9). Even if the rings appear healed, the hands must be kept under frequent observation because of the likelihood of increasing complications distal to the ring. In our series, 58% of the patients showed increasing severity in the constriction ring prior to surgery, the usual changes being in depth, distal edema, and cyanosis of the extremity. In none did the ring actually threaten the viability of the part.

TREATMENT

There is general agreement in the literature that staged Z-plasty around the circumference of the digit or limb is the operation of choice. We have used this method exclusively in our unit at the University of Iowa for over 30 years and in our review of 45 patients found only two with unsatisfactory results.

Simple constriction rings

Simple constriction rings are usually very shallow, are frequently incomplete, and may not need any treatment. Treatment is more often indicated for cosmetic than functional reasons. It should be remembered that as the infant's fat is absorbed from the hands during growth, the shallow grooves appear less obvious. Simple excision of the groove and the use of everting sutures for the skin and subcutaneous tissues will probably not be adequate because the

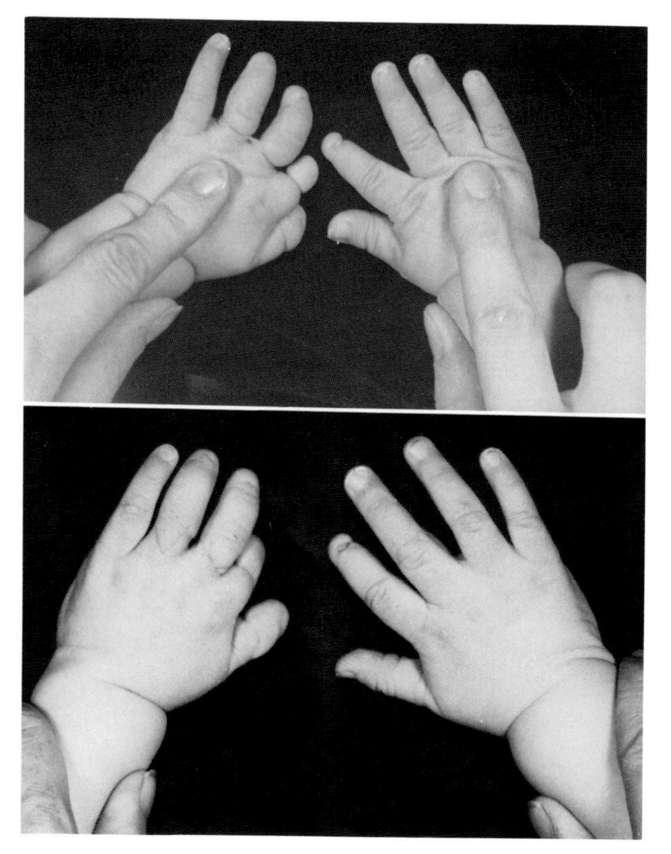

Fig. 11-10. *Simple constriction rings—Z-plasty.* Before and 2 years after the use of Z-plasties to relieve the constriction rings on all the left-hand digits except the small finger. On the right hand note the incomplete ring on the tip of the index finger, which is affecting nail growth.

resultant circular scar can contract to produce an hourglass-like waist in the digit. A much better technique is to use Z-plasty transposition flaps (Fig. 11-10).

The Z-plasty flaps should be planned to be as large as possible and to have an angle of about 60 degrees. Total excision of a circumferential ring in one stage risks grave interference with the distal blood supply, and it is wiser to do no more than one half of the circumference at one operation. In the infant's finger, one large Z-plasty may be sufficient for one side of a digit. In older children it is often wise to make two or more pairs of transposable Z-flaps.

When closing the flaps, it is best to use everting mattress sutures, and No. 6-0 ophthalmic catgut is suitable suture material.

Technically, it is important to excise rather than incise the constricted ring. There is breadth to the scarred or contracted ring. Simple incision will merely carry the abnormal portions of the ring onto the edges of the raised Z-plasty transposition flaps. An important point is made by Riordan when he stresses that although the ring has to be excised, the depth of the excision should be carefully judged. The nerve supply to the portion of the digit distal to the ring runs in the subcutaneous tissues. If the excision is carried boldly down through the deep fascia, the small cutaneous nerves are likely to be cut and distal anesthesia may well result.

Constriction rings with distal involvement

The common distal involvement is a varying degree of lymphedema, although sometimes the tourniquet effect may be so marked that there is distal cyanosis. Occasionally the circulatory embarrassment may escalate rapidly with increasing cyanosis and swelling to such a degree that early release of the ring is mandatory.

The surgical principles are the same as for the simple rings—excision of the ring tissue and relief of future constriction by breaking up the circumferential tightness with Z-plasty flaps (Fig. 11-11). The subcutaneous tissues are usually edematous, and I trim the excess swollen fat off the flaps and often from beneath the adjacent proximal and distal skin before closing. The fat excision makes it easier to oppose the flaps, and the subsequent healing is very satisfactory.

The Z-flaps will aid in reestablishing the lymphatic drainage of the swollen tissues, and the swelling will begin to recede even when only one half of the circumference has been relieved. Cyanosis and circulatory embarrassment are also usually relieved after the first operation. I usually do the second operation to complete the ring excision about 2 to 3 months after the first.

When doing the dorsum of a digit, one will often find it very difficult to locate the small subcutaneous sensory nerves, but on the palmar aspect, the proper digital nerves and vessels can usually be found. These nerves and vessels are often very closely adherent to the undersurface of the ring, and the greatest care must be used in excising the ring over the line of these neurovascular bundles. Occasionally the ring is very broad, and it has been reported that a cross-finger flap can be used to replace the abnormal area. I have never done such a procedure.

It is wise to warn the parents that although the swelling will be greatly reduced some months after the second operation, there will always be some residual degree of cosmetic deformity (Fig. 11-10). Reduction in the swelling does not necessarily occur in a symmetrical manner. The underlying skeleton may also show various reductions or malformations.

Constriction rings and acrosyndactyly

The surgical problem in treating children with constriction rings and acrosyndactyly is often the proper separation of the digits rather than the relief of the constriction bands (Fig. 11-12). There is a tendency for the ends of the fingers to be drawn together so that their tips resemble a small bunch of grapes. It is often extremely difficult to allocate the appropriate tip to each finger (Fig. 11-13). Simon is quoted by Fischl as having pointed out that the tip of the long finger is usually the most volar. It is not essential to allocate the tips in an

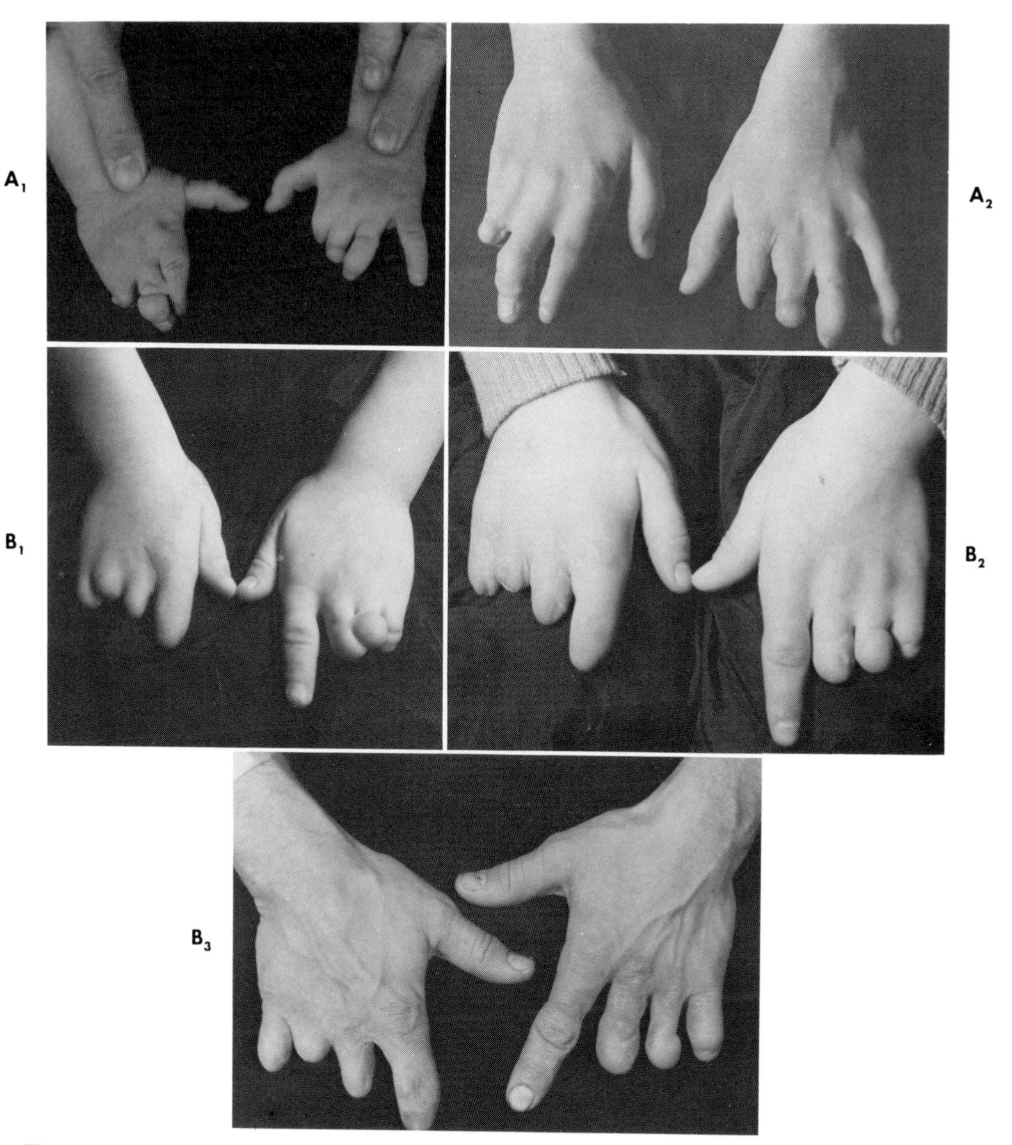

Fig. 11-11. *Acrosyndactyly and constriction rings.* Long-term follow-up of two patients showing how early release of the bands allows good growth of the fingers. **A1,** Before surgery and, **A2,** 11 years after surgery. **B1,** Another patient before surgery and 2 years, **B2,** and 27 years, **B3,** after surgery.

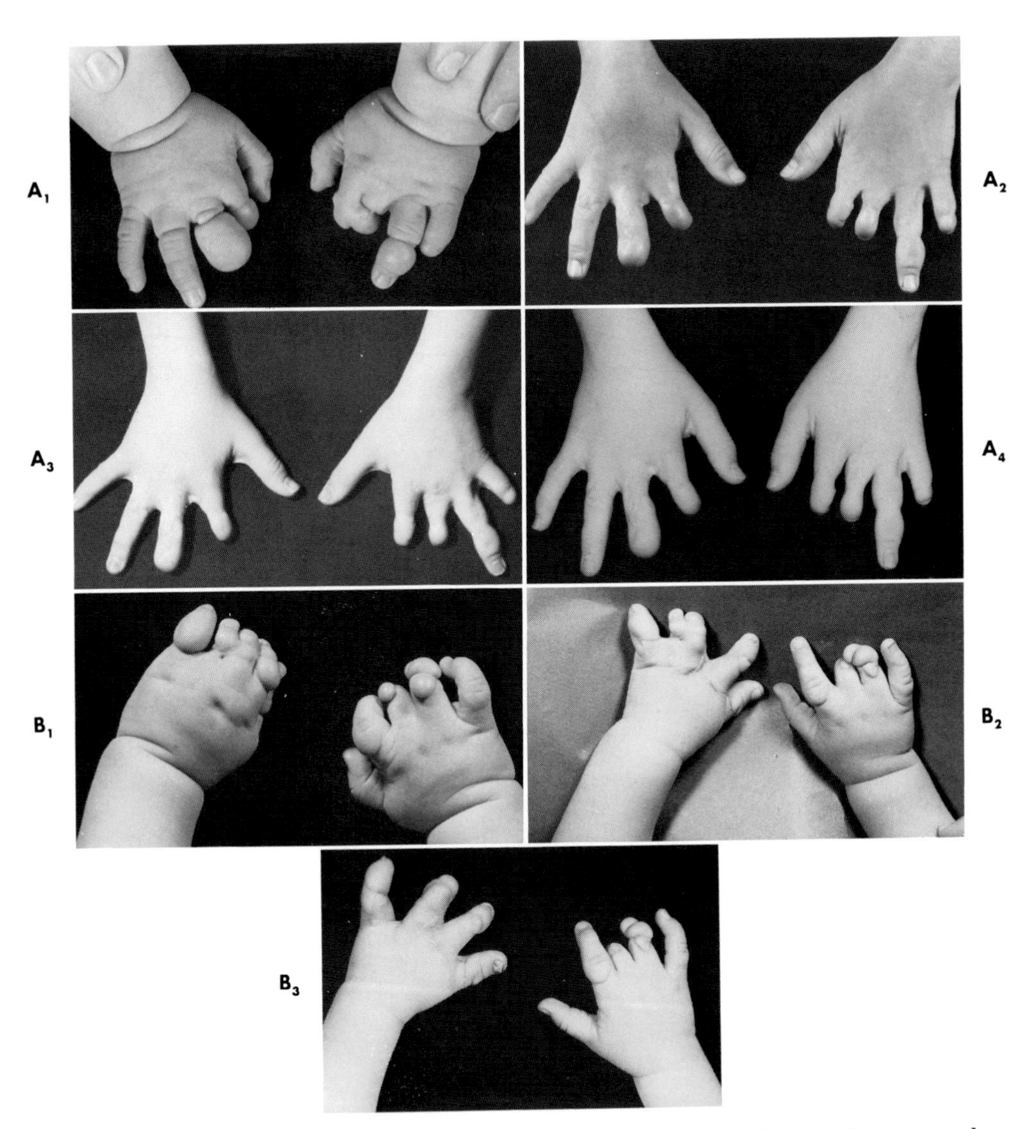

Fig. 11-12. *Severe constriction rings.* If digits are to survive and subsequently grow, early removal of rings causing lymphedema is essential. **A,** A 12-year follow-up showing how digits slowly return to near normal diameter after relief of the vascular congestion. **A1,** Before surgery. **A2,** Eight years later. **A3,** Ten years postoperatively. **A4,** Twelve years postoperatively. **B,** A 15-month follow-up showing staged surgical relief. Note how a three-fingered hand was made for the left hand. Further surgery is necessary on both hands. **B1,** Before surgery. **B2,** Eleven months later. **B3,** Fifteen months postoperatively.

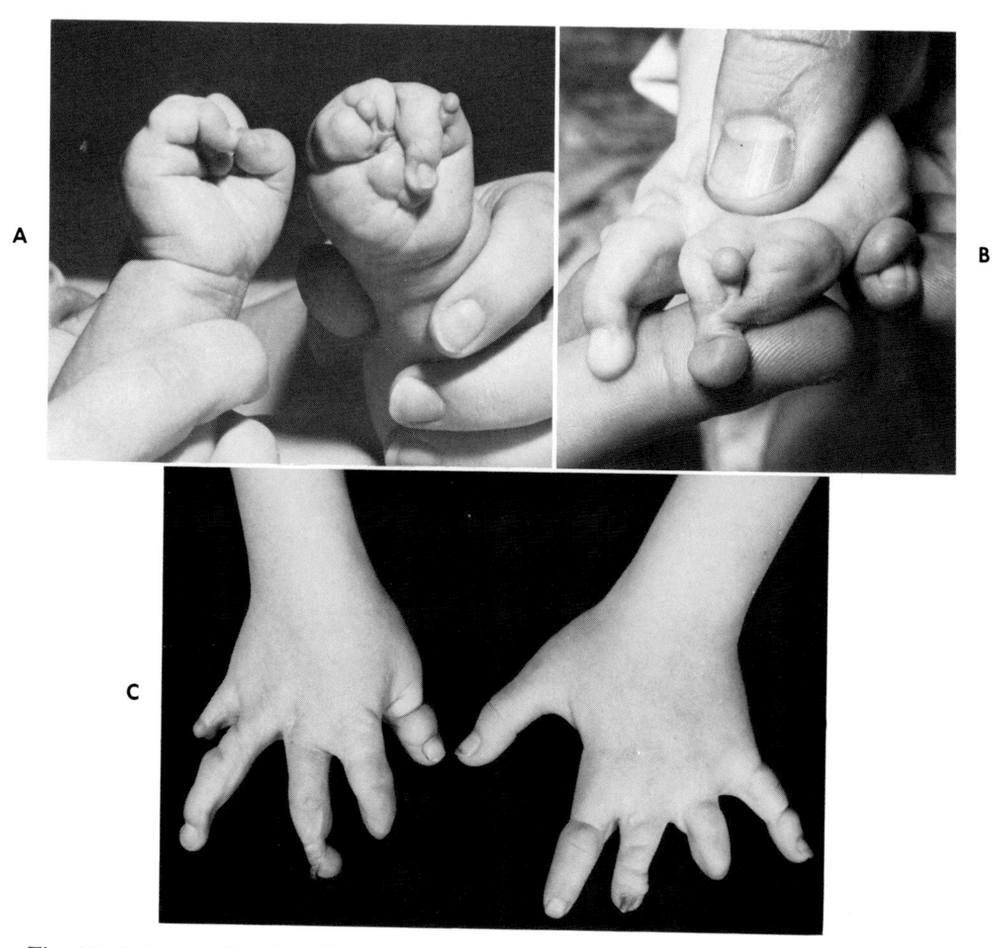

Fig. 11-13. *Acrosyndactyly and constriction rings.* **A,** In severe involvement it is often extremely difficult to allocate the appropriate tip to each finger. **B,** The left hand. **C,** Result 6 years after the first operation.

anatomically correct position since length is best preserved by retaining the tip on the end of the finger on which survival is most likely.

Because the fingers are often drawn together, they must be released early to allow proper parallel longitudinal growth. This will often entail the use of split-thickness skin grafts as well as appropriate Z-plasty flaps. I believe that in severe cases release must be done in the first 6 months of life, and it should certainly be completed by 1 year of age. There are a number of technical problems associated with acrosyndactyly, and these are discussed in appropriate detail in the section on acrosyndactyly (page 192).

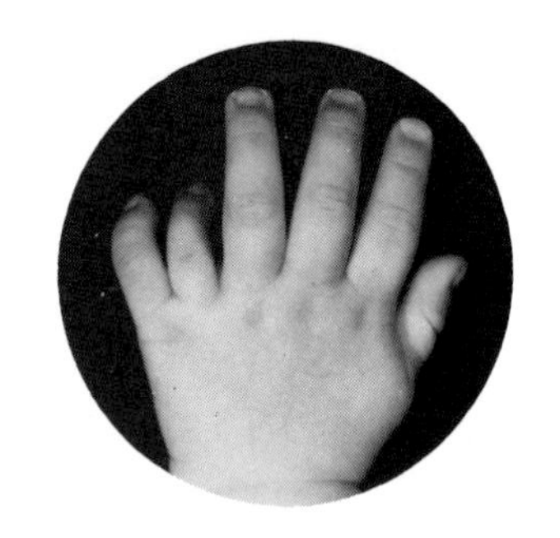

CHAPTER 12

EXTRA FINGERS

Polydactyly occurs in horses and cows, dogs and cats, rats and guinea pigs, and even chickens. It also occurs in humans; in fact, it is usually regarded as man's commonest single malformation of the extremities. In the United States, between 9,000 and 10,000 new cases are recorded each year. Polydactyly is said to be more common in females. It has intrigued the lay mind for centuries and is mentioned in several places in the Bible. In some societies it has been a sign of virtue and good breeding, while in others it was so despised that infants with polydactyly were killed. Henry VIII's queen, Anne Boleyn, is described as having "on her left hand a sixth finger" that "was to her an occasion of additional grace by the skillful manner in which she concealed it from observation."* To this day people are sensitive about their additional digits, and very few adults are found with more than the normal complement of fingers. The lay public rarely if ever counts the digits of people they encounter, but one cannot convince parents of this, and I personally believe that reduction amputations are functionally and socially justified.

The range of duplication is enormous, varying from small protuberances on the border of the hand through mirror hands with seven or eight digits to the rare "circus freak" who has several hands or partial hands attached to a grossly abnormal upper limb.

INCIDENCE

Polydactyly has been reported in all countries of the world, and no race is immune. It shows racial preference in that it is seen more frequently in blacks

---

*Strickland, A.: Lives of the Queens of England, vol. II, London, 1840-1848, H. Colburn, pp. 589-590.

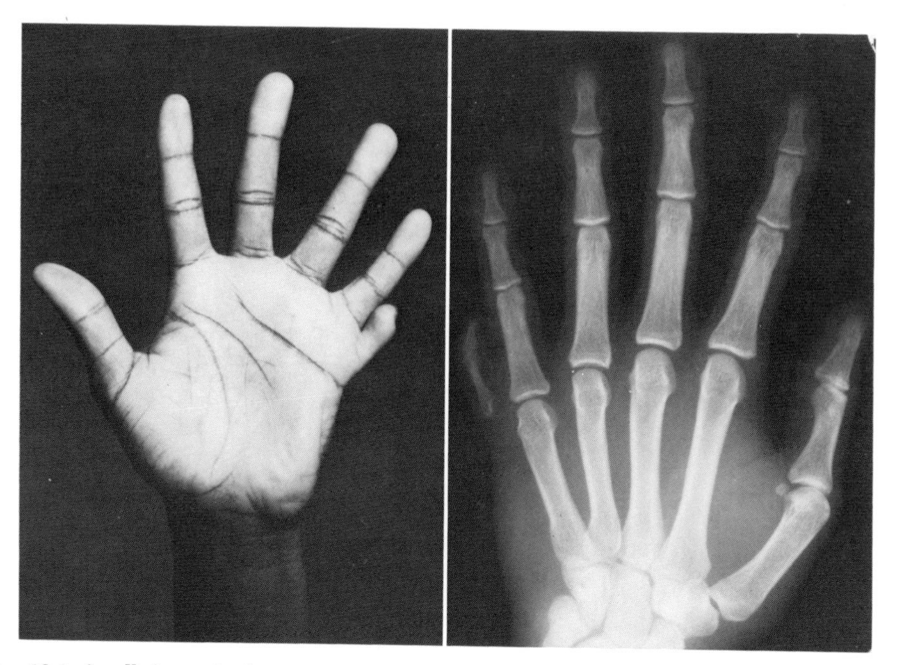

Fig. 12-1. *Small finger duplication.* Occasionally accessory small fingers are retained into adult life. They are rarely full size or functional. (From Flatt, A. E.: Problems in polydactyly. In Cramer, L. M., and Chase, R. A., editors: Symposium on the hand, vol. 3, St. Louis, 1971, The C. V. Mosby Co.)

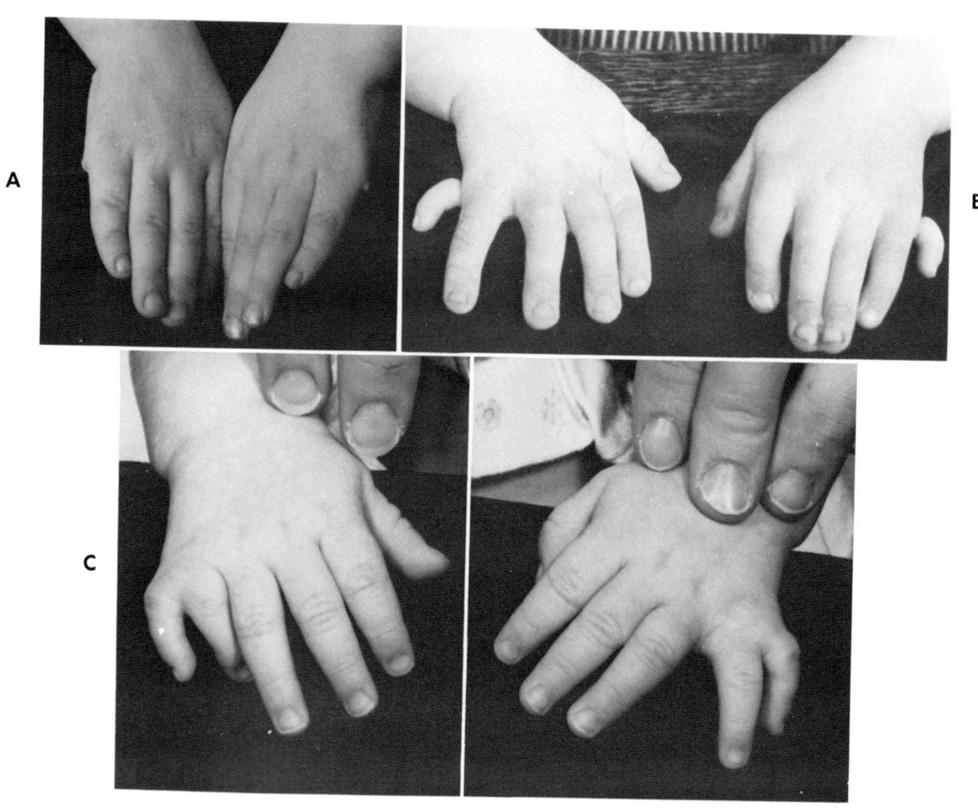

Fig. 12-2. *Small finger duplications.* Duplication can be represented by mere nubbins on the side of the hand, **A;** by floppy useless digits, **B;** or occasionally by virtually fully developed fingers, **C.**

than in whites or Orientals. The incidence in blacks is thought to be about 1 in 300 and in whites about 1 in 3,000. Duplications of the small finger are common in blacks, but in whites and Orientals, it is the thumb that is most commonly reproduced (Fig. 12-1).

Although each of the five digits of the hand can be duplicated, it is generally agreed that the most common extra digit is the small finger, perhaps with a proportion of eight occurrences to one of any other digit (Fig. 12-2). The great majority of infants with extra small fingers have the digits removed while

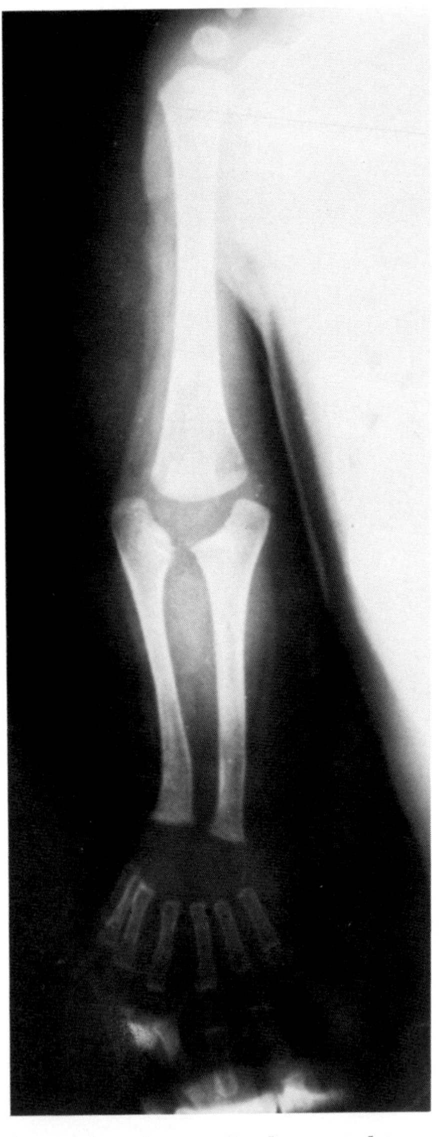

Fig. 12-3. *Mirror hand.* The number of mirror hands reported is small, and this poor quality x-ray film is the only record in the University of Iowa files.

they are in the newborn nursery. True figures are therefore extremely difficult to obtain since many extra small fingers are lost to statistical analysis.

We at the University of Iowa have reviewed 320 hands recorded as having duplications and have found that there were 143 thumb duplications and 135 small finger duplications. The remaining 42 hands showed central duplications.

Conditions such as mirror hands or multiple hands excite clinical curiosity and are usually published as isolated cases in the literature (Fig. 12-3). Even the number of mirror hands recorded is small, and the largest published series is three cases reported by Harrison, Pearson, and Roaf. Reconstructive principles have not been worked out, but Entin has stressed that the functional problem for these individuals is that the hand is flat and neither border can oppose since there is duplication of the ulna and its associated digits.

SYNDROME ASSOCIATIONS

In infants with multiple malformations caused by unknown degrees of genetic influence, polydactyly is usually a constituent part. It is also part of a significant number of syndromes determined by a single gene as well as of those caused by chromosomal abnormalities. Because of this association, an extra digit on the hand of a newborn should signal the need for a complete and thorough physical work-up since it may indicate concealed malformations or the possibility of anomalies arising later in life. Over 40 abnormalities have been reported as associated with polydactyly, and the most common local association is syndactyly.

Duplication of the small finger is commonly part of a syndrome. If small finger duplication occurs in isolation it can be caused by a dominant trait, but it usually is a result of an autosomal recessive trait when it occurs in a syndrome. Thumb polydactyly is only rarely part of a syndrome, but when it occurs, it is usually transmitted as an autosomal dominant trait.

Duplications of the thumb are usually described as preaxial polydactyly; those of the index, long, and ring fingers as central or axial; and those of the small finger as postaxial. Preaxial duplications are dealt with in Chapter 7, and the remainder of this chapter is devoted to central and postaxial polydactyly. These duplications are associated with many syndromes, most of which are common. They can be grouped as:

1. Chromosomal abnormalities
2. Syndromes with eye abnormalities
3. Syndromes with orofacial abnormalities
4. Syndromes with skin manifestations
5. Syndromes with bone dysplasias
6. Syndromes with mental retardation

Among all these possibilities, the most frequent associations are with Ellis-van Creveld (bone dysplasia), Laurence-Moon-Bardet-Biedl, Biemond (eye abnormalities), Meckel (orofacial abnormalities), and trisomy 13 syndromes.

TYPES

The extent of duplication varies from a completely developed digit to doubling of a single phalanx or to a skin tag. Stelling and Turek have classified the anomaly into three main types:

Type I is an extra soft tissue mass not adherent to the skeleton and frequently devoid of bones, joints, cartilage, or tendons.

Type II is a duplication of a digit or part of a digit that has normal components and articulates with an enlarged or bifid metacarpal or phalanx. Usually an entire digit contains all the elements of the normal digit.

Type III is rare and consists of a complete digit with its own metacarpal and with all the soft tissues involved.

PRINCIPLES OF TREATMENT

It might seem that removal of an extra digit would be an easy procedure and one done at no functional cost to the patient. Several publications have implied that the surgical treatment of polydactyly is not a problem, and one has even suggested that ablation "requires no ingenuity and creates no problem." Our follow-up studies have shown this to be far from the truth, and we have encountered several significant complications. I believe there are very definite indications for and methods of surgery applicable to the different digits.

Principles of early treatment used in caring for other congenital anomalies of the hand also apply to the treatment of polydactyly with few exceptions. In general, surgery should be completed before school age; often, in the more complicated problems, to wait until school age is to wait too long. By this time the supernumerary component may have displaced normal tissues and marked radial or ulnar deviation may have developed. Simple ablation is not the answer for such problems. Additional surgery, such as reconstruction of collateral ligaments, osteotomy of phalanges, or fusion of a joint, may be necessary.

There is increasing clinical evidence that surgery at an early age allows the maximum development of cerebrocortical patterns consistent with the parts that are present in the hand. If the operation is delayed too long, functional patterns established with the abnormal hand will have to be erased and new cortical patterns established following corrective surgery.

While the x-ray examination may show duplication of bony elements, it must be remembered that soft tissue duplication or even absence of soft parts may complicate the problem. A detailed and very careful functional assessment must be made of the hand before surgery is undertaken.

It must also be realized that premature surgery carries liabilities. Operating on a tiny digit makes it technically difficult to remove adherent supernumerary components without damaging the epiphysis of the remaining normal part. Premature epiphyseal closure and a shortened digit may be the result of such surgery. In relatively simple problems, a compromise age of around 3 years allows a good result to be obtained in tissues that are sufficiently large to make operation relatively easy.

CENTRAL DUPLICATION

Duplication of the three central digits increases in incidence toward the ulnar side, the ring finger being the most commonly duplicated while the index is the least commonly involved. The incidence in the patients in our study group is shown in Table 12-1. Twenty-three of the 28 patients had complex syndactyly with Type II polydactyly in which the extra digit was hidden in the web between two normal digits (Fig. 12-4). A different 23 patients showed involvement of the long or ring finger, and five patients had a total of seven duplicated index fingers. Three patients showed combinations of involvement between the ring and small fingers in one hand and the long and ring fingers in the other.

Barsky has commented that the complicated type of syndactyly with polydactyly is almost always seen when the long and ring fingers are involved. The extra digit is often atypical in form, the development of the nerve and blood

Table 12-1. Incidence of central polydactyly in 28 patients

	Ring	*Long*	*Index*
Male			
Right	7	2	4
Left	2	3	2
Female			
Right	8	3	1
Left	9	1	0
Total digits	26	9	7

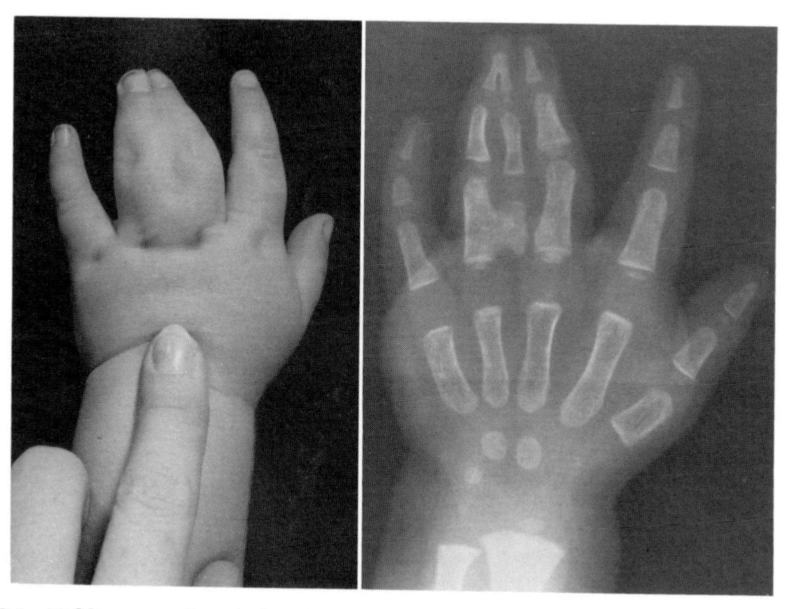

Fig. 12-4. *Hidden central polydactyly.* An apparently complete syndactyly often conceals a Type II polydactyly between the adjacent long and ring fingers. (From Flatt, A. E.: Problems in polydactyly. In Cramer, L. M., and Chase, R. A., editors: Symposium on the hand, vol. 3, St. Louis, 1971, The C. V. Mosby Co.)

vessels may be anomalous, and the tendon attachments may also show variations. The joints are usually abnormal, and the parent digit, while usually complete, frequently shows some joint distortion.

Index finger

In our series at the University of Iowa, duplication of the index finger represented 2.2% of our patients with polydactyly and only 0.3% of patients with congenital deformities of the hand. Thus index duplication is rare. There is no direct account of this condition in the English literature, but Burman has recorded some 15 references from a variety of journals, some over 100 years old. He also indirectly recorded a double index finger in a report on congenital radioulnar synostosis.

Kanavel believed that there was much confusion concerning polydactyly of the index finger. He felt that the confusion arose because hyperphalangia of the thumb is not uncommon, and a careful survey of the cases published up to 1932 convinced him that in some instances triphalangeal thumbs had been called duplications of the index fingers and that the converse was also true.

Wood has reported our five individuals with seven index finger duplications. Two of these patients represented the Type I polydactyly of Stelling and Turek's classification, two showed the Type II, and one represented the Type III. Of the patients with Type II polydactyly, one possessed a duplicated digit containing all normal elements, which was attached to the base of the middle phalanx, and the other patient showed a duplicated digit attached to a bifurcated index metacarpal. The patient with Type III polydactyly had a duplication of the index finger, with three index fingers present in the same hand. Because a normal thumb

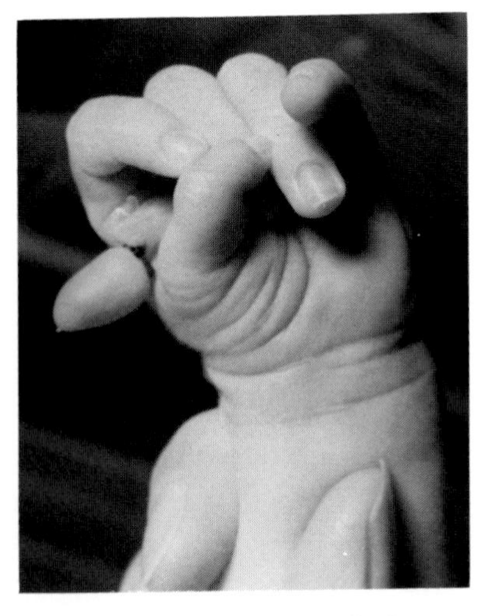

Fig. 12-5. *Index duplication—Type I.* This type of duplication can be treated by simple ablation of the pedunculated mass. It is common practice to tie off the pedicle, but it has been recorded that infants have bled to death from this procedure. (From Wood, V. E.: Duplication of the index finger, J. Bone Joint Surg. [Am.] **52**:569-573, 1970.)

was present, we felt that the finger with the extra metacarpal of Type III polydactyly definitely arose from the index finger.

Type I polydactyly can be treated by simple ablation of the pedunculated mass (Fig. 12-5).

The Type II duplication needs excision of the extra tissue, but the deviation in the digit left behind must be corrected. The patient illustrated in Fig. 12-6 had separation of the index-long partial syndactyly, removal of the distorted extra ulnar elements, and reconstruction of the ulnar collateral ligament of the proximal interphalangeal joint. This operation was done when the patient was 11 months old, and separation of the long-ring partial syndactyly was done considerably later. Follow-up showed that 30 degrees of radial deviation was still present and that both the proximal and distal interphalangeal joints remained stiff.

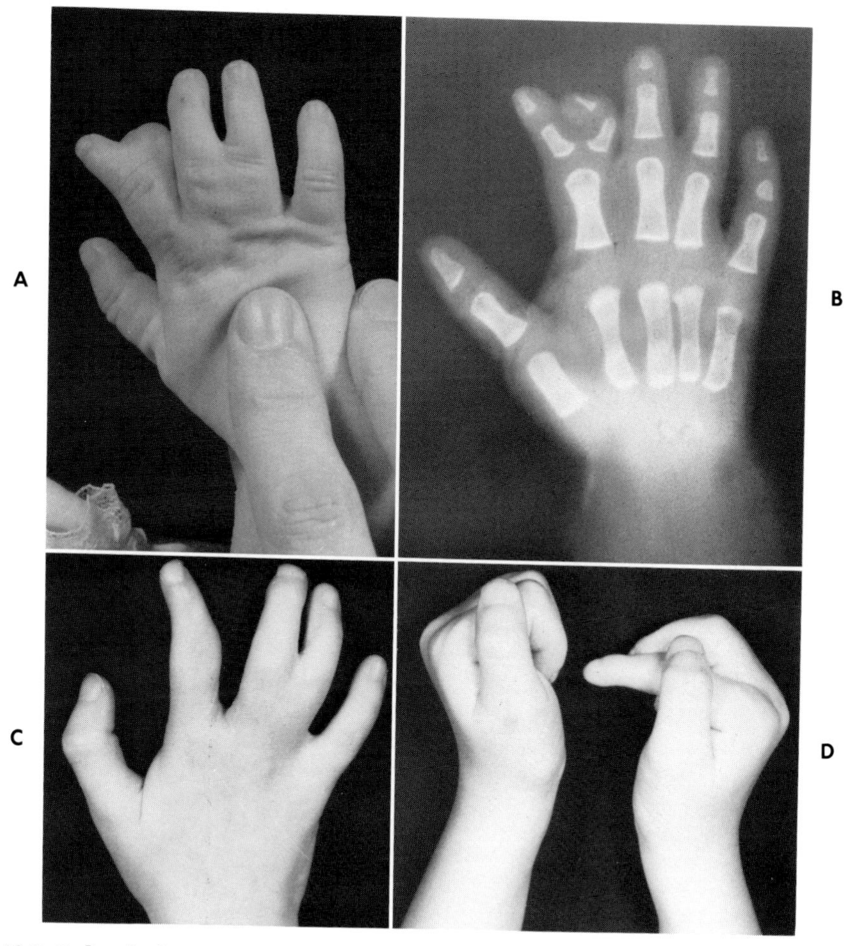

Fig. 12-6. *Index duplication—Type II.* **A** and **B,** Preoperative state showing an ulnar duplication of the middle and distal phalanges. It is probable that the width of the proximal phalanx represents a fusion of two duplicated phalanges. **C** and **D,** Postoperative appearance of the index finger showing the persistent radial deviation and restricted interphalangeal joint motion. (**B** from Wood, V. E.: Duplication of the index finger, J. Bone Joint Surg. [Am.] **52**:569-573, 1970.)

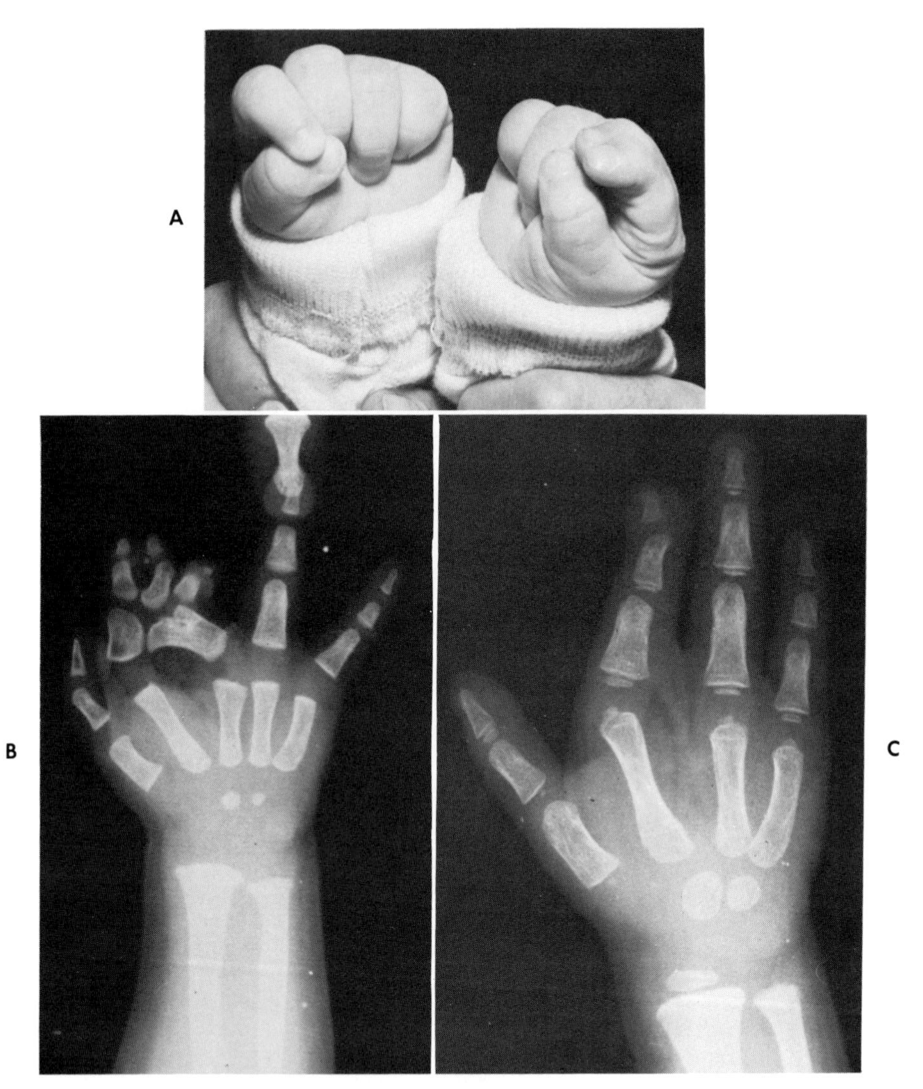

Fig. 12-7. *Index duplication and complex syndactyly.* **A,** The right hand shows what appears to be a simple syndactyly of the index-long fingers. **B,** The preoperative x-ray film shows an extremely complex index-long finger syndactyly with duplication of the index finger. **C,** Surgery at 22 months of age established a normal looking index finger after amputation of the long finger and its metacarpal. Careful longitudinal trimming of the wide index proximal phalanx did not disturb its proximal epiphysis. **D,** At the age of 13 years, the epiphysis of the index proximal phalanx continues to grow normally. **E,** The state of the hand at age 13 years. This patient is right handed and plays the violin. (**D** from Wood, V. E.: Duplication of the index finger, J. Bone Joint Surg. [Am.] **52**:569-573, 1970, and other figures from Flatt, A. E.: Problems in polydactyly. In Cramer, L. M., and Chase, R. A., editors: Symposium on the hand, vol. 3, St. Louis, 1971, The C. V. Mosby Co.)

D

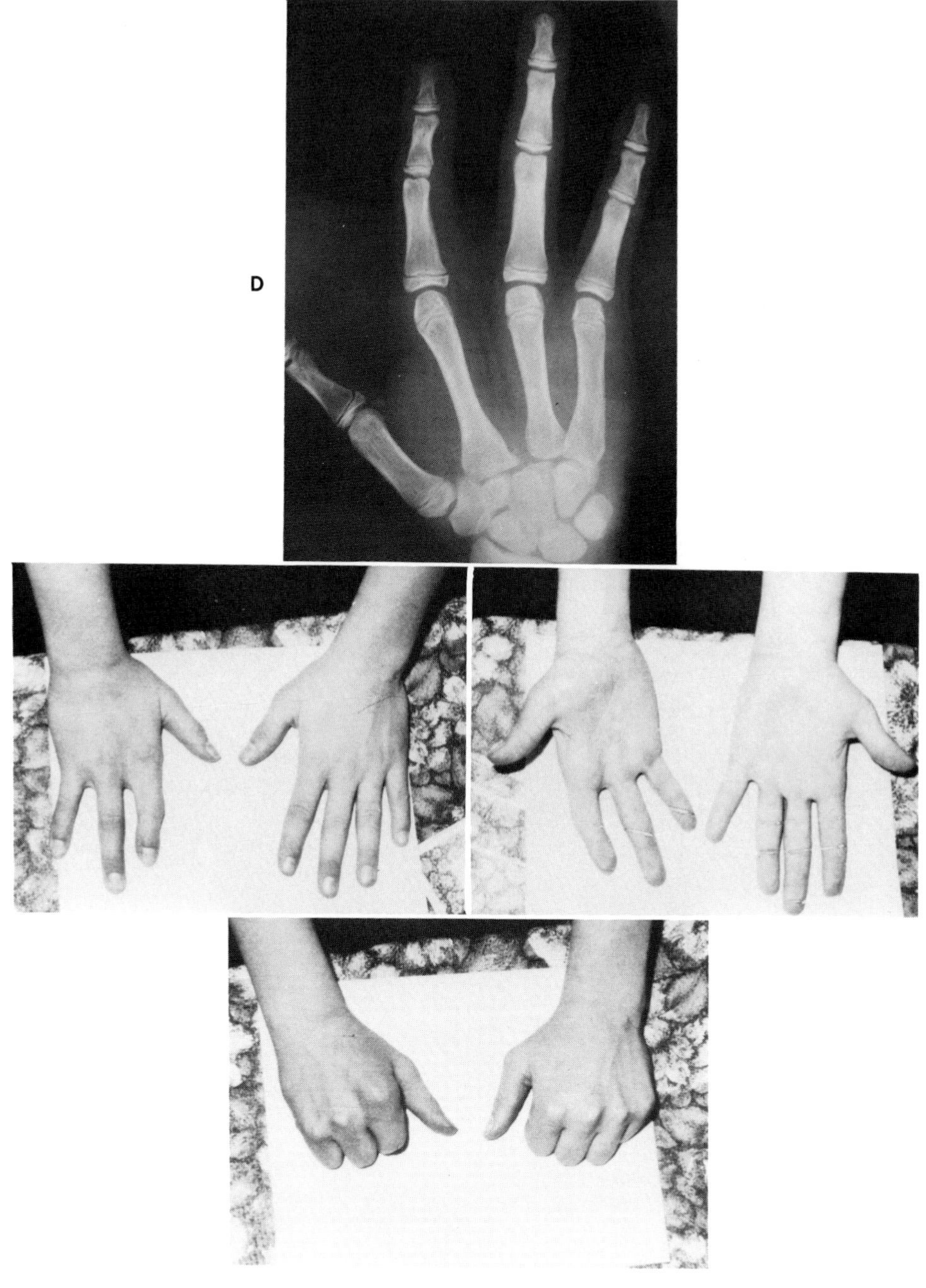

E

Fig. 12-7, cont'd. For legend see opposite page.

Occasionally Type II duplication of the index will be associated with syndactyly to the long finger. In order to provide maximum function one may be forced into creating a three-fingered hand from the jumbled components of the radial fingers. Fig. 12-7 shows the right hand of a boy who was operated on at age 22 months. The basic principle of early establishment of as normal a skeleton as possible was followed by the excision of the third metacarpal, longitudinal trimming of the wide proximal phalanx of the index, and the separation of the syndactyly. Great care was taken not to damage the epiphysis at the base of the index finger, and a scalpel was used to divide the wide phalanx. Follow-up x-ray examinations showed that the epiphysis grew normally, and at the age of 13 years the patient had a very functional hand. He plays the violin and is considered to be right-handed by his parents.

Type III duplications are usually somewhat easier to treat because the duplicated digit is a distinct entity, and amputation is relatively easy. Fig. 12-8 shows a patient with duplication and partial reduplication. A normal appearing five-digit hand was produced by selective amputation at 21 months of age. Follow-up 8 years later showed a decrease in pinch grip between thumb and index compared with the dominant left side. Fine coordination of the index finger was diminished, and precise prehensile activities had been transferred to the long finger.

A

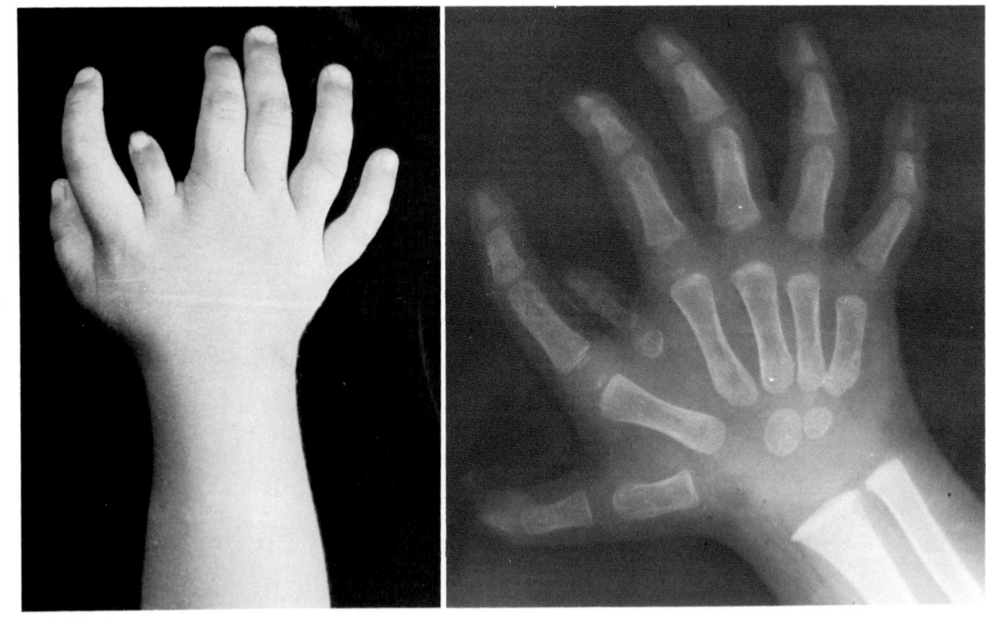

Fig. 12-8. *Index duplication—Type III.* **A,** Preoperative state of the hand. **B,** X-ray film showing a Type III duplication and, enclosed between it and the normal index, a partial index reduplication. **C** and **D,** Operative photographs. **E,** Postoperative x-ray film showing a normal skeletal appearance of the hand. **F** and **G,** The slightly wider thumb span can be seen on the operated right hand. (**B, E,** and **F** from Flatt, A. E.: Problems in polydactyly. In Cramer, L. M., and Chase, R. A., editors: Symposium on the hand, vol. 3, St. Louis, 1971, The C. V. Mosby Co.)

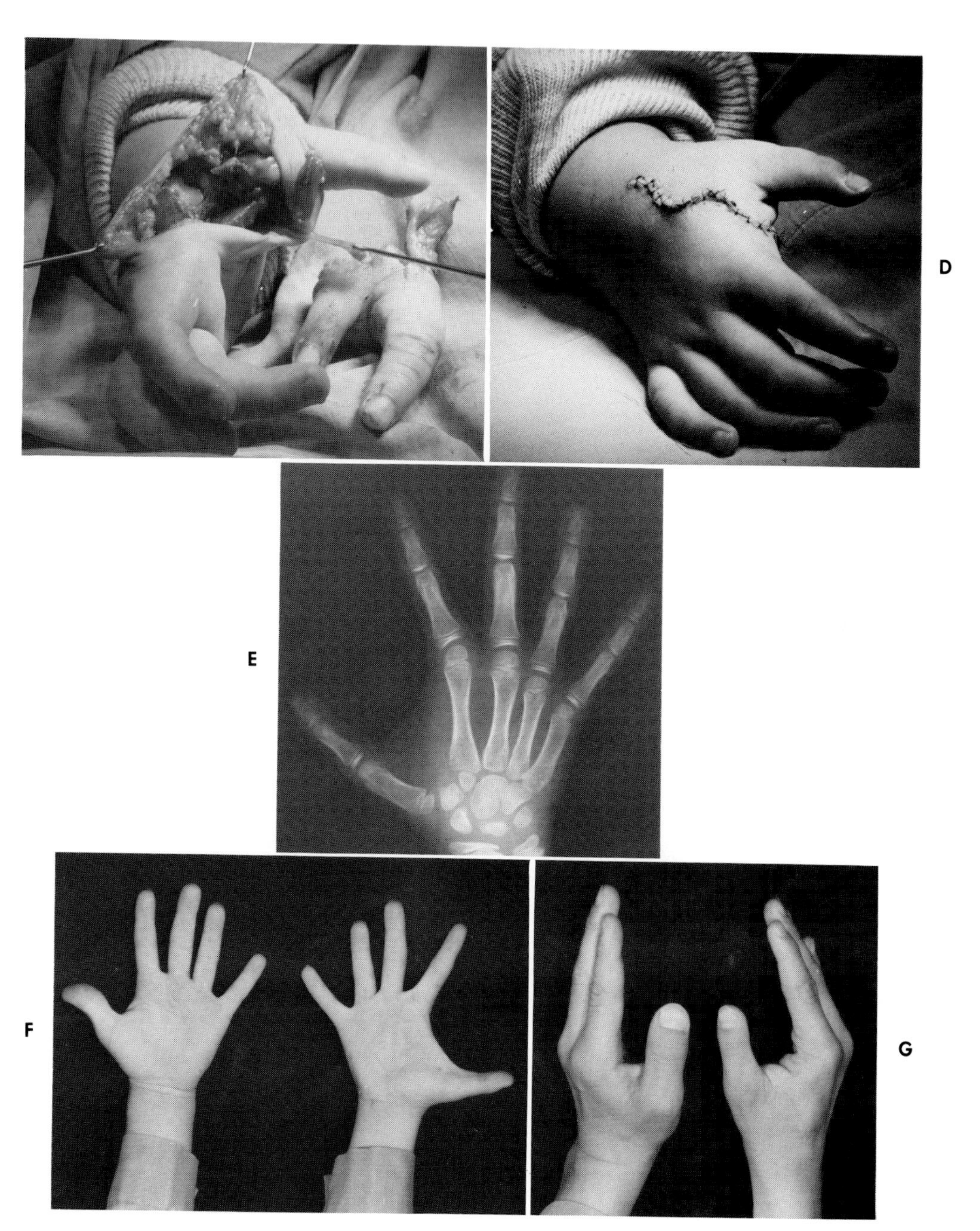

Fig. 12-8, cont'd. For legend see opposite page.

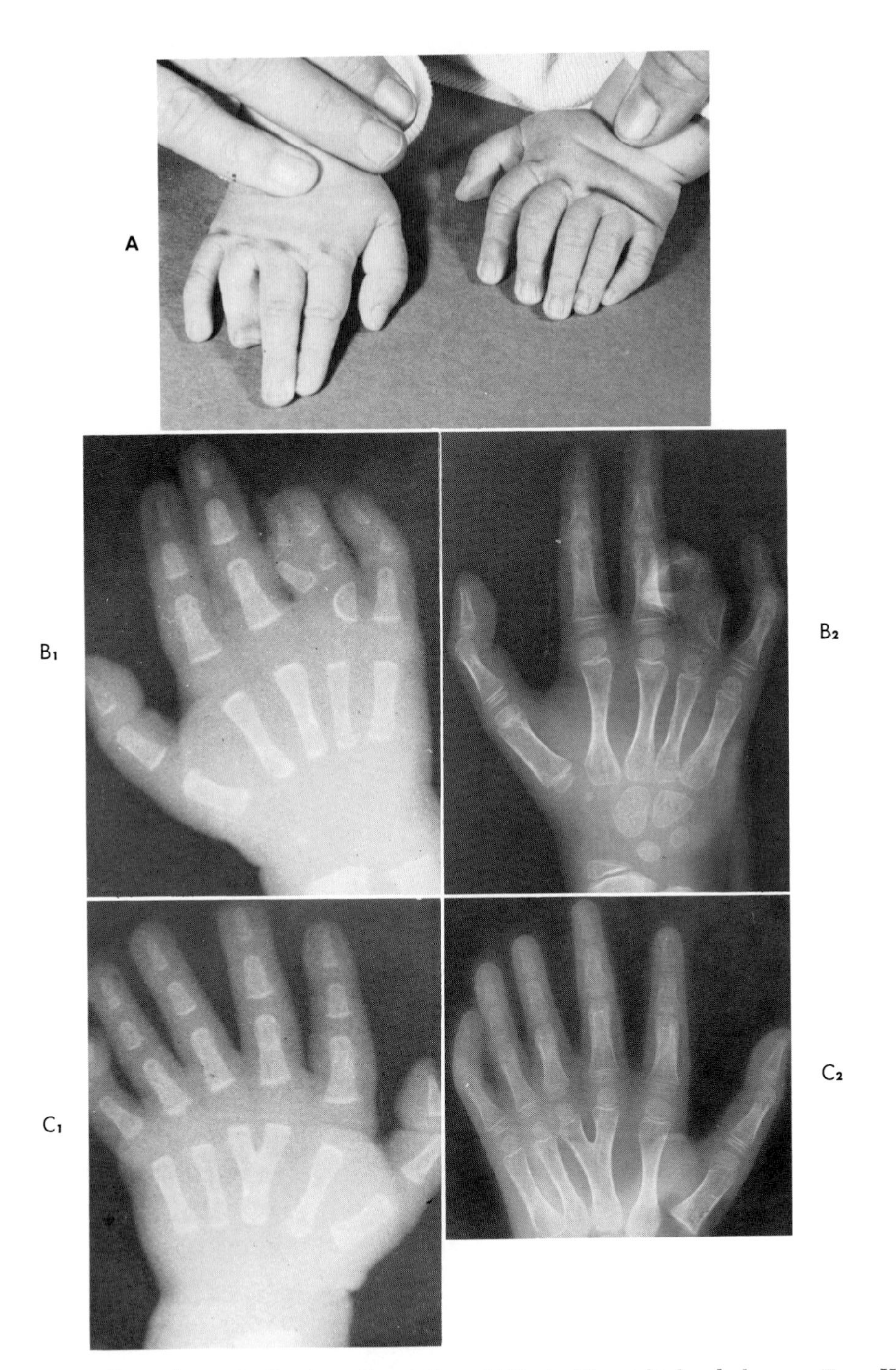

Fig. 12-9. *Long finger duplication—Types II and III.* **A,** The right hand shows a Type II polydactyly and the left a Type III. This child is severely mentally retarded, and no surgery has been performed. **B1** and **2,** Eight years of growth in the supernumerary components and aberrant epiphyses have caused a grotesque skeletal deformity. **C1** and **2,** During the same time the Type III duplication has grown uneventfully. (**C2** from Wood, V. E.: The treatment of central polydactyly, Clin. Orthop. **74:**196-225, 1971, and **A, B1, B2,** and **C1** from Flatt, A. E.: Problems in polydactyly. In Cramer, L. M., and Chase, R. A., editors: Symposium on the hand, vol. 3, St. Louis, 1971, The C. V. Mosby Co.)

Table 12-2. Anomalies associated with central polydactyly

Anomaly	*Number of patients*
Syndactyly of hands	20
Polydactyly of toes	3
Syndactyly of toes	3
Congenital amputation of thigh	1
Short fifth toe	1
Multiple hemangiomas	1
Hernia	1
Cryptorchidism	1
Mental retardation and internal hydrocephalus	1
Synostosis of carpal bones	1
Optic atrophy	1

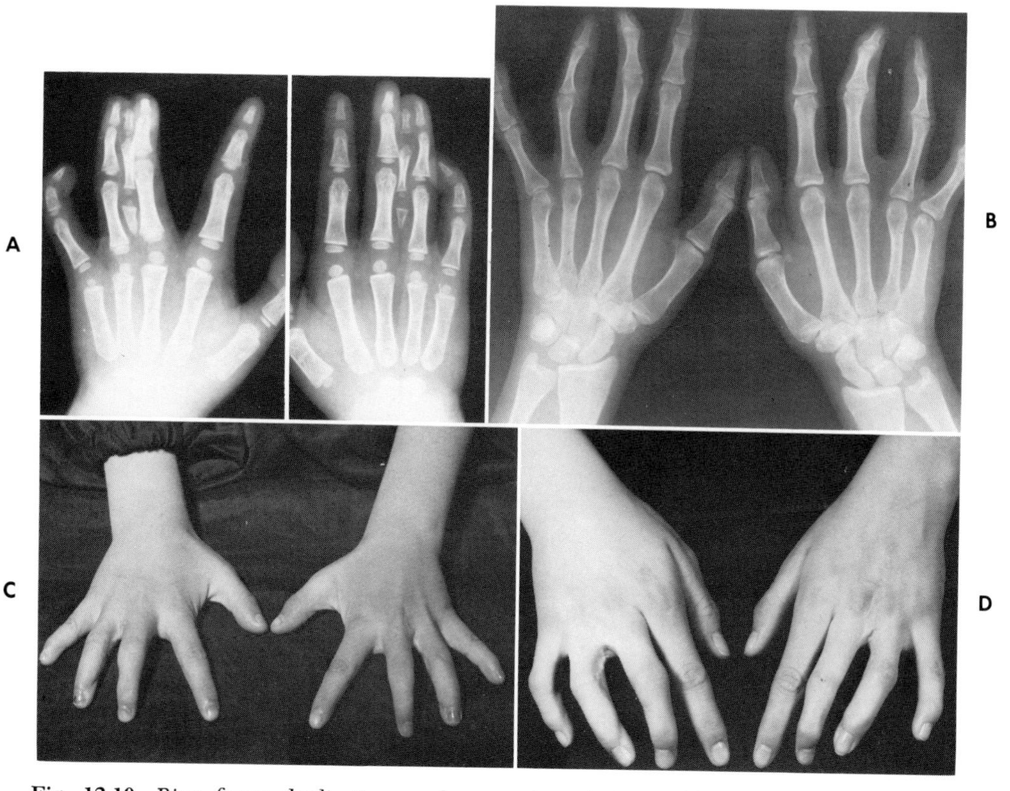

Fig. 12-10. *Ring finger duplication—early operation.* **A,** X-ray film taken prior to removal of an extra digit at age 6 months. **B,** X-ray film after bone growth had ceased. **C,** Appearance of the hands in teenage and, **D,** in adult life. (**A, B,** and **D** from Flatt, A. E.: Problems in polydactyly. In Cramer, L. M., and Chase, R. A., editors: Symposium on the hand, vol. 3, St. Louis, 1971, The C. V. Mosby Co.)

Long and ring fingers

Most patients with duplicated long or ring fingers have a Type II polydactyly in which the extra digit is hidden within the web between two normal digits (Fig. 12-4). The extra digit is usually atypical in form, in the development of its nerve and blood vessels, and in its tendon attachments and joints. The parent digit, while usually complete, frequently shows some joint distortion. Type III cases are rare and usually go unnoticed by the general public; patients with Type III extra digits are quite often resistant to the idea of amputation (Fig. 12-9).

A great variety of anomalies have been found to be associated with central polydactyly, but they are not usually life threatening. Those found associated with our patients are recorded in Table 12-2.

Treatment of the hidden Type II polydactyly is not simple. Although there is usually sufficient skin to allow good closure after separation of the conjoined digits, the interposed phalanges can prove very troublesome.

Each extra phalanx has its own epiphysis, which does not necessarily lie

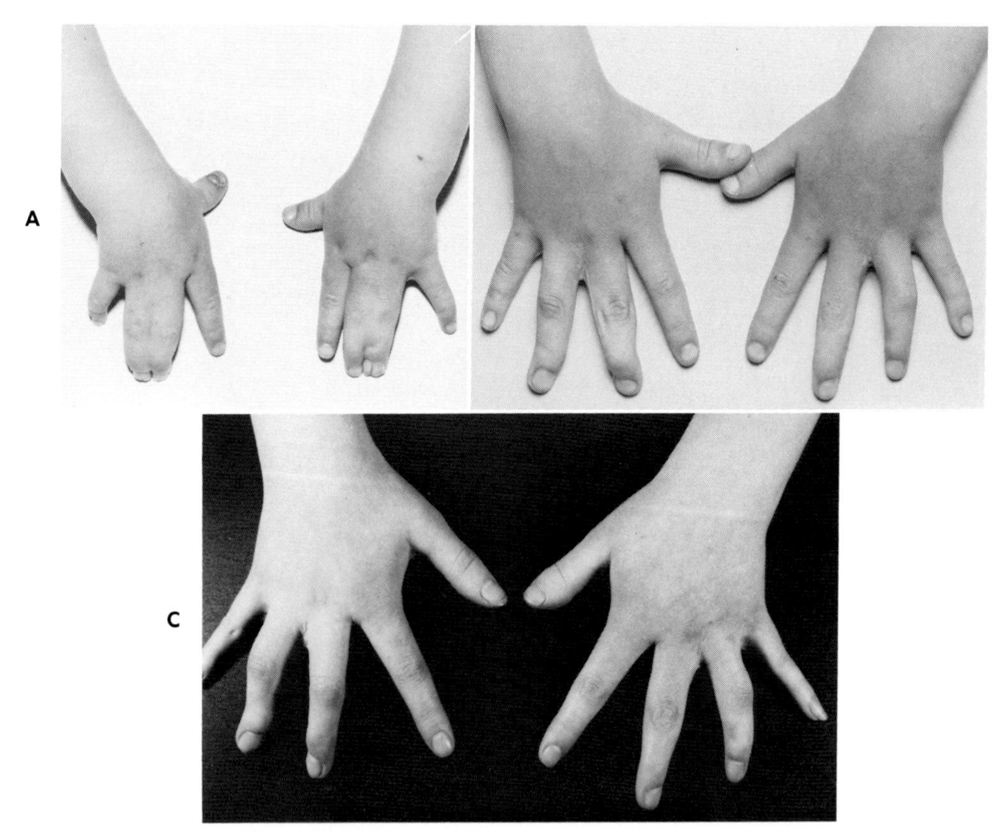

Fig. 12-11. *Ring finger duplication.* Another patient with a problem similar to that in Fig. 12-10 but not operated upon until 2 years of age. **A,** Before surgery. **B,** Six years later. **C,** Note how in adult life her hands show a shortening and twisting of the affected fingers similar to that in Fig. 12-10, *D.*

in the normal line of growth; if left in place these phalanges can produce distinct aberrations of shape in the other fingers. I believe these complications can be best avoided by amputating the extra phalanges and establishing a normal longitudinal skeleton before the patient is 6 months old (Figs. 12-10 and 12-11).

Duplicated distal phalanges usually diverge from the longitudinal axes and, therefore, early surgery is especially important. When surgery is delayed beyond 1 to 1½ years of age, the supernumerary components and abnormal epiphyses tend to displace the normal components into marked radial or ulnar deviation, and growth subsequently continues in this abnormal direction (Fig. 12-9). If this deviation has occurred, simple ablation is not the answer. More complicated surgery will be needed, and it may entail reconstruction of collateral ligaments, osteotomy of the phalanx, or fusion of an interphalangeal joint because of the continued divergence of the components from the longitudinal axis. It may also be necessary to excise the associated metacarpal to narrow the hand and bring the adjoining fingers together.

I have noticed that when all four fingers are retained there is often a ten-

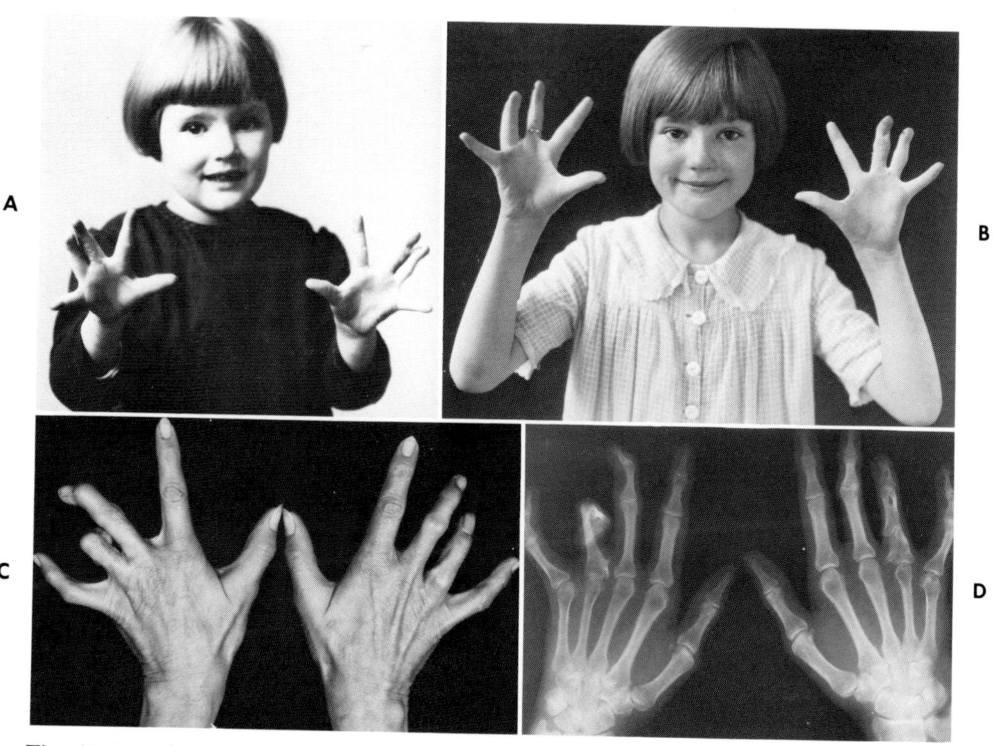

Fig. 12-12. *Bilateral long finger polydactyly—Type III.* **A,** This patient had removal of her duplicated long fingers at the age of 3 years. **B,** By pre-teenage the result appeared to be satisfactory. **C** and **D,** At 50 years of age it can be seen that the ring fingers are stunted and distorted. This patient had a strong family history of duplications, and she bore the child illustrated in Fig. 12-13. (**B, C,** and **D** from Flatt, A. E.: Problems in polydactyly. In Cramer, L. M., and Chase, R. A., editors: Symposium on the hand, vol. 3, St. Louis, 1971, The C. V. Mosby Co.)

dency for a lack of longitudinal growth in one or another of the involved fingers. This has been particularly noticeable in the ring finger. I have found, on follow-up of our long-term cases, that the ring finger frequently does not continue to grow, even though in childhood it may be proportionally the right length. By the time the patient is in the late teens or has reached adulthood, the ring finger is noticeably smaller (Figs. 12-12 and 12-13).

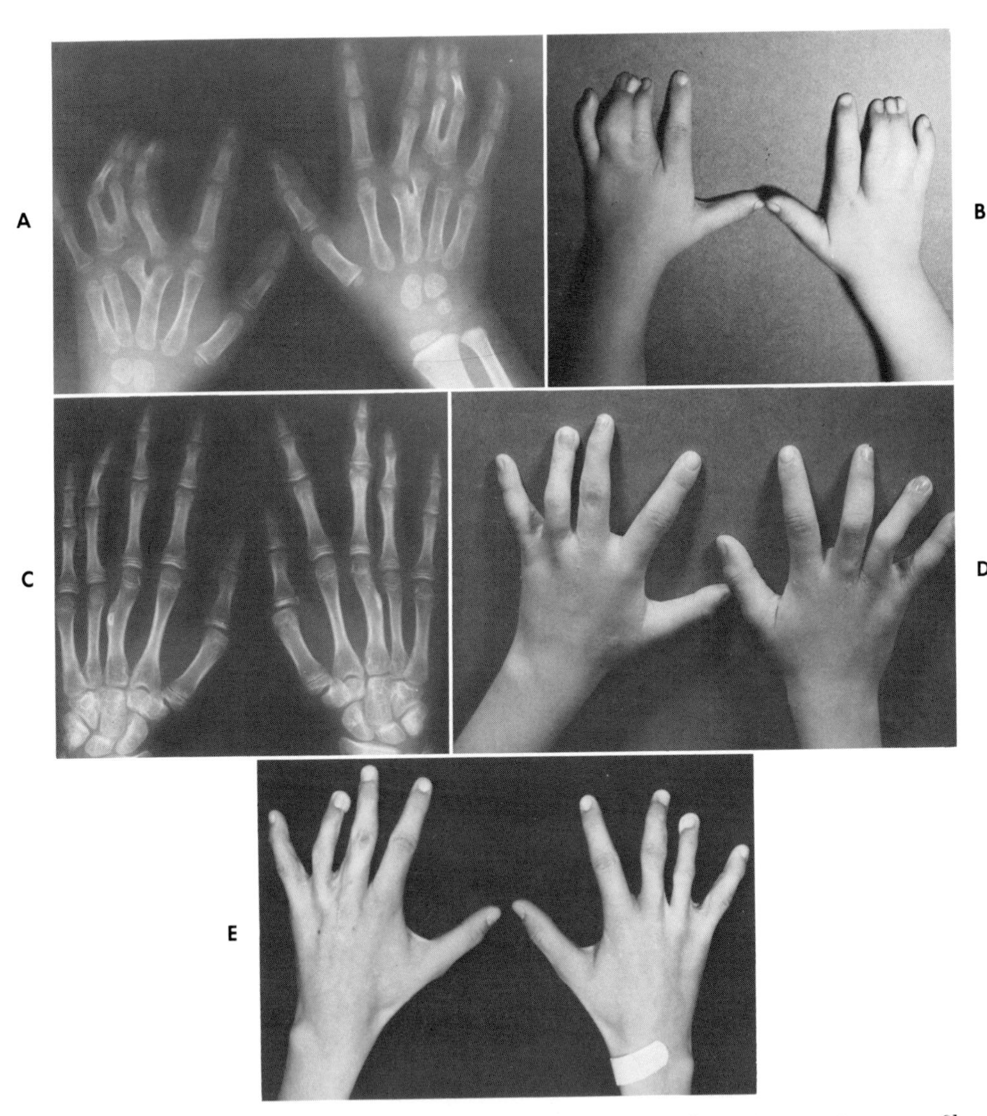

Fig. 12-13. *Bilateral long finger polydactyly—Type III.* **A** and **B,** Preoperative x-ray film and clinical appearance of the hands. **C** and **D,** Five years after removal of the duplicated digits. **E,** Nine years after the original surgery, development of the ring finger has failed to keep pace with the other fingers. (From Wood, V. E.: Treatment of central polydactyly, Clin. Orthop. **74:**196-225, 1971.)

THE THREE-FINGERED HAND

Although it may seem paradoxical, it is sometimes necessary to seriously consider the production of a functional three-fingered hand from what is in effect a five-fingered hand. In the past I have often attempted to retain four fingers, but this is no longer my routine surgical plan. Many parents are reluctant to accept this advice, but my experience with these types of complicated central polydactyly has been depressing (Fig. 12-14). Great judgment is necessary in deciding whether all four fingers should be preserved or whether a three-fingered hand would be more functional and perhaps even more cosmetically acceptable. I was led to the view that a three-fingered hand is socially

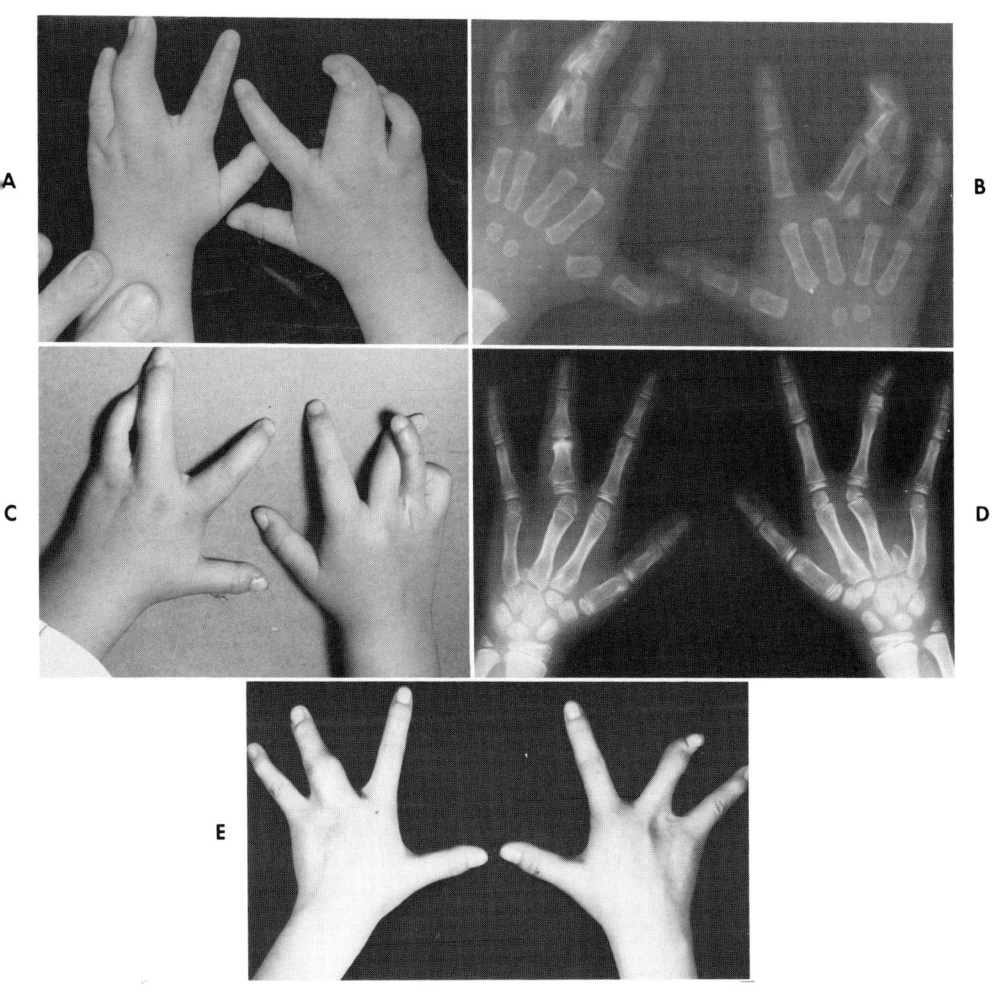

Fig. 12-14. *Three- or four-fingered hand?* **A** and **B**, Preoperative appearance of a bilateral concealed Type II duplication of the long finger. **C**, After four operations, a reasonable three-fingered left hand and a bad four-fingered right hand were produced. **D** and **E**, Final state of the hands after nine operations spaced through a period of 10 years. (From Wood, V. E.: Treatment of central polydactyly, Clin. Orthop. **74**:196-225, 1971.)

and functionally acceptable to many patients by being forced to amputate rigid and useless fingers that had been produced by earlier attempts to make four-fingered hands.

Nowadays, when faced with a complicated central duplication, I make a thorough explanation of the potential problems to the parents and obtain their consent to produce a three-fingered hand if I feel that it will produce the best functional and cosmetic result for their child.

Before surgery it is often found that there is limitation of movement in the interphalangeal joints. While this may be produced by the presence of accessory digits, it is more often caused by an inherent stiffness, and occasionally even true symphalangia, of the apparently normal fingers. If the anatomic state of the tendon attachments is doubtful in the slightest degree, a thorough exploration at surgery is mandatory. It may be found that a tendon will have to be transferred from the accessory digit or even from another part of the hand to produce the best functional result. In many patients there are split tendons or a central tendon with a wide split in the two segments. These variations must be checked at surgery, and the best functional structure should be left in the digit that remains. It is impossible to determine the full extent of any neurovascular abnormalities until the area is thoroughly exposed at surgery.

ULNAR DUPLICATION

The small finger is probably the digit most often duplicated. Dominant inheritance of the trait is common; in some families it is regularly transmitted, but others can show great variation in occurrence and extent of duplication. Polydactyly is the commonest congenital hand anomaly of blacks, and duplication of their small or postaxial finger is ten times more common than of their preaxial finger.

Temtamy and McKusick have divided the postaxial duplications into Types A and B by genetic differences. Type A digits are fully developed extra digits, while Type B digits are frequently pedunculated or rudimentary (Fig. 12-15). Children born to individuals with Type A accessory small fingers can produce children with either Type A or B digits, but people with Type B deformities can only produce more Type B progeny.

The Type A group may show duplication or thickening of the metacarpal, and occasionally carpal fusions or accessory carpal bones are found. It is the Type B, small, incomplete pedunculated digit that is more frequent.

Duplicate small fingers in the newborn should not be lightly dismissed because their presence is associated with a variety of syndromes. They are a typical component of trisomy 13 and also occur in such syndromes as Ellis-van Creveld, Laurence-Moon-Bardet-Biedl, and Jeune, in chondroectodermal dysplasia, and in several other conditions that are considerably less common.

The duplicated small finger is commonly undersized and is usually attached by a very narrow pedicle. More often than not it is tied off in the nursery, producing a blackened stump that drops off leaving an almost invisible

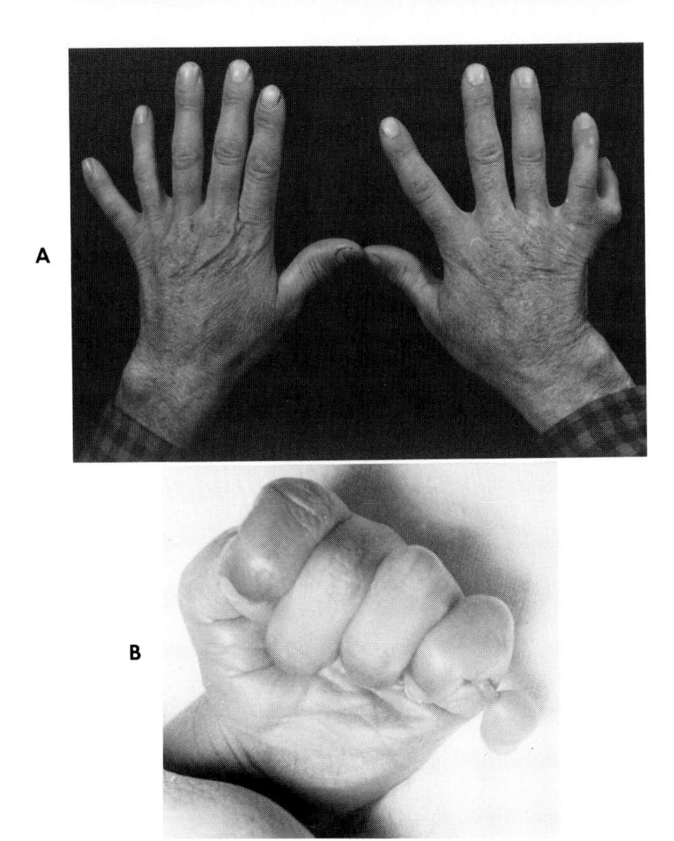

Fig. 12-15. *Postaxial duplication.* The two types described by Temtamy. **A,** The fully developed extra digit. **B,** Type B digits are frequently pedunculated or rudimentary.

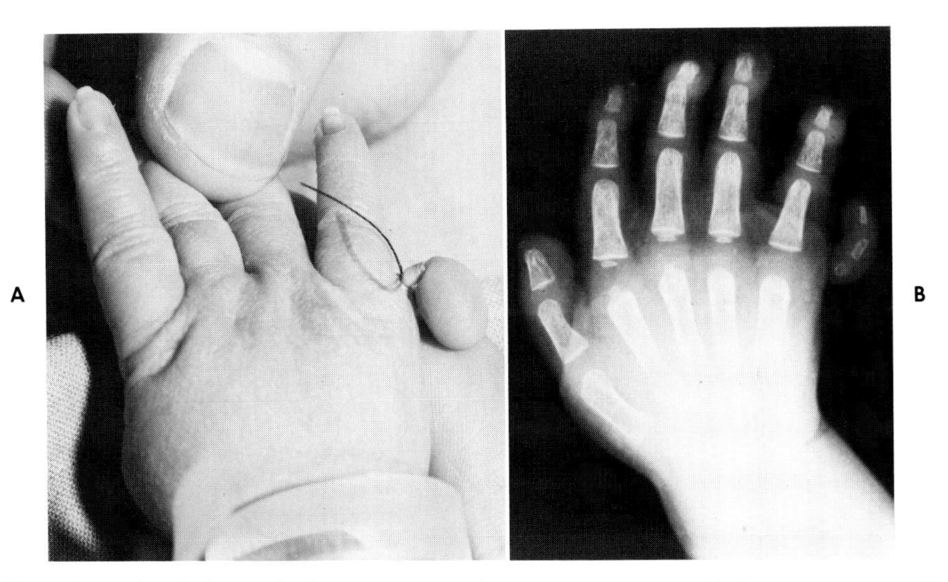

Fig. 12-16. *Small finger duplication—tying off.* **A,** Most extra small fingers are removed in the newborn nursery by tying off the pedicle. **B,** The presence of incomplete phalanges is no hindrance to such removal. The same risk of hemorrhage mentioned in Fig. 12-5 applies to this procedure. (From Flatt, A. E.: Problems in polydactyly. In Cramer, L. M., and Chase, R. A., editors: Symposium on the hand, vol. 3, St. Louis, 1971, The C. V. Mosby Co.)

scar (Fig. 12-16). It has been recorded that infants have bled to death following this procedure, but it must be an extremely uncommon occurrence.

Occasionally an accessory small finger that contains portions of phalanges is left on, and the patient grows to adulthood with an extra finger. These can be of varying lengths, but I have never seen a normal looking, fully duplicated small finger. Technically there is little trouble in terms of removal, even of the more formally duplicated digit.

CONCLUSIONS

The treatment of polydactyly is not simple. Unthinking amputation of an accessory digit, with no regard for the function that it contains or the function in its adjacent digits, will do a disservice to the patient. The problem is to decide which digit is the most functional and has the most normal potential for growth. Such a decision cannot be made in one visit to the newborn nursery. Very careful and prolonged preoperative assessment of all the anatomical structures in the accessory digit and its neighbors is vital before a proper surgical plan can be made. The infant or child will have to be watched both in sleep and at play before a full functional assessment can be made and the operative plan developed.

It is essential to produce as normal a skeleton of the hand as possible early in life so that growth may proceed unhindered. Long-term follow-up is also essential, since our studies have shown that there is a tendency for gradual increase in deviation deformities. It is my feeling that these deformities may show a significant increase at the time of the teenage growth spurt. I therefore believe these patients should be kept under clinical observation until skeletal growth has ceased.

CHAPTER 13

LARGE FINGERS

Differential diagnosis
Etiology
Clinical types
Neurological involvement
Treatment

Children are supposed to be born symmetrical, and it is taken for granted that as they grow their corresponding limbs, fingers, and toes will match in size. On rare occasions, something goes wrong; a localized enlargement of all tissues occurs, producing gigantism. It most commonly affects the extremities, and the upper limb is affected more often than the lower limb. Involvement may consist of a single digit, several digits, the whole limb, or even one half of the body.

Localized symmetrical involvement of one or more digits has been called clubfinger, macrodactyly, or megalodactyly. Edgerton prefers the term proposed by El-Shami, digital gigantism, because it encompasses "congenital pathological enlargement of any soft parts of the body with associated enlargement of the skeleton."* This useful definition excludes digits enlarged by conditions such as hemangiomas, arteriovenous fistulae, lipomatosis, neurofibromas, and neurilemmoma of bone.

The condition is one of the least common congenital deformities; in our study of 1,476 patients we have found only 19 hands so affected, giving an incidence of 0.9%. Twenty digits were involved—seven index fingers, five long fingers, and four ring fingers, and the border digits, thumb and small finger, were each involved twice.

Analysis of the literature shows that in 90% of patients the deformity is unilateral and that it more commonly occurs on the radial side of the hand with the index finger being most frequently involved. When more than one digit is enlarged, it is adjacent digits that are affected; there is no recorded case of enlarged digits separated by a normal digit (Fig. 13-1). Multiple digit involvement

---

*Edgerton, M. T., and Tuerk, D. B.: Macrodactyly (digital gigantism): its nature and treatment. In Littler, J. W., Cramer, L. M., and Smith, J. W., editors: Symposium on reconstructive hand surgery, vol. 9, St. Louis, 1974, The C. V. Mosby Co., p. 157.

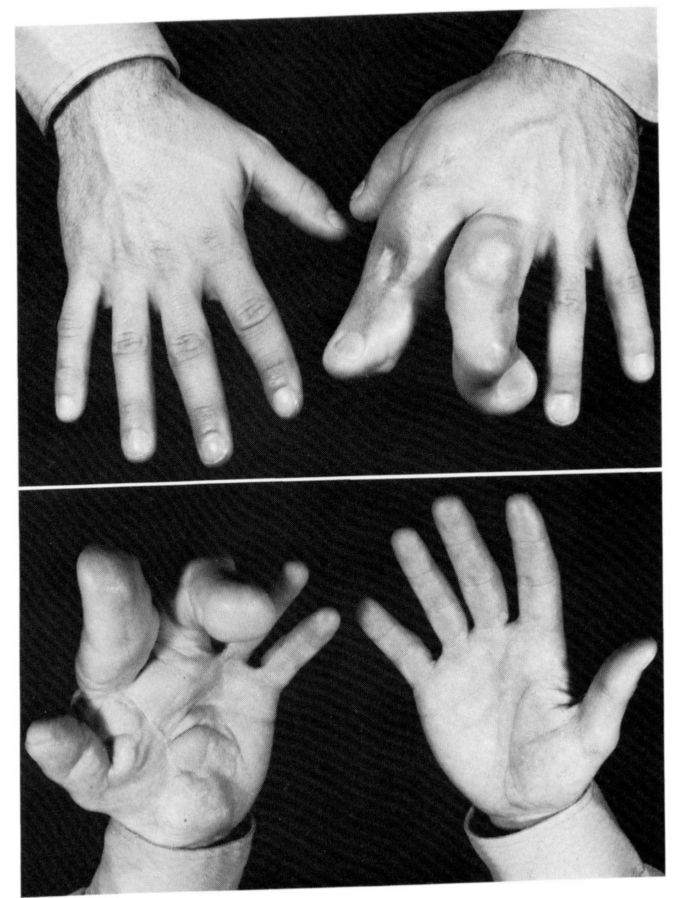

Fig. 13-1. *Typical digital gigantism.* This hand of a school teacher shows the common unilateral radial involvement. When more than one digit is enlarged, it is adjacent digits that are affected.

is said to be three times more common than single digit enlargement.

There is probably an equal sex distribution. Hereditary factors are not thought to be involved although two patients with family histories have been reported. Chromosomal studies have been reported by several authors, and all have been normal.

DIFFERENTIAL DIAGNOSIS

Differential diagnosis for these swollen fingers should include such conditions as Albright polyostotic fibrous dysplasia, osteoid osteoma, lymphedema, arterio-venous fistula, lymphagioma, hemangioma and associated Klippel-Trenaunay-Weber syndrome, and von Recklinghausen neurofibromatosis (Fig. 13-2). Associated anomalies have been recorded, but none shows any high incidence. Syndactyly occurs in about 10% of cases, and polydactyly, cryptorchidism, nevi, and other congenital defects have been recorded.

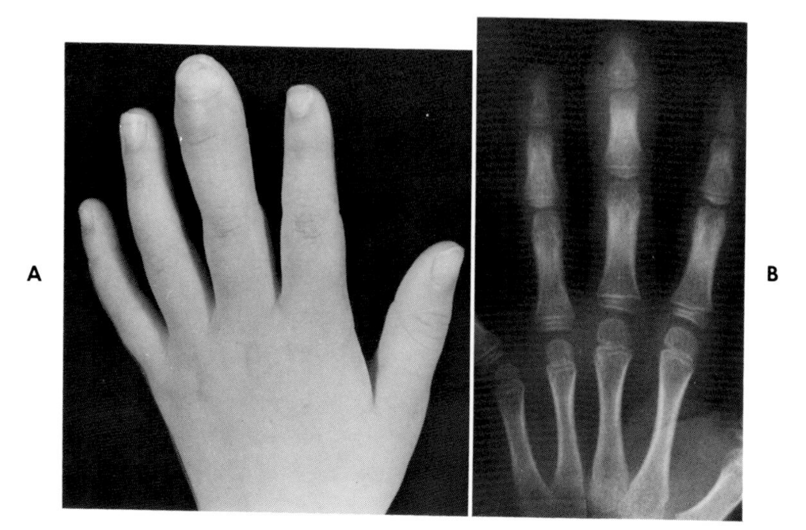

Fig. 13-2. *Differential diagnosis.* Osteoid osteoma of a phalanx must be considered. **A,** The enlargement is usually localized rather than generalized and associated with symptoms. **B,** The x-ray film should be diagnostic.

ETIOLOGY

No good explanation exists for the excessive growth. Barsky, who has studied this condition in detail, writes, "My present concept of the cause of macrodactyly is that during fetal development some disturbance of the growth-limiting factor occurs in the local area affected and that in the presence of this lack of inhibition the part continues to increase in size. This would account for the progressive overgrowth in later years."* He quotes Streeter, who in a 1930 letter to Rowe wrote, "Local gigantism . . . is not bilaterally symmetrical but is exhibited in haphazard focal areas. It does not appear to be a mutation or to be a gene-determined character. Rather, it is a consequence of germ plasm abnormality or pathology. By accident in cell cleavage or organogenesis certain areas in some instances become supersensitive to growth stimuli. Such an overresponse is at the basis of local gigantism, as I see it."†

CLINICAL TYPES

Whatever the explanation, there is general agreement that two clinical forms of the condition exist. In the more static type the infant is born with the enlargement and as the child grows, the enlarged digit keeps pace proportionately with the growth of the rest of the hand. In the other, progressive, type there is a disproportionate growth of the affected digit so that its size increases at a faster rate than does the rest of the hand.

*Barsky, A.: Macrodactyly, J. Bone Joint Surg. (Am.) **49**:1255-1266, 1967.

†Rowe, E. W.: Local gigantism, Radiography and Clin. Photography **20**:20-21, 1944.

The soft tissue involvement is usually greater on the palmar aspect. There is sometimes thickening of the skin and hypertrophy of the nail. Generally the involvement is greatest distally, and the enlargement gradually tapers off in the more proximal tissues (Fig. 13-3).

The phalanges are invariably involved, and in very extensive cases, the metacarpals may also be enlarged. As the phalanges enlarge, the finger tends to curve, developing an ulnarward curvature with the convexity on the radial side (Fig. 13-4). Occasionally, if two adjacent fingers are affected, the tips will curve toward each other with the more radial digit pointing in an ulnar direction. Infrequently, a finger may be involved only on one side in which case it will curve, with the excessive growth being on the convex side.

As the child matures, the finger gradually begins to lose motion. Growth of the whole digit tends to cease when the epiphyses close, and by this time motion is usually severely restricted. This is probably caused by several factors; the bony enlargement is somewhat irregular and may cause secondary joint limitation. I believe most of the lack of motion is caused by deliberate lack of use of the hand

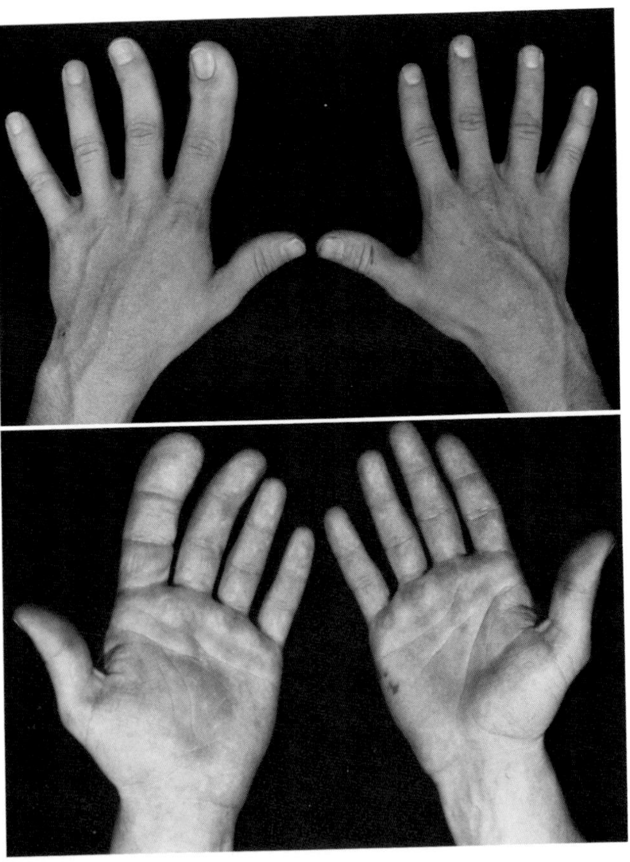

Fig. 13-3. *Index gigantism.* This index finger shows the characteristics of digital gigantism: enlarged nail, involvement greatest distally tapering off into proximal tissues, and soft tissue involvement greatest on palmar aspect.

and its frequent concealment from public gaze. Certainly the psychological impact of the deformity on children is great. They are subjected to frequent peer abuse and to remarks like "monkey hand," "banana fingers," and other less savory allusions (Fig. 13-5).

One still finds adults who have survived this tormenting and who now claim that it mattered little to them, but on private questioning they will readily admit to the difficulties they encountered while growing up. Occasionally they can put the finger to good use—such as the bartender, whose hand is shown in the circu-

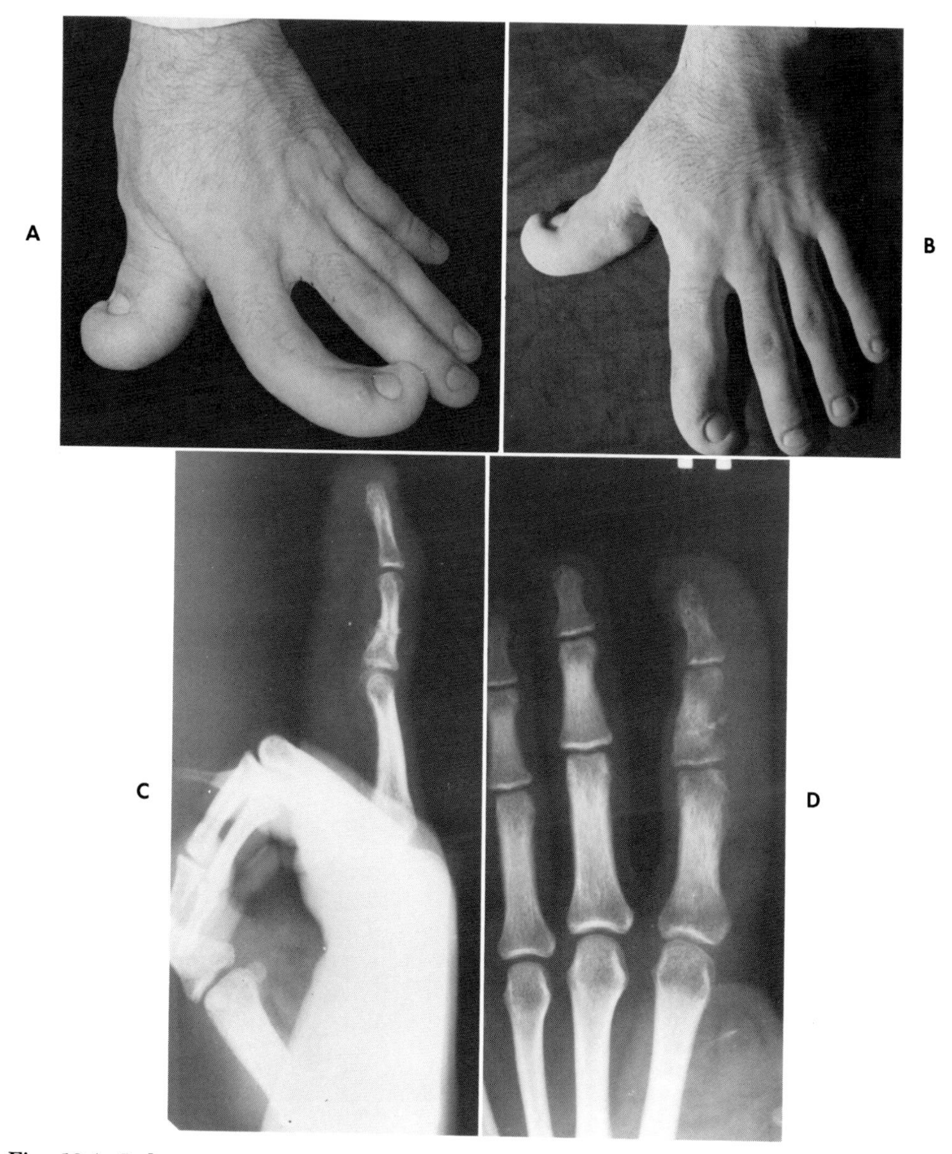

Fig. 13-4. *Index gigantism—curvature.* **A,** The index usually develops an ulnarward curve with the convexity on the radial side. **B** and **C,** Corrective osteotomy straightened this finger. (From Wood, V. E.: Macrodactyly, J. Iowa Med. Soc. **54**[10]:922-928, 1969.)

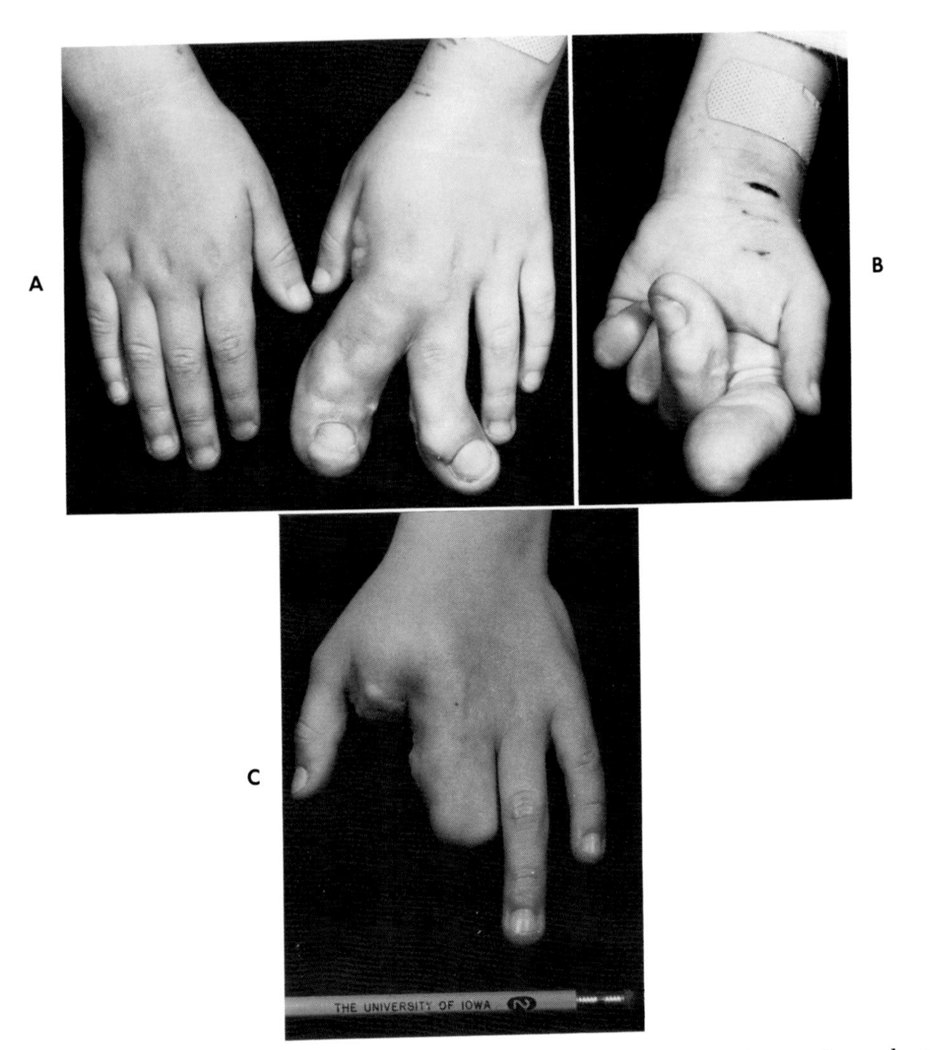

Fig. 13-5. "*Banana fingers.*" **A,** Children are frequently subjected to cruel peer abuse about their fingers. **B,** Lack of flexion is common in the affected fingers. **C,** Selective amputation is often the only treatment that is helpful.

lar illustration at the beginning of the chapter and whose large finger gave him uncanny leverage with bottles and glasses. Some make light of the deformity, such as the patient in Fig. 13-1, who is an athletic coach in a high school. But the significant fact is that they still seek advice to "see whether anything can be done" short of amputation.

NEUROLOGICAL INVOLVEMENT

Adults may seek help because of the cosmetic deformity or because of lack of motion, and some will come for examination with a characteristic history of carpal tunnel syndrome. A few will have developed trophic ulcers on the involved

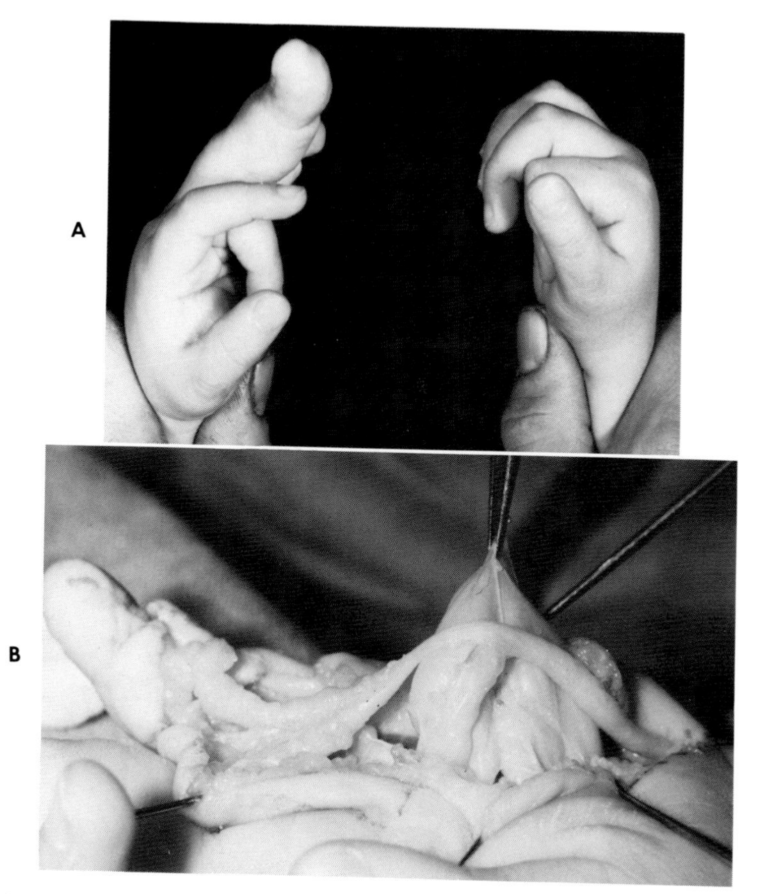

Fig. 13-6. *Neurological involvement.* **A,** An enlarged long finger in a child's left hand. **B,** Exploration showed gross involvement of the digital nerves.

digits. These neurological symptoms point up the greatest problem in the etiology of the condition. The greater part of all the cases in the literature—and my few personal cases—show significant enlargement of the nerve supply to the affected digits (Fig. 13-6). Children may complain of paresthesias or hypesthesia. By the time they reach adult life the enlargement of the median nerve is so great that its compression beneath the transverse carpal ligament leads to slowing of conduction and the onset of symptoms. Because of this frequent association of neural enlargement with macrodactyly, many physicians consider the primary cause to be a form of neurofibromatosis. Edgerton has summarized the rather impressive evidence for the connection between macrodactyly and neurofibromatosis by pointing out that:

1. Adipose tissue and fibrous tissue proliferation in the peripheral nerves are found in both.
2. Both conditions show similar neurogenic patterns of involvement.
3. Some macrodactyly patients have definite neurofibromas, lipomas, and cafe au lait spots.

4. The digital nerves that supply the enlarged macrodactylic digits are usually enlarged.
5. Some (but not all) of the patients with macrodactyly may be found to have minute plexiform neurofibromas on special nerve staining of the excised fatty tissues.
6. Periosteal nerve neurofibromas are associated with localized enlargement of bone.
7. Neurofibromatosis of splanchnic nerves is associated with localized gigantism of those portions of the gut.

Associated with these findings is the fact that the tissues involved in the enlarged digit are invariably those that respond late in development to neurogenic influence, that is, nerves, fat, skin appendages, and bones. The tendons and blood vessels are not usually enlarged (Fig. 13-7).

Most authorities have linked the neural enlargement and the macrodactyly as cause and effect and use such terms as "localized neurofibromatosis" to describe the condition. However, macrodactyly can occur as a true hypertrophy of all tissues without selective enlargement of the nerves, and gross enlargement of the median nerve and its branches can occur without associated digital enlargement.

Another term currently in vogue that links neural enlargement and macrodactyly is nerve territory oriented macrodactyly (NTOM). This is a useful term because it emphasizes the anatomical relationship between nerve distribution and the site of enlargement. Enlargement is certainly more common in median nerve

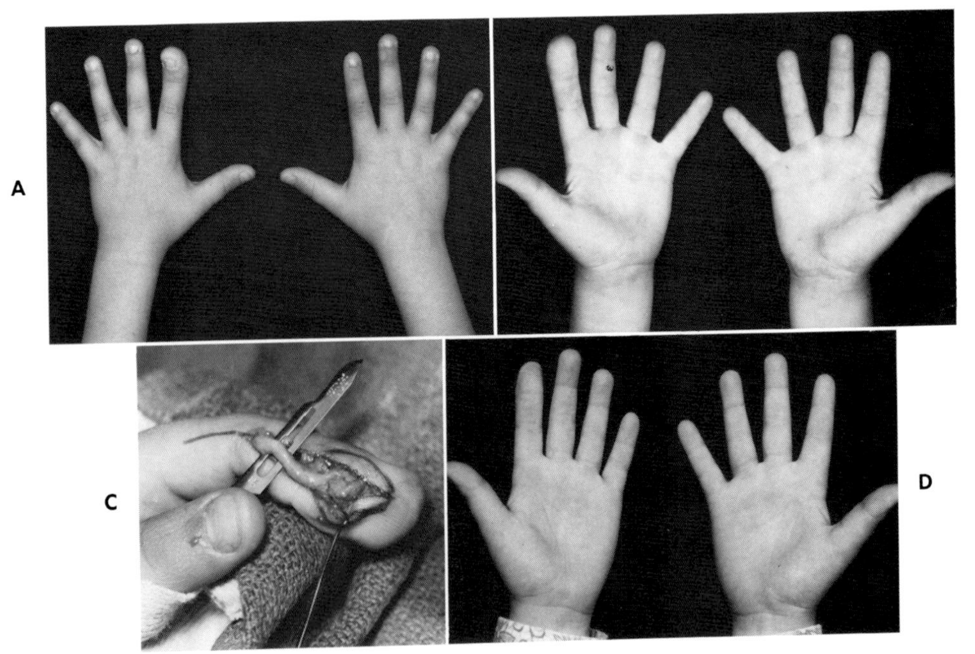

Fig. 13-7. *Digital nerve involvement.* **A** and **B**, Terminal enlargement of the left index finger. **C**, The gross size of the distal end of the digital nerve. **D**, Bulk reduction was done and 3 years later the size remains acceptable.

territory, but it has also been recorded in ulnar nerve supplied areas (Fig. 13-8).

Grossly, the epineurium of the enlarged nerves has a ground-glass appearance, has lost its smoothness, and is thickened. The lobules of fat that infiltrate and separate the nerve bundles are dark yellow, large, and hard to remove because of the many fine vessels that traverse the fat. Microscopic studies show normal axons with dense collections of neurilemmal cells in the center of the nerves and dense peripheral fibrosis.

Some pathologists regard these areas as representing a solitary plexiform neuroma. There is said to be about a 13% risk of malignant change in generalized neurofibromatosis, but no data regarding the risk of malignant change in macrodactyly have been published. I believe children with this condition should certainly be observed to maturity.

TREATMENT

Deformity, with its functional and psychological handicaps, is the reason for treatment, which must be planned on an individual basis for each patient. There is no medicinal treatment for the condition; radiotherapy is valueless, and surgical treatment is ablative. One is reluctant to suggest amputation, but for an adult with a single, stiff, ugly finger, it is probably a kind and practical suggestion. If the finger involved is either the long or ring, then ray resection and narrowing of the hand by moving the border digits over is the operation of choice. Barsky and Tsuge are the two surgeons who have published the most

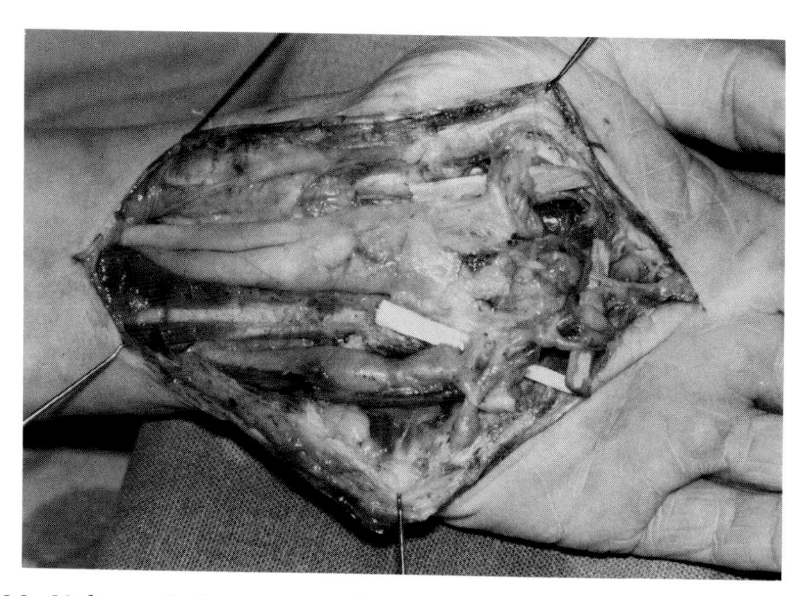

Fig. 13-8. *Median and ulnar nerve involvement.* The hand of a 12-year-old boy without digital gigantism but with gross enlargement of both median and ulnar nerves. The median nerve, with its median artery, passes across the center; the ulnar nerve lies parallel and below the median; the connecting branch between median and ulnar nerves lies over the piece of white tongue depressor.

complete operative reports and whose surgical plans are feasible and practical.

For treatment of the enlarged thumb or fingers of a child, reduction operations rather than amputations are the procedures of choice. The technical problems are great, and although bulk reduction may be achieved, the cosmetic result is often far from pleasing (Fig. 13-9). Thumb reduction has one practical advantage over finger reduction. If the carpometacarpal and interphalangeal joints of the thumb have an acceptable range of motion, considerable length reduction can be obtained by bone resection and fusion of the metacarpophalangeal joint. When this is accompanied by soft tissue reduction, the functional result is often good.

For the infant with mild hypertrophy of a digit, inhibition of growth should be planned. Treatment should be started early in infancy by attempting to gain proportional lengths of the affected and nonaffected digits. There is no "correct" age at which this surgery should be started. Timing will depend on the extent and severity of the deformity and the experience of the surgeon. Initially, surgery should consist of multiple defatting procedures on one side of a digit at a time. It is probably best to wait at least 3 months before operating on the second side of a digit, and caution is necessary in the dissecting and defatting of the flaps. The blood supply of the skin in these enlarged digits is relatively poor, and Edgerton has emphasized that these flaps run more than a usual risk of necrosis. He recommends removing the skin and subcutaneous tissues in one piece, defatting the skin to the dermal level, and returning it as a full-thickness graft. This is excellent advice, and it works, but I would caution those not used to employing full-thickness grafts that survival of the graft is sometimes precarious on the uneven bed left after excision of bone, fat, and even neural elements.

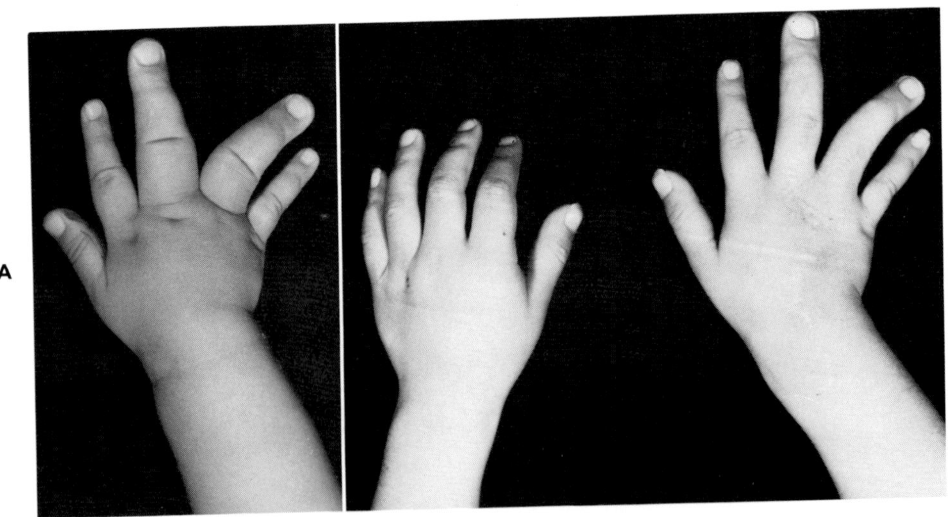

Fig. 13-9. *Dorsal bulk reduction.* **A,** Occasionally there is considerable bulk present on the dorsum of the hand and digits. Bulk reduction is more easily done there than in the palm. **B,** Three years after reduction. Osteotomy of the ring finger will be done.

Tsuge recommends combining the defatting procedure with complete separation of the digital nerve branches. He dissects all the branches of the digital nerve in conjunction with removal of the adipose tissue. He carefully preserves the digital vessels and the digital nerve trunk itself after completely excising all its branches. I have no personal experience with this procedure. Tsuge believes that his method of stripping the nerve inhibits or checks growth in the defatted digits. He comments that this procedure will cause a certain degree of sensory disturbance, but if finger growth can be inhibited, it should be considered an acceptable procedure. Follow-up shows a gradual improvement in sensibility over a period of time.

As the child grows, the length of the involved digits should be monitored against the hand size of the patient's parent of the same sex. When the external length is the same, epiphyseal arrest at the proximal end of each phalanx should be done. Stapling or wiring these small epiphyses is very difficult to do; I use a fine burr on a high-speed drill to destroy the growth area. Epiphyseal destruction controls only growth in the longitudinal direction; the circumferential bone growth can be expected to continue and will require further correction. Longitudinal excisional osteotomies of approximately one quarter to one third of each side of each phalanx will need to be done. I usually do some of this side-to-side narrowing before the epiphyseal arrest. Even though one takes great care to preserve the articular surfaces of each phalanx, joint motions inevitably decrease following surgery.

In older children, adults, and younger children with extreme macrodactyly it will be necessary to do staged bulk reduction operations, which are designed to produce a finger as normal looking as possible. If the finger has curved during its enlargement, it should be straightened by wedge or rotational osteotomies. Length reduction should not be done by terminal amputation, but rather the nail should be preserved by Barsky's method. He recommends obtaining length correction by shortening the middle phalanx and fusing the distal interphalangeal joint (Fig. 13-10). Shortening of the distal phalanx would risk interference with the growth of the nail root. The length of the distal phalanx should also be preserved so that the proximal end can be hollowed to receive the distal end of the middle phalanx, thereby ensuring good bone-to-bone contact. It is possible to combine both shortening and curvature correction by excision of the distal interphalangeal joint and arthrodesis in the corrected position.

The operation is done from one side of the finger at a time. A longitudinal narrowing and excision of nail, nail bed, and nail root can be done from a lateral approach. Skin adipose tissue and even excess nerve tissue can be excised before the finger is shortened and the fusion site then immobilized by a Kirschner wire. Skin over the dorsum of the distal interphalangeal joint will have to be excised, but the skin over the nail root should be left intact. Sometimes the profundus tendon is so slack that it should be shortened and reattached to the terminal phalanx. Small degrees of laxity can be taken up over a period of time, and if the insertion of the superficialis into the middle phalanx has not been dis-

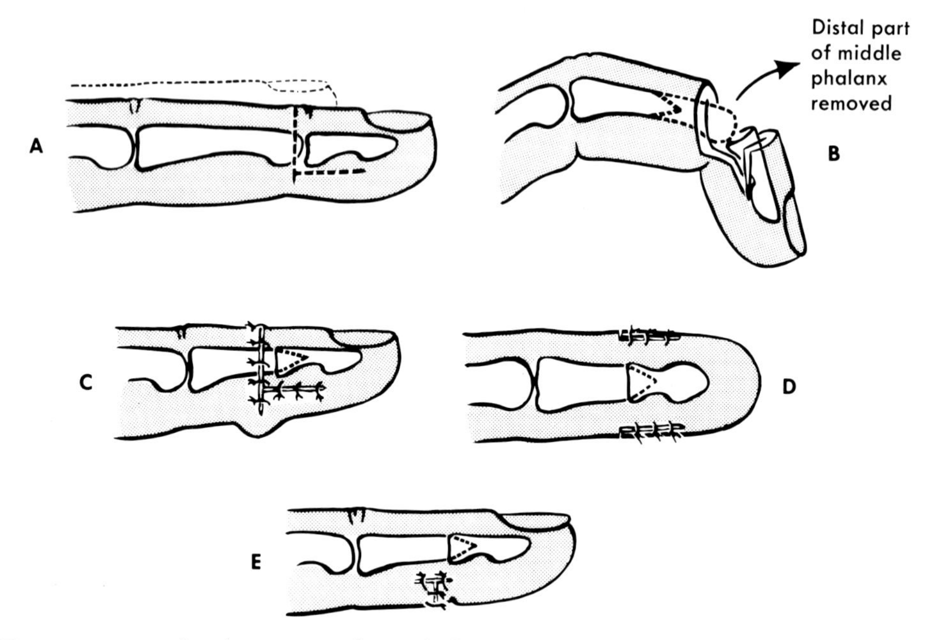

Fig. 13-10. *Digital reduction—Barsky method.* **A,** An enlarged digit contrasted with one of normal size. The skin incision is outlined. **B,** The distal part of the middle phalanx is removed, the proximal portion is spiked, and the base of the distal phalanx is hollowed out. **C** and **D,** Impaction of the bones and closure of the skin wounds produces a palmar hump. **E,** The hump can be excised some months later.

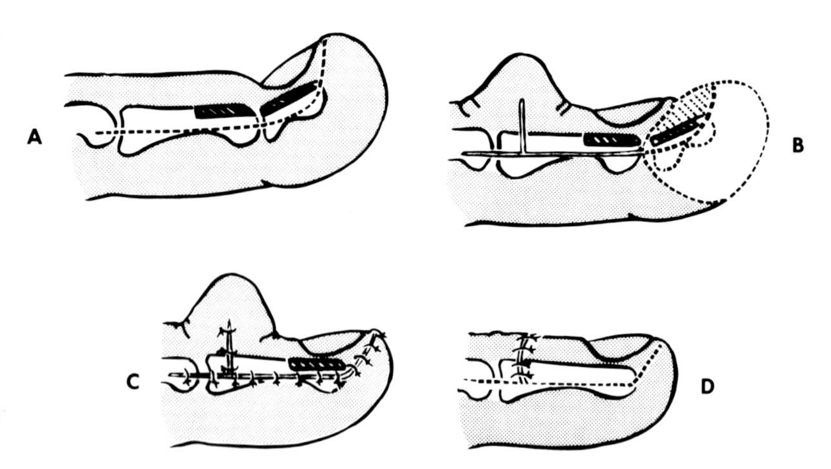

Fig. 13-11. *Digital reduction—Tsuge method.* **A,** Incisions on either side of the finger create a dorsal flap carrying the nail and the dorsal one third of the terminal phalanx. **B,** The terminal pulp and remainder of the terminal phalanx are removed. **C,** The dorsal nail-carrying flap is moved proximally, creating a dorsal hump. **D,** The hump can be excised some months later.

turbed, reattachment of the profundus is not mandatory. This method of shortening from the top and one side is the first of three stages. It will leave a significant palmar bulge, which is removed together with additional adipose tissue at a second operation about 6 weeks later. The third operation, which is carried out about 3 months later, repeats the lateral excision of skin, fatty tissue, and nerve and includes another narrowing of the width of the nail if this is indicated.

Two other procedures for length and bulk reduction have been recorded; I have tried neither, but both appear to have much to recommend them.

Tsuge has devised an "opposite method" to that of Barsky, performing the shortening from the dorsal rather than the lateral side. He recommends this method for toe shortening but has also used it for fingers (Fig. 13-11). Incisions are made on either side of the finger on the neutral border line, creating a dorsal flap that will carry the nail and its root together with the dorsal one-third thickness of the terminal phalanx. A similar sized area is excised from the dorsum of the middle phalanx and the nail-bone complex moved back and fixed in the new site. This produces a dorsal skin bulge that can be excised 6 weeks later. The terminal pulp and distal terminal phalanx are excised, and the nail is shortened. Finally, the palmar skin is brought up and sutured to the terminal nail bed. Tsuge comments that this procedure is simple, but it does have the disadvantage of not providing as good a blood supply in the skin-nail flap as the Barsky method.

Millesi recently published an ingenious but rather intricate scheme for reduction in size of a thumb (Fig. 13-12). The distal half of the terminal phalanx is amputated transversely after two lateral skin flaps have been raised. The proximal half of the terminal phalanx and its overlying skin and nail are split into three parts by two sagittal cuts. The central third is discarded and the two lateral portions joined by Kirschner wire fixation. The insertions of the extrinsic flexor and extensor tendons usually remain intact on these two lateral portions. Shortening of the proximal phalanx is accomplished through a neutral border incision on the radial side. Two oblique saw cuts through the bone allow removal of the central

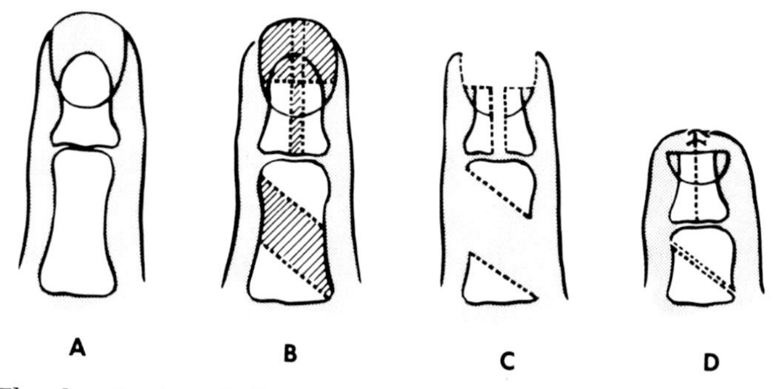

Fig. 13-12. *Thumb reduction—Millesi method.* **A,** The enlarged thumb. **B,** Two lateral flaps are raised and the central portion of the distal terminal phalanx is removed. Cross-hatching shows the other areas of bone removed. **C,** All tissue to be excised has been removed. **D,** The bones are reconstituted, and the lateral flaps sewn together over the end of the thumb.

portion together with adjacent adipose tissue. These two portions are also joined with Kirschner wire fixation. Millesi records that satisfactory functional and cosmetic results are achieved except for the appearance of a longitudinal ridge in the middle of the nail.

Significant cosmetic improvement may be gained by these procedures, but it may still be necessary to do bulk reductions of the proximal phalanx area, including narrowing of the width of the phalanx.

The parents of these unfortunate children must be honestly told of the potential number of operations necessary and of the limitation of function produced by the surgery, and they must accept that at the end of it all the finger will look better but will still seem abnormal to strangers seeing it for the first time. I agree with Edgerton's plea that treatment for this monstrous deformity should be generally more aggressive and undertaken at an earlier date than has been reported in the past. Hopefully, by operating early, excessive growth can be minimized by epiphyseal arrest, and some of the important secondary problems, such as psychological disturbances, joint stiffness, and curvature of the fingers, may be avoided.

SECTION FOUR

THE HAND

CHAPTER 14

CLEFT HAND AND CENTRAL DEFECTS

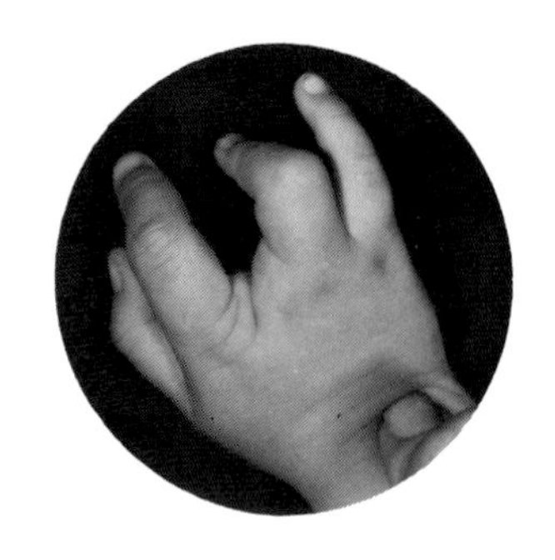

Absence of the central ray of the hand constitutes the true split or cleft hand. Such cases are rare. The many other varieties of absence of rays and their associated anomalies have been usefully grouped by Barsky into two main types—typical and atypical.

TYPES

The classical typical cleft hand consists of a deep V-shaped defect in which the entire long finger ray is missing. This divides the hand into radial and ulnar components (Fig. 14-1). The defect is usually bilateral and frequently familial. Equivalent foot involvement is common. In more severe involvement the two remaining digits in each component may be webbed, with ring-small syndactyly probably being slightly the more common. If true syndactyly of the thumb-index is present, it is the rule to find significant adduction contracture of the thumb. More extensive involvement than is present in the typical cleft hand may result in absence of the index and long finger rays; only very rarely is the ring finger ray alone absent.

The atypical cleft hand shows absence of more than one central ray and has a U-shaped defect between the remaining border digits. It is characteristically unilateral, and there is usually no foot involvement (Fig. 14-2).

ETIOLOGY

The exact mechanism by which these defects are produced is not known. Certainly the insult must occur to the developing hand plate around the seventh week of fetal development. It has been suggested that the predominance in males

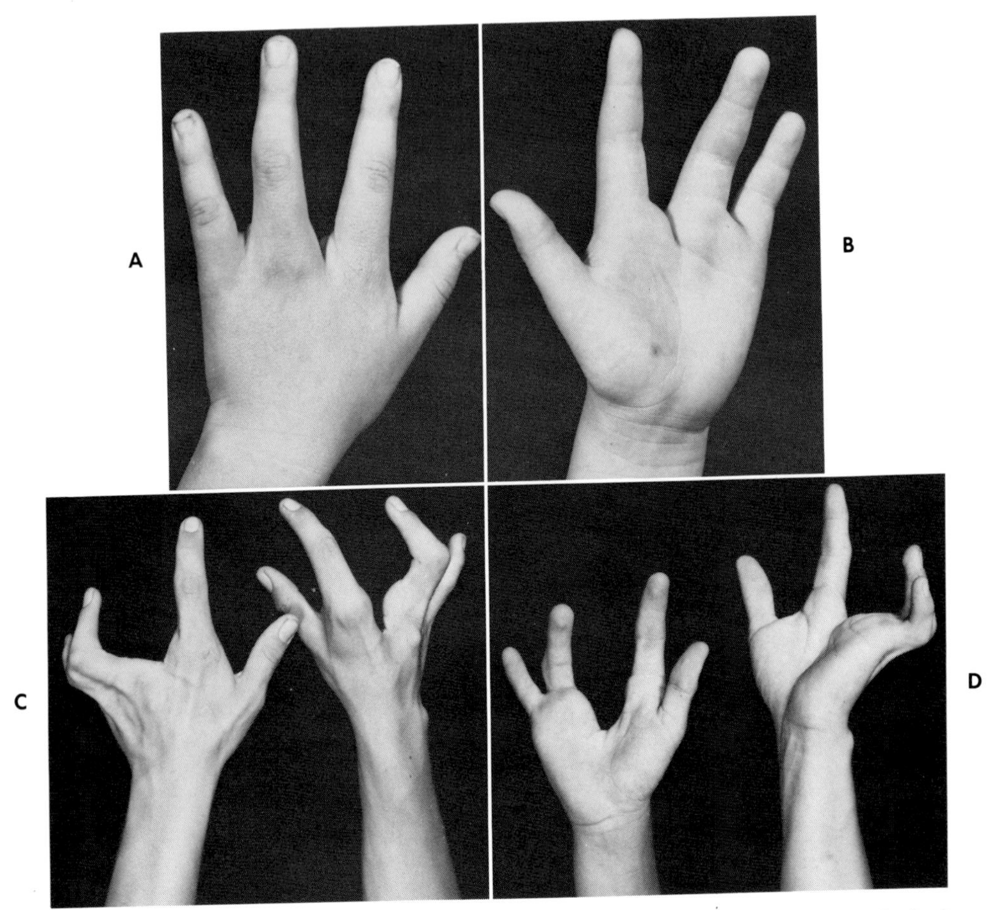

Fig. 14-1. *Typical cleft hand.* The typical cleft hand has a V-shaped defect from which the entire long finger ray is missing. **A** and **B,** Mild type. **C** and **D,** Severe type.

can be explained by the fact that hand plate development in the female is further advanced than in the male during the seventh week.

Maisels has illustrated the centripetal suppression theory as a progression of insult to the developing hand plate (Fig. 14-3). The mildest form shows a simple split of the hand with no absence of tissue. Next in severity is the classical absence of the long finger ray followed by progressive suppression of the radial rays. Eventually the ulnar rays are also affected. Sometimes the remaining rays are fused together to give a lobster claw appearance. Unfortunately, this theory cannot explain the great variety of absences and combinations that occur.

It would certainly be simplistic to picture the atypical cleft hand developing from a greater involvement of the central mass because this deformity is morphologically different and its appearance is sporadic without occurrence in other members of the family.

Many pedigrees have been published demonstrating the dominant transmission of the typical cleft hand. Irregular dominance does occur, and recessive in-

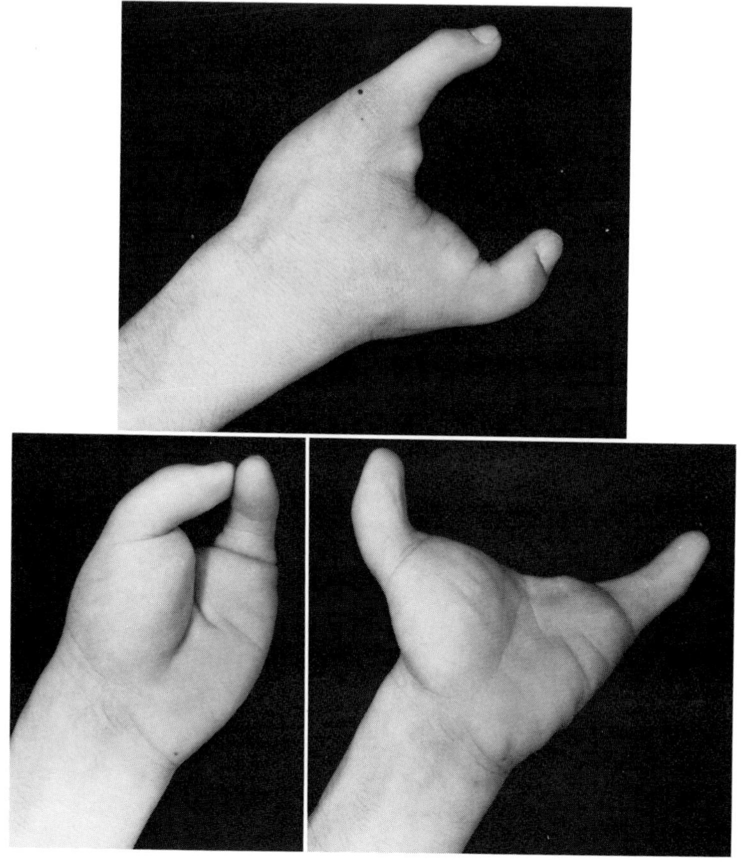

Fig. 14-2. *Atypical cleft hand.* An atypical cleft hand lacks more than one central ray. It has a U-shaped defect between the remaining border digits, which retain good pincer function.

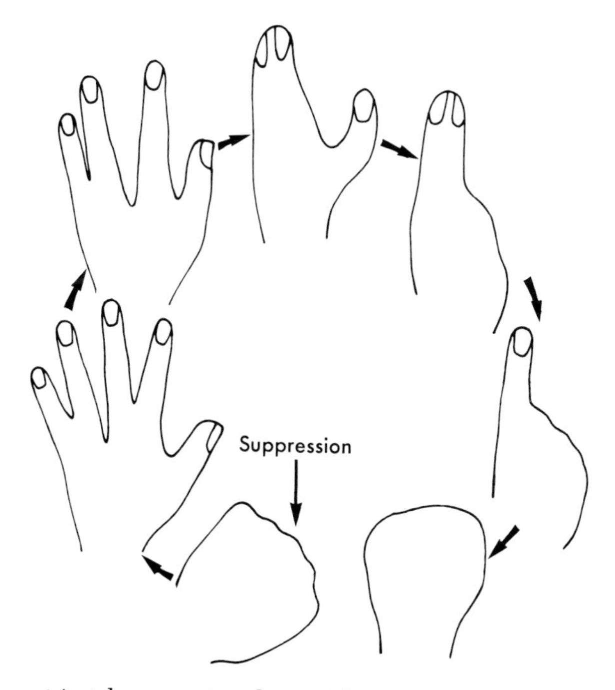

Fig. 14-3. *The centripetal suppression theory.* The progressive reduction in tissue from an insult to the developing hand plate starts with a mild cleft between normal fingers and ends with total suppression of all digits. (From Maisels, D. O.: Theory of pathogenesis of lobster claw deformities, The Hand **2**:79, 1970.)

heritance has also been noted. Even in families with clearly dominant inheritance the genetic prognosis for unaffected offspring must be guarded since skipping of a generation has been shown to occur occasionally.

The incidence of cleft hand is not high. Comparison of the Iowa study with the reports of other large series shows an average incidence of less than 2.3% of the total number of hand anomalies.

A significant number of anomalies are associated with cleft hand. Cleft feet occur in at least half the cases, and cleft lip and palate show a high association. Walker and Clodius have developed statistical evidence to show that this association between cleft lip and palate and splitting of the hand should be considered a separate syndrome. Other associated anomalies reported include congenital heart disease, imperforate anus, anonychia, cataract, and deafness.

I do not believe that the terms cleft hand and lobster claw hand are synonymous and prefer to reserve the latter for the fully developed atypical type. I also do not believe it etiologically correct to include in the lobster claw group those hands that possess border digits with an intermediate web of skin bearing finger nubbins. Frequently these hands contain portions or all of the central

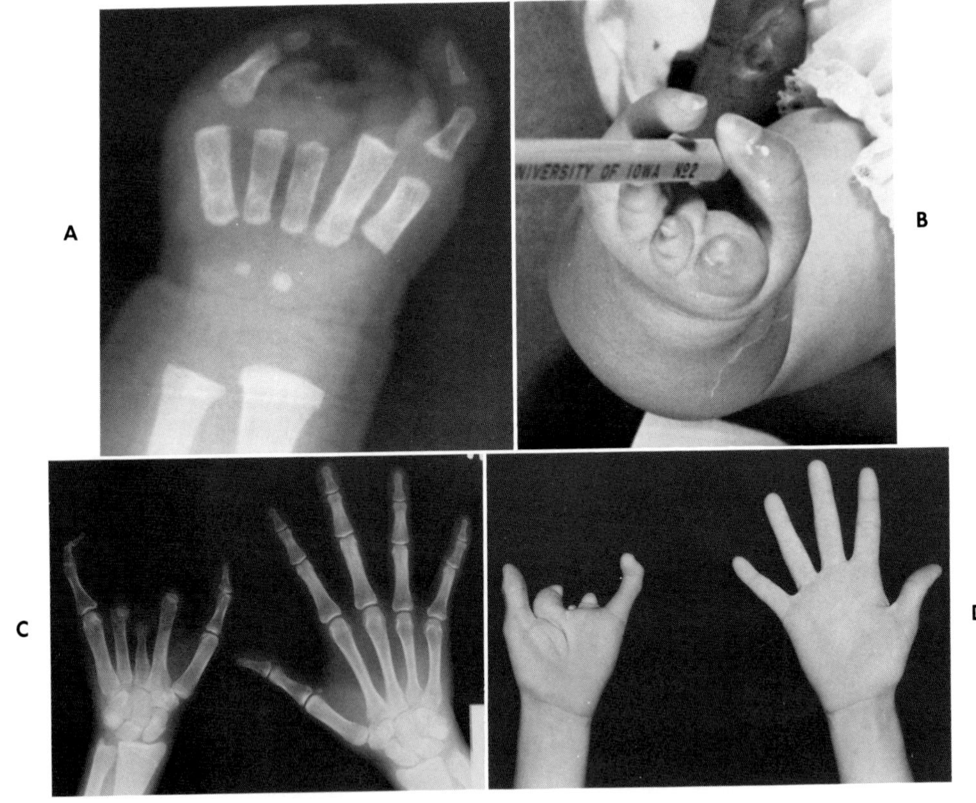

Fig. 14-4. *Central absence.* Hands that possess border digits and intermediate finger nubbins are not examples of cleft hand, principally because they have metacarpals. **A** and **B,** A baby with central absence. **C** and **D,** Same patient in adult life. **E** and **F,** Two patients demonstrating the good grasp possible with this type of hand.

metacarpals and are more properly considered examples of symbrachydactyly with the primary deficiency being in phalangeal, and not metacarpal, development (Fig. 14-4). Functionally, however, these hands have exactly the same problems as the true lobster claw hand, and they share the same reconstructive surgical procedures.

OPERATIVE TREATMENT

General principles

Typical cleft hand is a functional triumph and a social disaster. Patients with such hands can perform any function they wish but more often than not hide their hands because of their grotesque appearance. Careful examination often shows that there is an adduction contracture of the thumb and that function can be improved by addition of skin to the thumb web. Patients who show syndactyly between the border digits should have the webs released.

The atypical cleft or fully developed lobster claw hand also functions well

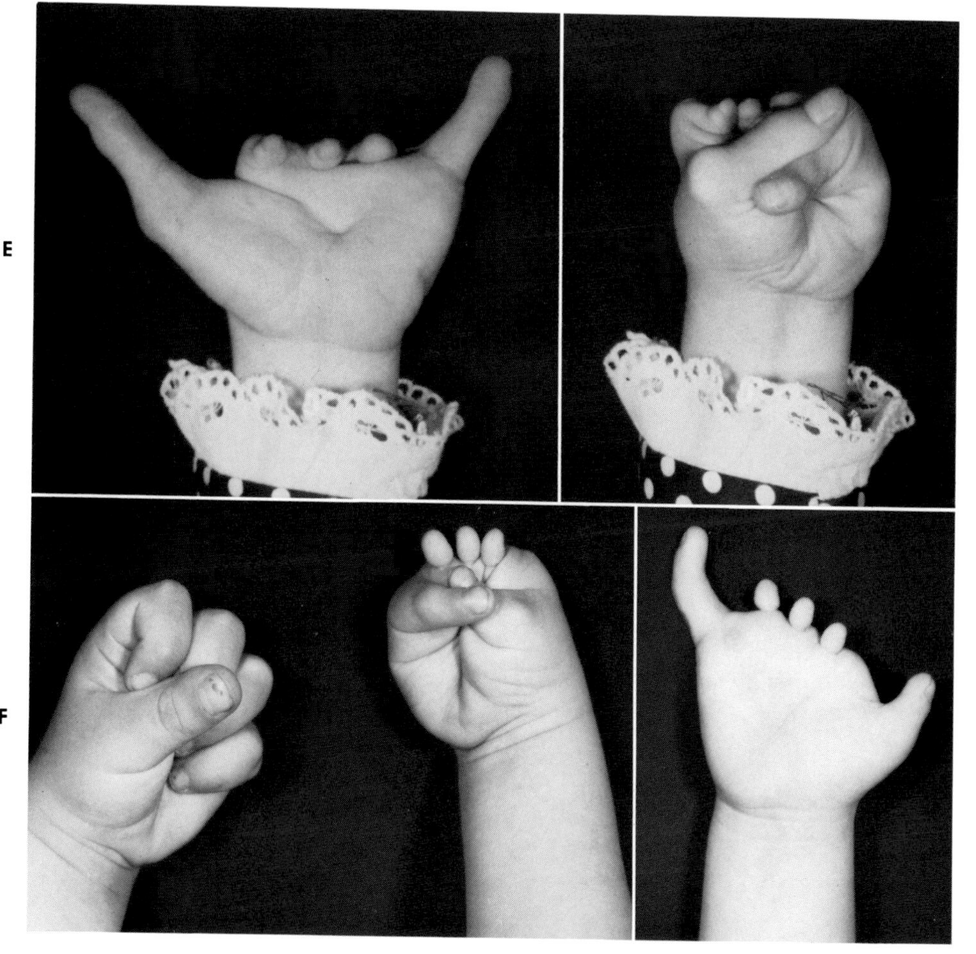

Fig. 14-4, cont'd. For legend see opposite page.

but is an equal social disaster. Patients so affected usually develop good grasp and can pinch accurately. Occasionally the extent of grasp can be increased by deepening the cleft. The stability of grasp or pinch may be hindered by hypoplasia or absence of phalanges in the two border digits. Usually the middle phalanges are involved, and fusions of the distal or occasionally the proximal interphalangeal joints can significantly improve function.

Typical cleft hand

I have found that the biggest problem in developing my operative plan for cases of typical cleft hand is deciding on the extent of thumb mobility necessary. Maximum mobility of the thumb is admirable and should always be the objec-

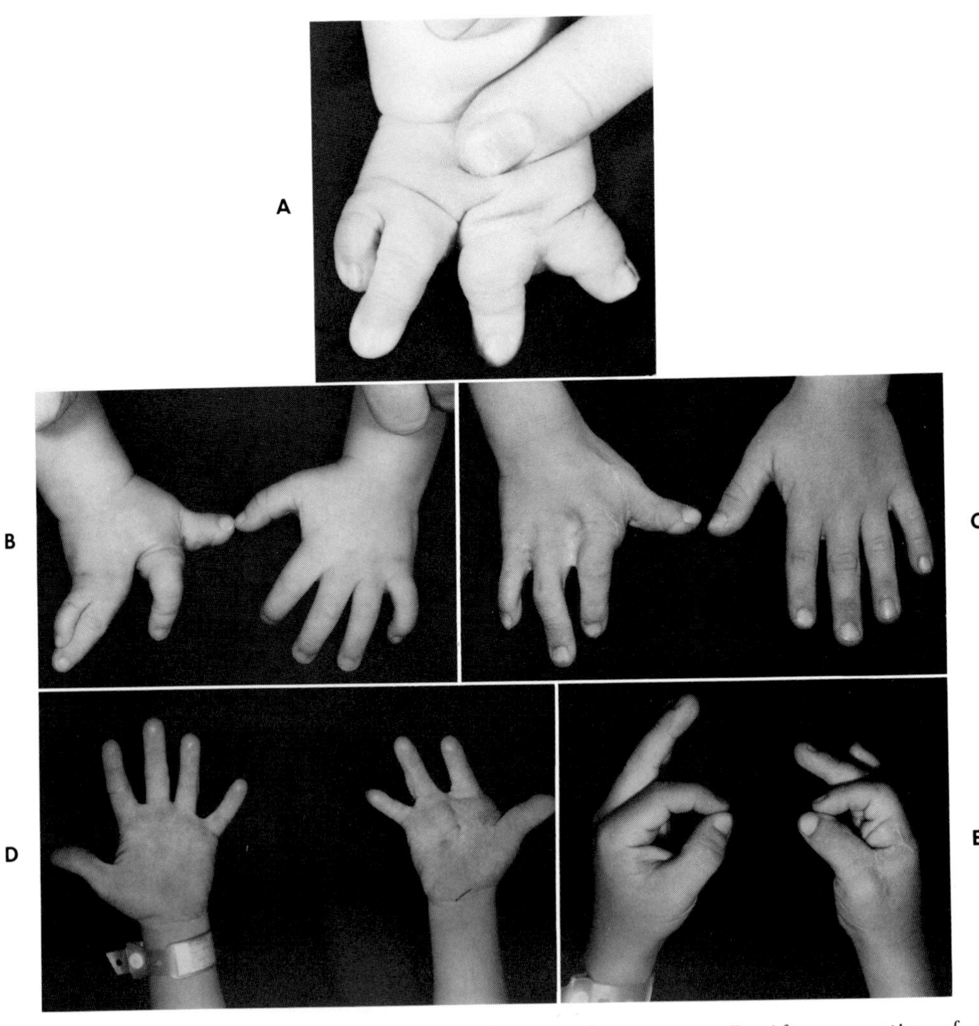

Fig. 14-5. *Cleft hand and border syndactyly.* **A,** Before surgery. **B,** After separation of the syndactyly, which must be done early in life. **C, D,** and **E,** Five years after closure of the cleft and 7 years after the syndactyly release.

tive. Achieving the maximum will often require an extensive operation and may in fact not be necessary. Minor degrees of thumb adduction can often be accepted and the central cleft closed without intruding on the first web space.

When either or both pairs of the border digits are joined in a syndactyly, the first stage of correction must be the separation of these digits using standard procedures (Fig. 14-5). After a minimum recovery period of at least 6 months, thumb mobility must be assessed and the decision made either to simply close the central cleft or to at the same time increase the mobility of the thumb. If the latter is necessary, the best operation is that devised by Littler in which the excess skin of the cleft is transposed on a proximal palmar pedicle into the thumb web and the index metacarpal is shifted to a more ulnar position adjacent to the ring finger.

Simple cleft closure. The problems in this operation are to obtain a satisfactory commissure and to prevent later spreading of the repair. Some surgeons advocate osteotomy of one or both of the adjacent metacarpals. I have no strong views on this; it is pointless to advise "never," but in general I do not usually find osteotomy necessary. It is, however, important to obtain good opposition of the distal ends of the metacarpals, and every effort must be made to reconstitute the transverse intermetacarpal ligament.

The skin incisions must be planned to yield a zigzag interdigitated line on both palmar and dorsal surfaces. The commissure is too far proximal for it to be rotated or transposed distally and should be excised. It is wrong to plan the incisions so that the distal adjacent edges lie in the new commissure site; subsequent healing will draw this junction proximally into a narrow V shape. A flap will have to be fashioned, based distally on one or another digit, to form the transverse skin of the commissure. Barsky uses a diamond-shaped flap, which sits well across the commissure and up onto the side of the adjacent digit. The diamond design gives good healing without linear contractures (Fig. 14-6).

This diamond-shaped flap is broadly based, and it should be defatted down to the subdermal plexus of vessels to ensure that it will fit satisfactorily into the adjacent finger. In siting the base of the flap, one should place it more dorsally than truly laterally because the former position will yield sufficient skin on the dorsal aspect to give the normal sloping character to the commissural skin.

To draw the metacarpals together one must excise any excessive soft tissue and even bony remnants that lie between them. Any tendency for the metacarpals to spring apart must be relieved. Incising the carpometacarpal joint capsule on the outer sides may be sufficient, but occasionally an incomplete transverse osteotomy on the outer side of one or rarely both metacarpals may be needed. This osteotomy should be done cautiously until the inner or cleftward cortex and periosteum can be bent sufficiently to form an intact hinge. It is not necessary, and is actually destructive, to completely transect the metacarpal. I do not think it wise to rely on the skin repair to hold the metacarpals together and believe they should be tethered. In the normal hand the intermetacarpal ligament passing between the palmar plates accomplishes this, and I always try

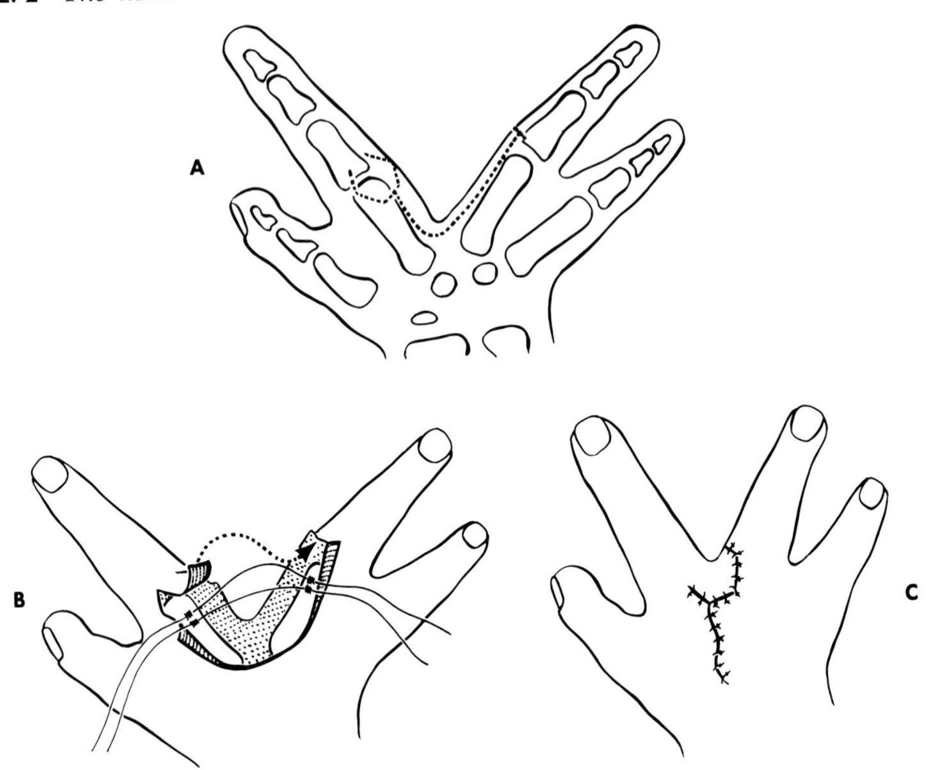

Fig. 14-6. *Closure of typical split—Barsky method.* **A,** The essence of this operative plan is to construct a good commissure using a diamond-shaped flap raised on one finger. **B,** The adjacent metacarpals must be tethered together. Strong sutures may be used. **C,** Skin closure with the diamond-shaped flap lying across the commissure.

to fashion some sort of ligament out of the adjacent soft tissues using fine absorbable sutures. The result usually looks unpromising, but follow-up studies show that this procedure does some good.

If I fail to make a satisfactory ligament, then I do not hesitate to tie the two metacarpals together with relatively strong catgut or even silk sutures. It may be necessary to drill holes in the metacarpal necks to stabilize the ties; these holes should be placed as far distal as possible without affecting the epiphyseal plates.

After the metacarpals have been approximated, the dorsal and palmar skin wounds are closed in a proximal to distal direction. Usually a significant amount of excess skin has to be excised to provide a good contoured closure. I usually fashion two curved interdigitating flaps and avoid the risk of contracture inherent in a straight-lined closure (Figs. 14-7 and 14-8). Healing often tends to draw the wound down into the line of the original cleft, and the final appearance can be improved with an additional layer of subcutaneous sutures before closing the skin. When the commissural flap is laid in place, it will often seem too large. The temptation is to trim it to fit, but it is better to retain the size of the flap and trim away excess dorsal skin, thereby gaining the dorsopalmar slope of a normal commissure.

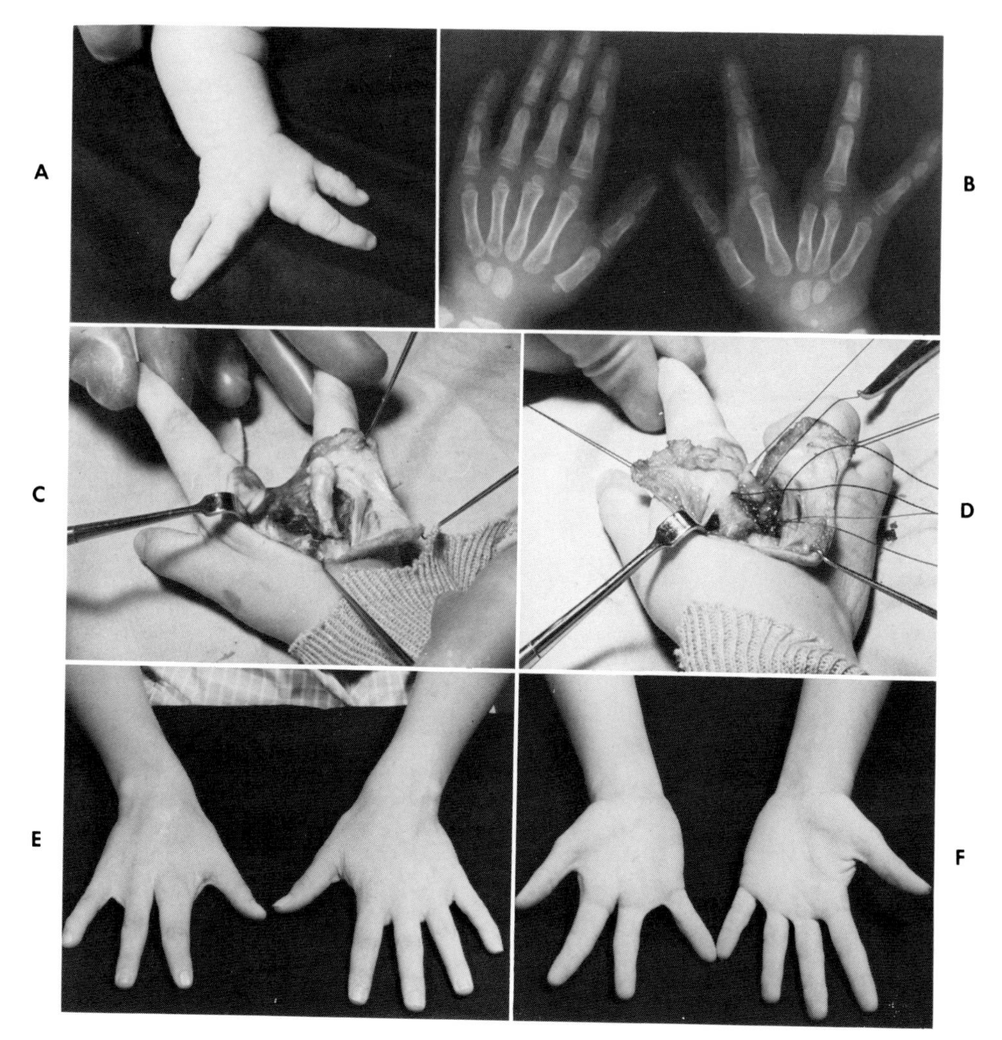

Fig. 14-7. *Typical cleft hand closure.* **A** and **B**, Before surgery. **C** and **D**, Surgery at age 9 months. The central metacarpal is removed and adjacent metacarpals tied together. **E** and **F**, Follow-up 12 years later.

In the postoperative period I protect the intermetacarpal ligament repair by putting a well molded plaster cast to the level of the metacarpal heads over a minimum amount of dressings. It is wise to extend the cast above the flexed elbow. The repair should be protected for at least 3 weeks.

Cleft hand with thumb adduction. There are many ways of providing supple skin to relieve adduction contracture of the thumb (see Chapter 5). It is possible, and acceptable, to close the cleft hand first by the method described previously and then to later mobilize the thumb. If this staging is chosen, the planning of dorsal rotation flaps into the thumb web may be compromised by the presence of the zigzag scar from the primary operation.

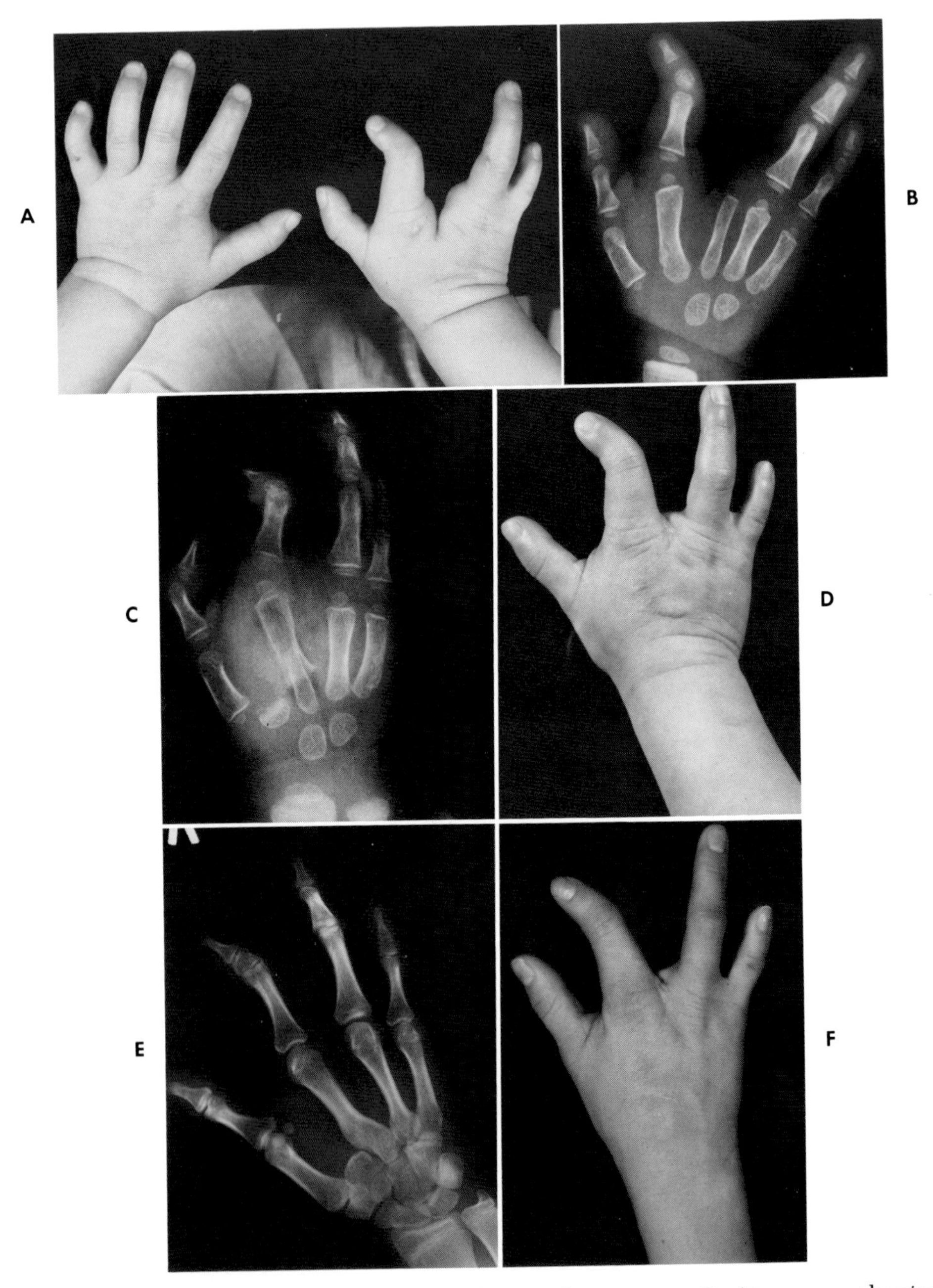

Fig. 14-8. *Typical cleft hand closure.* **A** and **B**, Before surgery. **C**, At surgery, showing shifting of index metacarpal onto base of central metacarpal. **D**, One year after surgery. **E** and **F**, Thirteen years after surgery, showing remodeling of the index metacarpal and persistent curvature of the finger because of the abnormal middle phalanx.

A satisfactory but technically somewhat difficult operation devised by Littler closes the split by ulnar transposition of the index ray and corrects the thumb adduction contracture by use of a palmar based flap fashioned from the skin of the cleft.

The skin incisions are complicated and are shown in Fig. 14-9. The essential features are as follows: First, one outlines the sides of the split on the dorsal surfaces of the index and ring fingers, forming a proximal V-shaped apex and extending distally to the level of the metacorpophalangeal joints. A small, straight extension is then made on the ulnar side of the index finger to later accommodate a small commissural flap that will be raised on the radial side of the index finger. As the incisions pass around the sides of the adjacent metacarpal heads they curve back onto the palmar aspect almost parallel and to the cleft side of the midline of the two fingers. Proximally they should extend no further onto the palm than to a point opposite the V-shaped proximal apex of the dorsal incision. These incisions outline the palmar based flap that will create the thumb web (Fig. 14-10).

One additional incision is necessary to release the adducted thumb. This incision starts on the dorsum of the thumb-index web at the same proximal

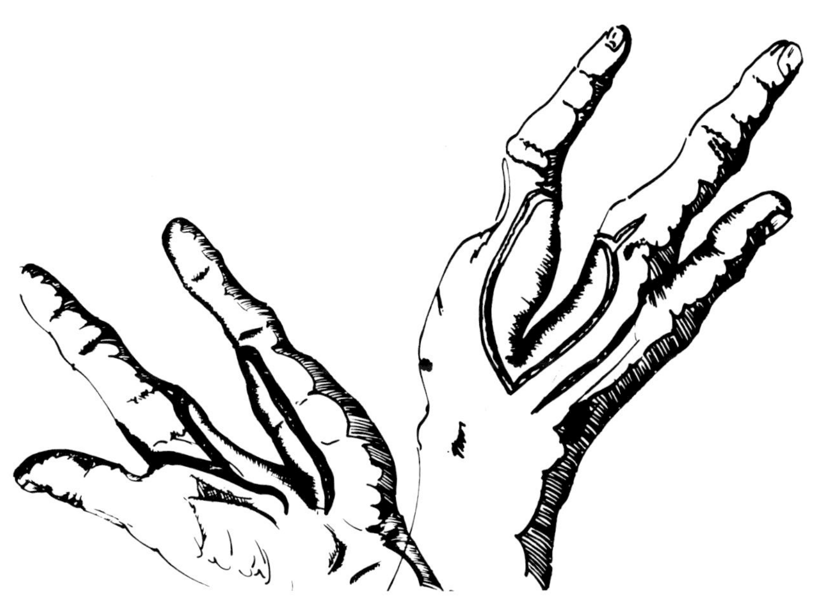

Fig. 14-9. *Cleft hand with thumb adduction.* The skin incisions are planned to aid in closure by ulnar transposition of the index. A small flap is designed on the ulnar side of the proximal phalanx of the index finger. This creates the new interdigital web space and places the suture line on the radial border of the ring finger. (From Snow, J. W., and Littler, J. W.: Surgical treatment of cleft hand, Transactions of the International Society of Plastic and Reconstructive Surgeons, 4th Congress, Amsterdam, 1967, Excerpta Medica Foundation, pp. 888-893.)

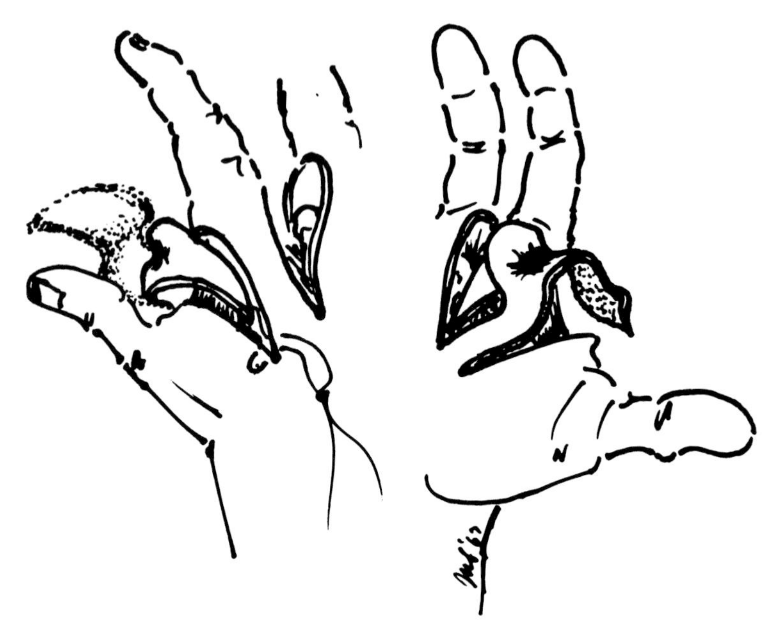

Fig. 14-10. *Development of the skin flaps.* The index ray being transposed ulnarward and the palmar base flap being brought over to form the new thumb-web space. The dorsal skin bridge over the index metacarpal assures good venous return and also contains the extensor tendons. (From Snow, J. W., and Littler, J. W.: Surgical treatment of cleft hand, Transactions of the International Society of Plastic and Reconstructive Surgeons, 4th Congress, Amsterdam, 1967, Excerpta Medica Foundation, pp. 888-893.)

level as the V-shaped cleft incision. It is extended distally parallel with the index split incision until it reaches the distal edge of the thumb-index web. By this means a strip of dorsal skin is left connecting the index to the dorsum of the hand. This strip should be made as wide as is feasible since it will cover and contain the dorsal venous drainage and the extensor tendons of the index finger. On the palmar aspect the incision is carried proximally and across in an ulnar direction to meet the palmar index split incision at its approximate midpoint, thereby leaving a tapered palmar index flap for later insertion into the closed cleft (Fig. 14-10).

The split flap is developed from the dorsum, proximodistally in the areolar plane overlying the extensor tendons. The dorsal veins that pass across the incisions must be carefully tied and not dissected off the flap. The flap is long, and this retention of the veins and their connections into the web space and the palm will ensure good venous drainage. As the dissection proceeds around the curve of the split and onto the deep palmar aspect, the branches of the median nerve supplying sensibility should be preserved.

Next the thumb-index web incision is developed and deepened. It is unnecessary and unwise to mobilize the skin on either side of the incision. Fibrous bands will have to be released between the two metacarpals, and part of the origin of the first dorsal interosseus from these bones must be detached.

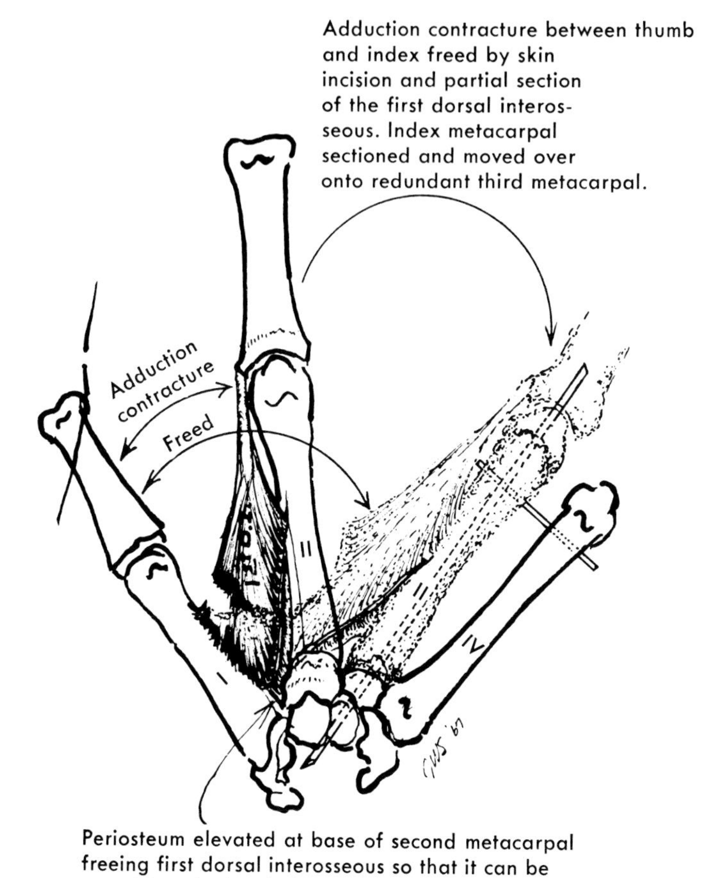

Fig. 14-11. *Transposition of the index.* The base of the third metacarpal has been smoothed to receive the base of the index metacarpal, which is fixed in place with Kirschner wires. The first dorsal interosseus has been freed subperiosteally at the base of the index and has been partially incised at its origin from the thumb metacarpal. (From Snow, J. W., and Littler, J. W.: Surgical treatment of cleft hand, Transactions of the International Society of Plastic and Reconstructive Surgeons, 4th Congress, Amsterdam, 1967, Excerpta Medica Foundation, pp. 888-893.)

In addition, the adductor pollicis and the radial belly of the flexor pollicis brevis may have to be elevated from their origin to allow full abduction of the thumb. As this dissection proceeds proximally, the radial artery should be protected as it passes between the two heads of the first dorsal interosseus muscle. Occasionally the dissection will have to be carried down onto the capsule of the carpometacarpal joint of the thumb and a relieving incision made on the radial side to permit full thumb abduction.

The index ray has now been mobilized sufficiently for its metacarpal base to be cut across and the ray transposed ulnarward (Fig. 14-11). Usually it is necessary to spread the tissues around the extensor and flexor tendons to allow transposition without tension. The method of fixation of the transposed index will depend on the extent of third metacarpal present. If the third metacarpal is small, it should be shaped into a peg and the index ray impaled upon it. Even if there is an extensive amount of bone present, I find this a satisfactory fixation method; others prefer to retain the metacarpal and fuse the index to its distal end to provide additional stability. When this is done, a triangular wedge of bone should be removed from the base of the index to allow its transposition. When the third metacarpal is completely absent, this wedge excision will allow the index to be angled ulnarward and to still maintain bone-to-bone opposition at its base.

I believe that the most secure fixation is obtained using intramedullary and transverse Kirschner wires, but care must be taken to assure proper alignment of the metacarpals (Fig. 14-12). The alignment of the transposed index ray must maintain the transverse and longitudinal arches of the hand and allow the fingers to flex normally into the palm without overlapping.

The final stage of closure is best done by first suturing the skin between the

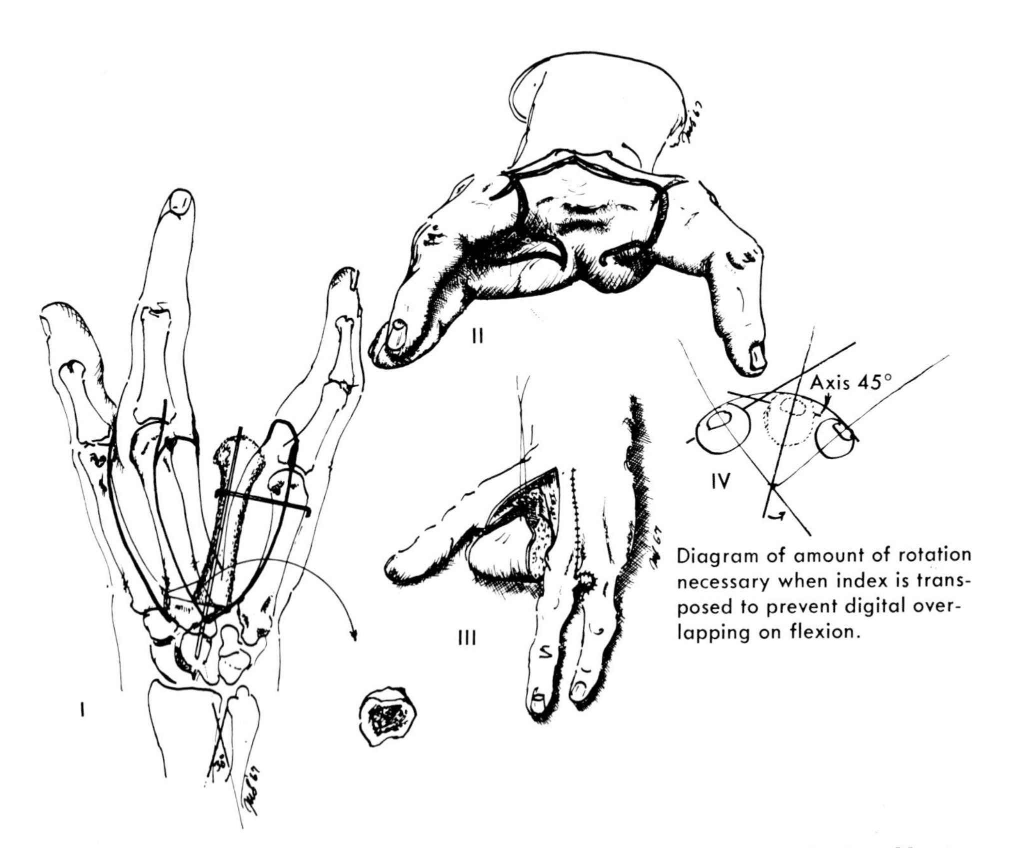

Fig. 14-12. *Setting in of thumb-web flap.* The index has been transposed, the adduction contracture of the thumb web has been freed, and the flap is being sutured into the new thumb-web space. (From Snow, J. W., and Littler, J. W.: Surgical treatment of cleft hand, Transactions of the International Society of Plastic and Reconstructive Surgeons, 4th Congress, Amsterdam, 1967, Excerpta Medica Foundation, pp. 888-893.)

ring and transposed index fingers. Care must be taken to inset the ring finger skin into the small longitudinal incision on the ulnar side of the index and thereby make an acceptable commissure. Closure of this wound establishes the lie and tension of the dorsal strip of skin over the index finger and allows the proper setting of the long palmar based flap into the new thumb-index web space (Fig. 14-12).

As in many reconstructive skin coverage procedures in the hand, even this large palmar based flap does not always carry the skin to the needy areas. Fig. 14-13 demonstrates the relationship between the palmar base of the flap and the newly abducted thumb. There is a greater distance from the proximal V-shaped apex of the dorsal thumb-index skin wound to the distal edge on the radial side of the index than to the distal edge of the ulnar side of the thumb. Sometimes the flap can be laid in slightly obliquely to accommodate this extra length on the index, but to avoid too much tension in the flap it may be necessary to split skin graft the distal portion of the index wound.

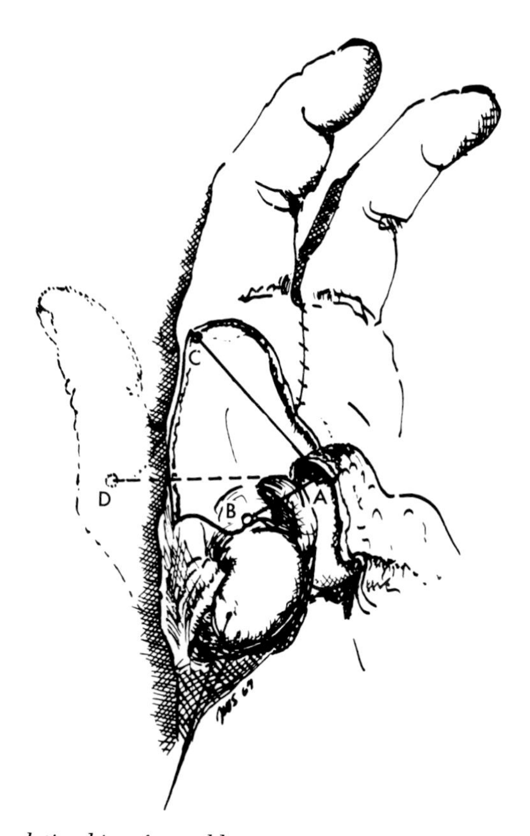

Fig. 14-13. *Thumb relationship after adduction contracture release.* The thumb portion of the cleft is moved closer to the base of the pedicle, therefore requiring less tissue for coverage. The most distant area for coverage is at point *C*, and here a split-thickness skin graft may be needed. (From Snow, J. W., and Littler, J. W.: Surgical treatment of cleft hand, Transactions of the International Society of Plastic and Reconstructive Surgeons, 4th Congress, Amsterdam, 1967, Excerpta Medica Foundation, pp. 888-893.)

I recommend placing a twist drain into the depths of the thumb-web space before closing the wound with interrupted skin sutures. The drain should be led out onto the dorsum and so placed in the dressings that it can be withdrawn in 24 or 48 hours. The hand should be bandaged in a position maintaining the two transverse and the longitudinal arches and allowing movement of the fingers and thumb. An outer plaster shell will protect and maintain this position.

The skin sutures are removed about the end of the second postoperative week, but the plaster should be reapplied for a total of 6 weeks before full use of the hand is permitted.

LOBSTER CLAW HAND

Surgery has little to offer patients with this malformation. Most patients show amazing dexterity, and nothing of significance can be done to increase the social acceptability of this deformity (Fig. 14-14). Some patients request obliteration of the unsightly cleft. Whenever possible, I advise against this because closure removes function and narrows the span of the hand and also because spreading of the cleft usually recurs.

In severe degrees of involvement one or both of the border digits may be rudimentary or flail. Fusion of adjacent bones or their augmentation by bone grafting can be useful. The key to grasp will be the mobility of the border

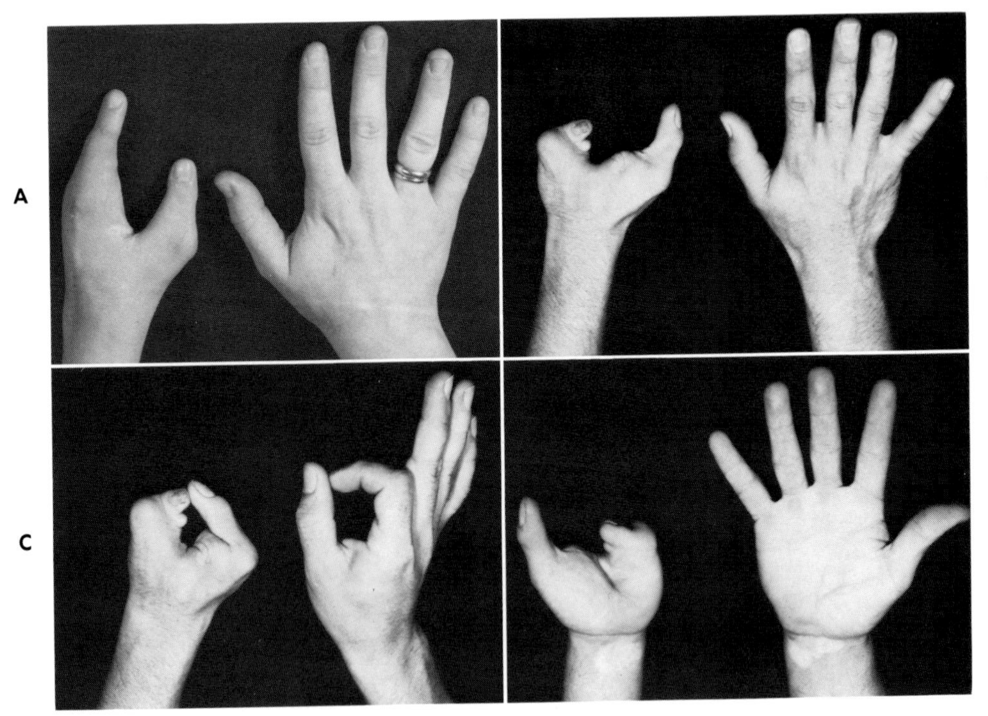

Fig. 14-14. *Lobster claw hand.* **A** and **B,** Typical deformities. **C** and **D,** The apposition possible; no surgery was done on this hand.

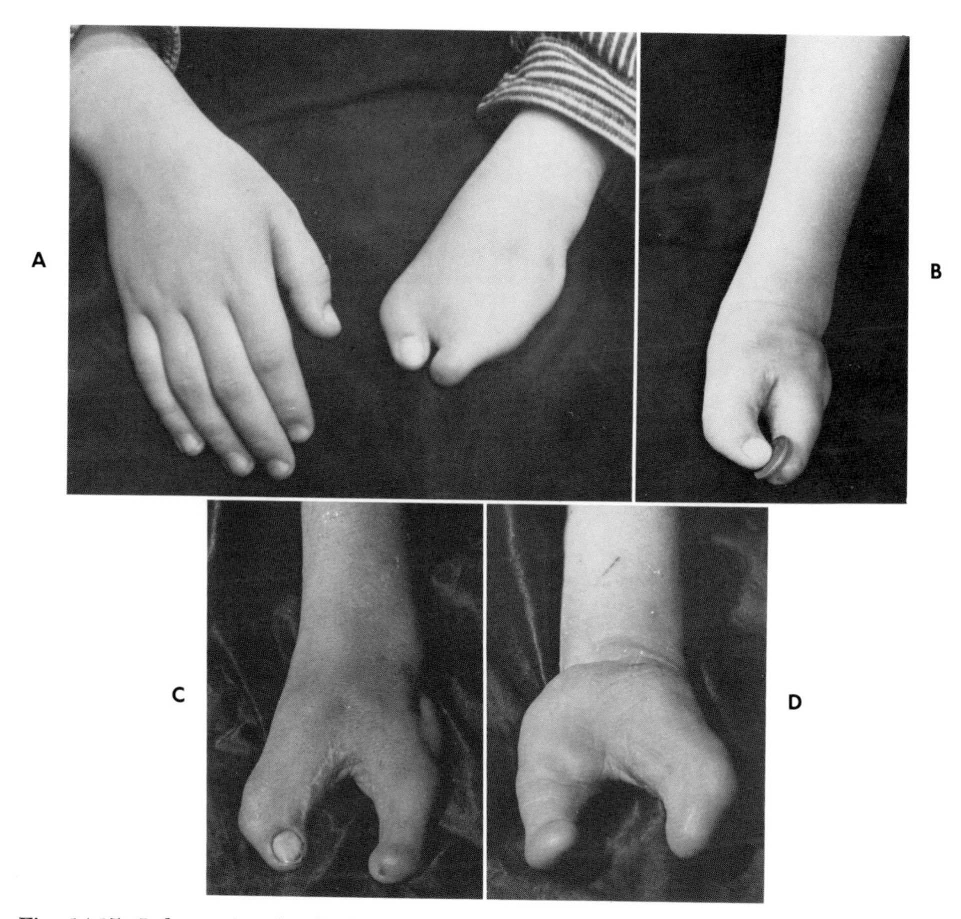

Fig. 14-15. *Lobster claw hand—deepening.* **A** and **B,** Boy's hand before and after deepening between the border digits. **C** and **D,** Thirty years later he retains good grasp in this hand.

A

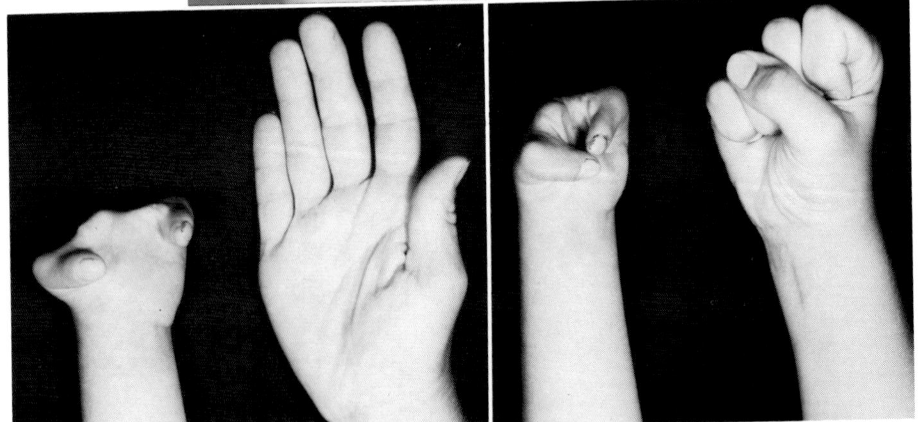

Fig. 14-16. ***Lobster claw hand—tendon transfers.*** **A,** Hand of a 3-year-old. This hand has no voluntary motion. Tendon transfers, prolonged by grafts, have been done. **B** and **C,** Range of motion possible 2 years after surgery; he still retains this motion in late teenage.

digits—if neither can be opposed, there is no purpose in providing two rigid claws; if one can be voluntarily controlled, then fusion or augmentation can serve a useful purpose (Fig. 14-15). The voluntary motion possible may not oppose one digit to the other. If this is so, rotational osteotomy of one or both metacarpals when the patient is around 3 years of age may considerably increase function.

If there is a good, stable passive range of motion in the border digits, active motion can be restored by tendon transfers. Wrist flexor and extensor tendons make suitable motors, but they will have to be prolonged by tendon grafts (Fig. 14-16). I usually use a plantaris tendon to ensure adequate length for both transfers. The distal ends of the tendons are secured by pull-out wires, which I remove after 3 weeks. It is usually wise to place the wrist in mild flexion during the 3 weeks of immobilization. It is probably best to delay surgery until about the age of 3 years to ensure some cooperation in the post-operative exercise period.

SEVERE SUPPRESSION DEFORMITIES

Function in these hands is severely reduced compared with normal hands, but with ingenuity, patients can produce function out of the most bizarre

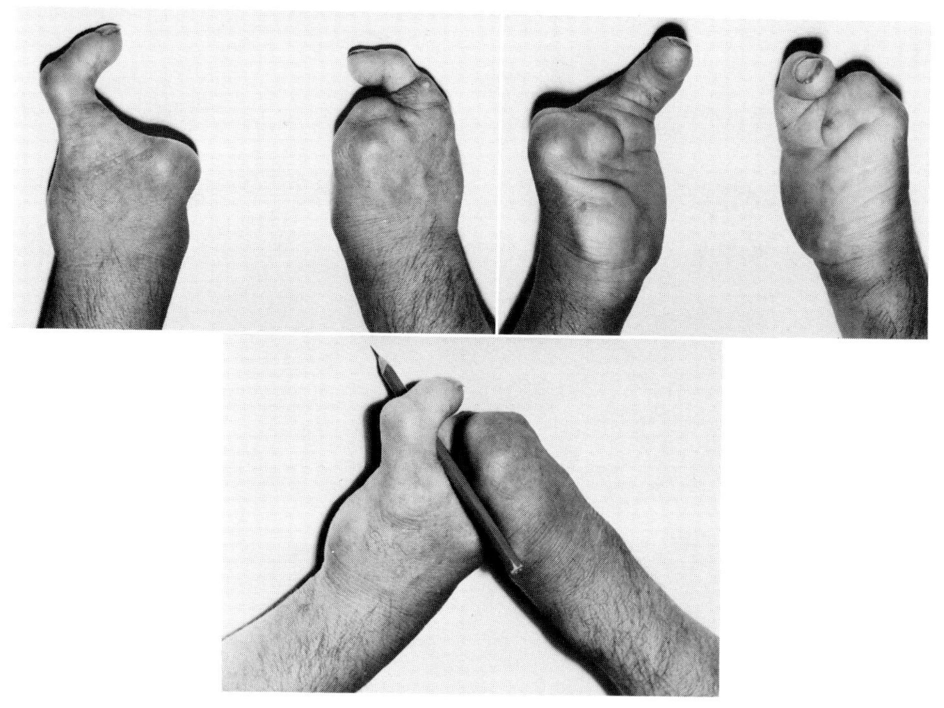

Fig. 14-17. *Severe suppression deformity.* Function in these hands is minimal, but by combining both extremities, considerable function is possible.

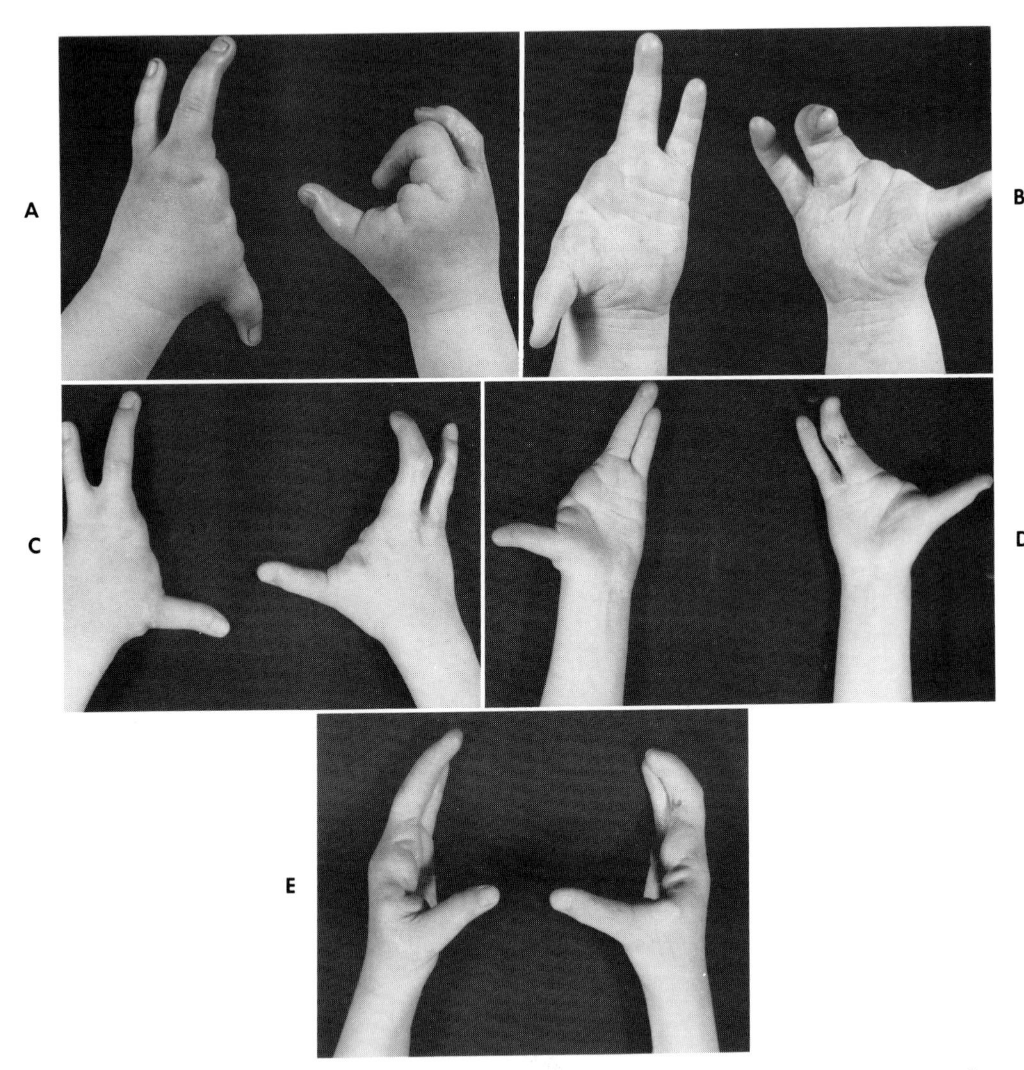

Fig. 14-18. *Severe suppression deformity.* Joint abnormalities sometimes occur in these deformities and corrective surgery can improve function. **A** and **B**, Instability of the left thumb and flexion deformity of the right long finger. **C**, **D**, and **E**, Five years after stabilization of the thumb and correction of the flexion contracture by skin grafting.

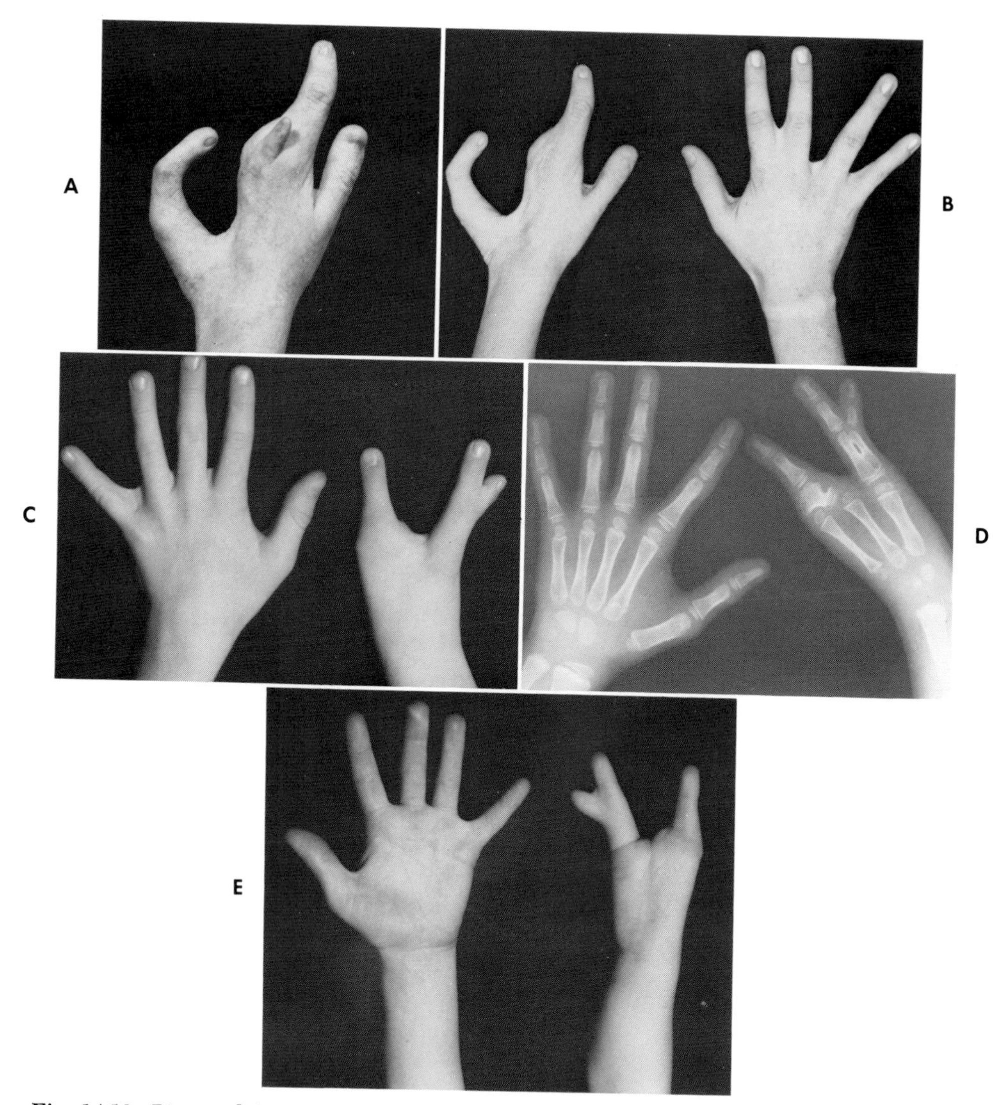

Fig. 14-19. *Bizarre deformities.* Vestigial digits are unsightly and may be removed if without function. **A** and **B**, This dorsal vestige was useless, and removal improved the cosmetic appearance. Note the incomplete syndactyly in the right hand, which the patient did not want corrected. **C**, **D**, and **E**, The small finger sprouting from the side of the ring finger is unsightly, but it possessed flexion and added considerably to grasp ability.

deformities (Fig. 14-17). There is so little tissue present in the involved hands that reconstructive possibilities are severely limited; however joint deformities can be corrected and function improved (Fig. 14-18). Malpositioned vestigial digits can be removed for cosmetic purposes, but very careful functional analysis must be made before removing tissue from these poorly equipped hands (Fig. 14-19).

CHAPTER 15

RADIAL CLUBHAND

"Radial clubhand" is a convenient but inaccurate label. It is commonly used to include a large variety of developmental abnormalities occurring along the preaxial border of the upper limb in association with a hand deviated at the wrist. The term "radial dysplasia" is useful to differentiate the deformity from an acquired clubhand but does not accommodate the significant anatomical deformities that are usually found in the forearm, elbow, shoulder, and the brachial plexus.

The spectrum of abnormalities varies from a minor degree of thumb hypoplasia to total absence of the thumb, its metacarpal, the scaphoid and trapezium, and the whole radius. In a small percentage of severe cases the developmental failure may trespass on the second ray and involve the scaphoid, trapezoid, second metacarpal, and index finger. The soft tissue structures of the hand and forearm are also involved, and variations of great surgical significance are common in the muscles, nerves, and vascular supply. Despite the fact that this condition is considered a preaxial affliction the ulna is almost invariably involved.

In moderate and severe cases the ulna is usually only about 60% of its normal length at birth, and this discrepancy is never corrected during subsequent growth.

ETIOLOGY

The etiology of this defect is obscure. The essential cause must be a dysplastic factor acting during the first few weeks of fetal life. Some theories claim that the deformity is produced by environmental factors such as compression during intrauterine life (Dareste theory), inflammatory processes (Virchow theory), maternal nutritional deficiencies, roentgen rays, and drugs such as insulin and thalidomide. Other theories explain the deformity as phylogenetic in origin, but most physicians consider a genetic factor more likely. No single theory is strong enough to resist reasoned criticism.

The genetic basis for the deformity is poor since hereditary tendencies are not common. Radial defects are unilateral in about half of the patients so affected, and the great majority of these defects occur sporadically. In unilateral cases the right side is affected almost twice as frequently as the left side. Bilateral involvement has been recorded in identical twins, and varying views have been expressed as to the degree of dominant or recessive heredity. In the University of Iowa series of 81 radial clubhands in 2,159 hands and in those reported in the literature, we have found no real occurrence of radial clubhand from one generation to the next.

Radial clubhand is frequently associated with other malformations, and there are a great number of syndromes in which radial clubhand usually occurs.

ASSOCIATED ABNORMALITIES

Cardiac abnormalities are particularly common in these patients, and the incidence of congenital heart disease has been recorded as between 10% and 13% in various series. There is a strong relationship between forearm anomalies and ventricular septal defects. It is interesting that the shaft of the radius begins development in the fifth week of fetal life, which is exactly the time that ventricular septum development begins. The Holt-Oram syndrome, which includes atrial septal defects, results from a dominant hereditary trait and shows a high association with radial aplasia. In our series there were only two patients with congenital heart disease and none with Fanconi syndrome, a recessively inherited trait with a reportedly high association of radial dysplasia. The most frequent abnormality we found was dysplasia of the hip joint to various degrees, followed by congenital scoliosis. Other conditions we found included Klippel-Feil syndrome; cleft lip and palate; dextracardia; imperforate anus; strabismus; Sprengel deformity; abnormalities of the lumbar and cervical spine; mental retardation; abdominal hernias; hammer toe; meningomyelocele; malformations of the ears, knees, and feet; and absence of a kidney. Additional conditions that we did not find in our patients but that have been reported as associated include tracheoesophageal atresia, polydactyly, syndactyly, triphalangia of the thumb, and thrombocytopenia.

Goldberg and Meyn have discussed the many syndromes that are involved and have grouped them under the following headings.

1. Syndromes with chromosomal abnormalities (6 syndromes)
2. Syndromes in which mental deficiency predominates (2 syndromes)
3. Syndromes in which craniofacial defects predominate (12 syndromes)
4. Syndromes in which cardiac anomalies predominate (2 syndromes)
5. Syndromes in which blood dyscrasias predominate (2 syndromes)
6. Syndromes in which vertebral anomalies predominate (3 syndromes)
7. Teratogenic syndromes (2 syndromes)

These are by no means all the syndromes that have been reported as associated with radial clubhand, but they are the more common. The clinically most important are the three associated with cardiac anomalies and blood dyscrasias: Holt-Oram, Fanconi, and TRA (thrombocytopenia-radial aplasia).

ANATOMICAL VARIATIONS

This section contains a somewhat detailed account of the anatomical variations associated with varying degrees of involvement of the preaxial border. I have included it because I believe a surgeon needs to be aware of these anomalies. The impatient may ignore it but perhaps at their patients' peril because scattered in these pages are practical comments on the surgical problems created by these alterations in normal anatomy.

Skerik and I have studied the anatomical variations associated with these deformities, and illustrated in Fig. 15-1 are the major anatomical defects seen in cross sections of the forearm and wrist.

The major differences in the anatomical relationships at midforearm level are seen between the normal and hypoplastic hand and between the hypoplastic and partially aplastic hand. The cross sections at wrist level accentuate the differences between partial and total aplasia of the radius. The flexor digitorum superficialis tendons to the index and small fingers and the extensor digitorum communis tendon to the index are frequently absent. Less frequently, but not uncommonly, the flexor digitorum profundus to the index is also absent.

Bones and joints

Both O'Rahilly and Heikel have published detailed descriptions of the skeleton in radial dysplasia, and only the major defects of the skeleton are reviewed here.

Bones normal in more than two thirds of cases. The capitate, hamate, triquetrum, and the ulnar four metacarpals and their phalanges are the only bones of the upper extremity that are present and free from defect in nearly 100% of cases of radial clubhand. The trapezoid, lunate, and pisiform are also usually normal; however, they are more frequently involved (10%) than the others. Abnormalities of these carpals tend to be hypoplasia, fusion, or delayed ossification rather than total absence.

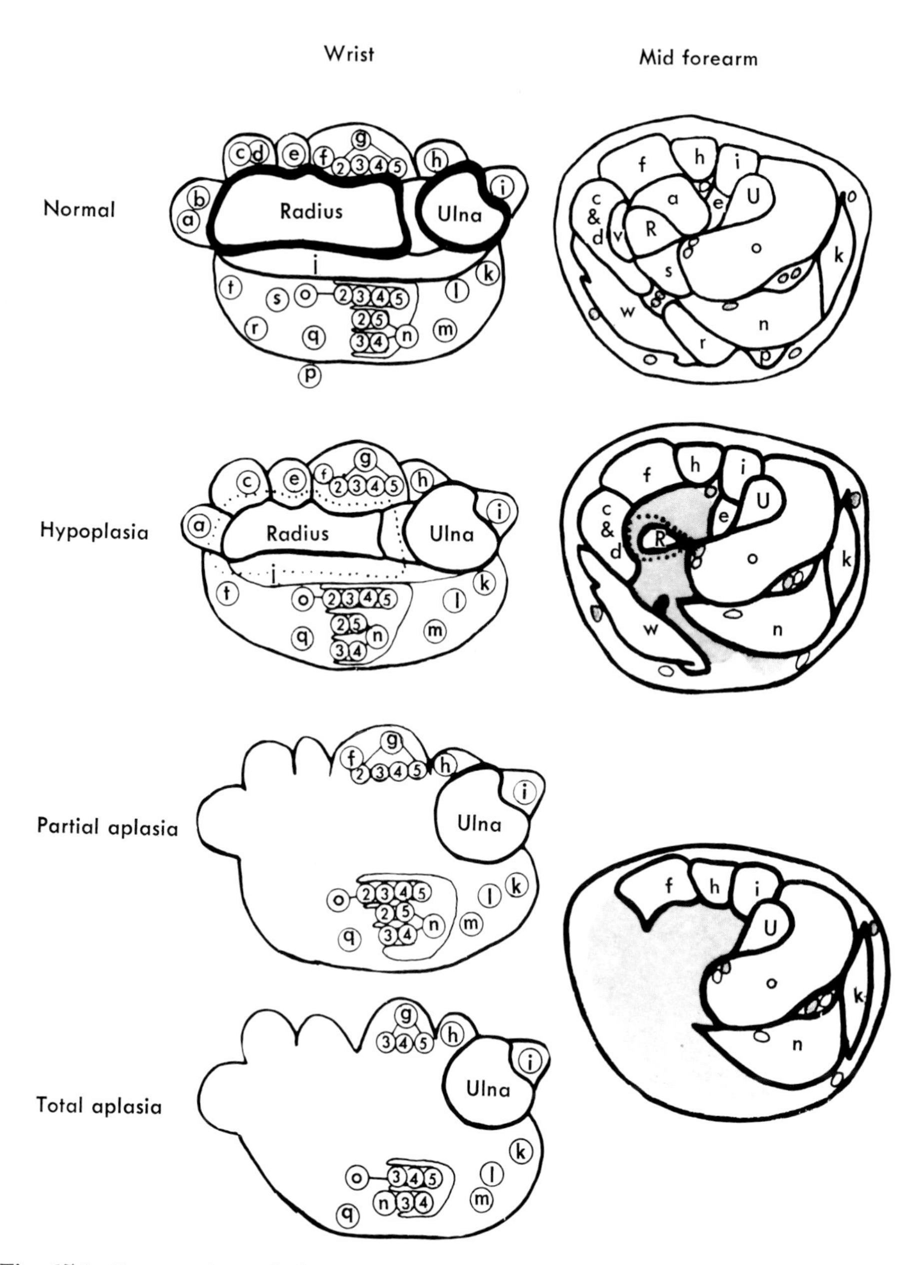

Fig. 15-1. *Cross-sections of the forearm and wrist in different degrees of radial dysplasia.* **R,** Radius; **U,** ulna; **a,** abductor pollicis longus; **b,** extensor pollicis brevis; **c,** extensor carpi radialis longus; **d,** extensor carpi radialis brevis; **e,** extensor pollicis longus; **f,** extensor indicis proprius; **g,** extensor digitorum communis; **h,** extensor digiti minimi; **i,** extensor carpi ulnaris; **j,** pronator quadratus; **k,** flexor carpi ulnaris; **l,** ulnar nerve; **m,** ulnar artery; **n,** flexor digitorum superficialis; **o,** flexor digitorum profundus; **p,** palmaris longus; **q,** median nerve; **r,** flexor carpi radialis; **s,** flexor pollicis longus; **t,** radial artery; **v,** pronator teres tendon; **w,** brachioradialis. (From Skerik, S. K., and Flatt, A. E.: The anatomy of congenital radial dysplasia, Clin. Orthop. **66**:125-143, 1969.)

Bones present but abnormal

Humerus. In radial clubhand the humerus is generally considerably shorter than in a normal limb. In bilateral cases the humeri are usually of equal length except in cases in which the degree of involvement is much greater on one side. Other humeral defects are absence of the capitulum, the coronoid fossa, the intertubercular sulcus, the medial condyle, or the entire distal end of the bone. A small or poorly formed trochlea or olecranon fossa is a common defect. The proximal epiphysis is rarely mentioned in radial dysplasia, but the distal humeral epiphysis may demonstrate varying degrees of involvement, which seem to relate to the severity of the radial dysplasia.

Ulna. The ulna is usually curved, shortened, and thickened. It may be displaced backward on the humerus, its olecranon tuberosity may be defective, and its styloid and coronoid processes may be missing. Heikel reports, "There seems to be a tendency toward a delay in appearance of the distal epiphyseal nucleus of the ulna and possibly also toward too early a fusion of the distal ulnar epiphyseal line in partial and total aplasia of the radius."*

Wrist joint. A well developed articulation between the ulna and carpus has been demonstrated in only a few cases. In most cases there is only a fibrous connection, although a flat cavity sometimes lined with hyaline cartilage has been observed on the radial aspect of the distal ulna. The bones are usually bound together by tough fibrous tissue and are not covered with articular cartilage.

The patients in our series demonstrated only half the degree of active wrist motion demonstrated by the patients in Heikel's series. Average wrist motion in flexion and extension was 45 degrees and average active wrist motion in radial and ulnar deviation was 17 degrees. Heikel reported averages of 83 degrees and 28 degrees respectively (Fig. 15-2, *A* and *B*).

Digital joints. The most significant clinical feature of the finger joints is their lack of complete range of motion. Heikel found the index and long fingers more limited in flexion at the metacarpophalangeal joints than the other fingers, but hyperextension of the joint was often possible. Dissection showed that the joint surfaces of the interphalangeal joints of the index and long fingers were flat and irregular. He found flexion contractures most frequently at the proximal interphalangeal joints.

In contrast to Heikel's findings, the metacarpophalangeal joints could not fully flex in nine of the 11 limbs in our series. Of the two who could flex their joints, one had normal motion and one lacked 10 degrees of full motion. These two patients with normal or near normal motion were the youngest in the study. Three patients were able to hyperextend at the metacarpophalangeal joints, and all patients were able to actively extend their fingers to neutral at the metacarpophalangeal joints.

---

*Heikel, H. V. A.: Aplasia and hypoplasia of the radius, Acta Orthop. Scand. (Suppl.) 39:1, 1959.

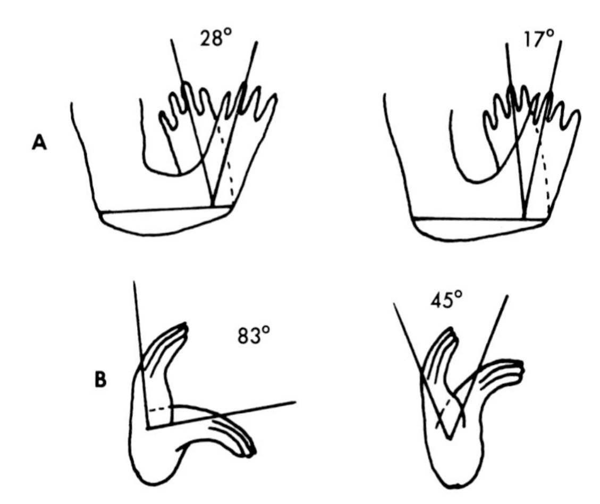

Fig. 15-2. *Active wrist motion.* **A,** Comparison between the range of radial and ulnar deviation reported by Heikel (left) and that recorded in the University of Iowa series (right). **B,** A comparison between the range of palmar flexion and dorsiflexion reported by Heikel (left) and that recorded in the University of Iowa series (right). (From Skerik, S. K., and Flatt, A. E.: The anatomy of congenital radial dysplasia, Clin. Orthop. **66**:125-143, 1969.)

Table 15-1. Average ranges of digital motion in all three finger joints

Finger	*Average degrees metacarpo-phalangeal flexion**	*Average degrees metacarpo-phalangeal flexion†*	*Average degrees proximal interphalangeal flexion**	*Average degrees distal interphalangeal flexion**
Index	35	51	47	17
Long	40	55	58	22
Ring	40	56	63	22
Small	37	54	58	25
Average	38	54	57	22

*From 0 degrees extension.
†From hyperextension.

The average ranges of digital motion in all three finger joints are recorded in Table 15-1.

Bones absent in more than one half of cases

Radius. If the radius is not totally absent, the proximal portion of the bone varies in shape and size and it can be represented by a fragment that may or may not be fused with the ulna. Several cases have been described in which the proximal portion was absent and the distal radius present. This condition is very uncommon and, according to Steindler, not prone to produce deformity. Surprisingly, the literature lacks detailed descriptions of the radius in cases of partial aplasia or hypoplasia, but several cases have been reported in which the lower epiphysis of the hypoplastic radius was intersected with fibrous tissue. This tissue band provided attachment for some of the muscles that ordinarily originate or insert on the radius.

Scaphoid. Four of 21 cases reviewed had a normal complement of carpal bones, while the scaphoid was missing in the remaining 17 cases. If the scaphoid is present, it is usually normal for it is rarely described as either rudimentary or fused with other carpal bones.

Trapezium. The trapezium, like the scaphoid, is frequently absent. In this review it was missing in 14 of the 21 cases. When it was present, it was rudimentary more often than the scaphoid.

First metacarpal and its phalanges. O'Rahilly states that the thumb, including the metacarpal and its phalanges, is absent in more than 80% of the cases of partial and total aplasia of the radius. It was absent in 15 of the 21 cases under review, but as previously mentioned, four of these cases had a hypoplastic radius. A rudimentary thumb is not uncommon. When this exists, the thenar eminence is usually absent and the thumb is a fleshy stump, sometimes having a small bone chip but no muscles.

In our series of 12 limbs there were two normal thumbs, six hypoplastic thumbs, two rudimentary thumbs, and two absent thumbs.

The hypoplastic thumbs were able to adduct and flex the first metacarpal but were severely limited in their ability to extend, abduct, and rotate the first metacarpal. All thumbs showed total absence of flexion and extension at the interphalangeal joint. The rudimentary thumbs had no functional capacity.

Surgical implications. The lack of normal skeletal support is the fundamental problem in the restoration of function to the limb. In the past, varied attempts have been made to create an internal support for the carpus and thereby retain some degree of wrist motion. In general, the follow-up studies have shown that these operations have not been successful. The surgical pendulum has now swung back to methods of stabilization of the carpus on the distal end of the ulna. Some surgeons maintain that this surgery should be delayed until growth has ceased in the distal ulna.

The amount of growth potential in the distal ulna varies greatly since its epiphysis always appears late and closes early in comparison with the normal limb. I believe that the prolonged wait advocated by some surgeons can cause profound secondary changes that ultimately are far worse than any disadvantage created by the premature closure of the distal ulna epiphysis caused by surgery (Fig. 15-3). The bones of the affected limb are always shorter than normal, and the majority are sufficiently short to require alteration of the sleeves of ready-made clothing. There is no significant functional difference between two short forearms differing in length by only 1 or 2 cm.

Lack of digital joint motion is one of the most severe handicaps suffered by these patients, and surgery has little to offer in the restoration of finger motion. Arthrogryposis is a poor diagnosis for these stiff fingers since the lesion lies in the joint and capsular tissues rather than in the extrinsic and intrinsic muscles. In some patients the stiffness can be extreme, but true symbrachydactyly is rare.

Those patients with a markedly curved ulna are sometimes subjected to

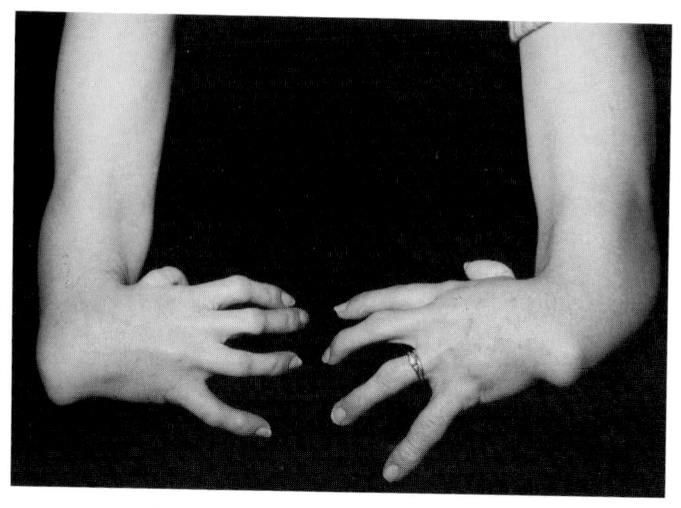

Fig. 15-3. *Untreated radial dysplasia in an adult.* The dislocation of the hand proximally along the ulna is profound. The distal ulna protrudes, and no length advantage has been gained in the limbs by delaying surgical implantation.

correction osteotomies. I have tried this procedure and have abandoned it except in extreme curvature because I was unable to demonstrate any significant functional improvement. Usually a closing wedge osteotomy is done, and rapid healing occurs; however, Wolff law continues to operate, and the curvature continues. Even multiple osteotomies have not helped, and I do not believe that even the cosmetic improvement is sufficient to justify the procedure.

Muscles

A large number of abnormalities in the musculature are recorded in the literature. These are discussed in some detail because the variations around the wrist can be utterly confusing to those not experienced with this problem.

Muscles normal in more than two thirds of cases. All muscles inserting on the humerus, except the pectoralis major and the coracobrachialis, are usually normal.

In the forearm the origin of the flexor carpi ulnaris is always normal, but occasionally the insertion is aberrant.

If the extensor carpi ulnaris is abnormal, its origin or insertion or both may deviate somewhat from normal or it may be fused with the extensor digitorum communis or with the flexor carpi ulnaris.

The extensor digiti minimi is occasionally abnormal. When anomalous, it is usually fused with either the extensor carpi ulnaris or the extensor digitorum communis. The insertion, then, may be normal, or it may be in common with the muscle to which it is attached.

Within the hand, the lumbricals have been described as part of the flexor digitorum profundus. Their innervation frequently varies. Total ulnar innervation is common, or the first lumbrical may be innervated by the median. If a lumbrical is missing, the one most frequently lacking is the lumbrical to the index.

Although the interossei are usually normal, the first dorsal interosseus, and occasionally others, can be absent. They can also be rudimentary, atrophied, or present but undifferentiated.

The hypothenar muscles are almost always normal, but like the interossei, they can occur as an undifferentiated mass in which the individual muscles can not be isolated.

Muscles frequently abnormal

Muscles absent in part. The abdominal origin of the pectoralis major from the aponeurosis of the external oblique is often missing. Less frequently, it may have an abnormal insertion, usually into the capsule of the shoulder joint, or it may have an abnormal connection with other muscles, particularly the deltoid.

The long head of the biceps is almost always absent, but if present it originates from the anterior diaphysis of the humerus, the crest of the greater tuberosity of the humerus, or the capsule of the shoulder joint. When it is present in total absence of the radius it usually inserts into the lacertus fibrosus of the biceps.

The short head of the biceps is always present but rarely normal. It is usually fused with another muscle, either the coracobrachialis at its origin or, more distally, the muscles of the forearm. The combined muscles usually originate from either the epicondyle or the forearm flexors. A terminal tendon is often absent and the insertion is into the joint capsule and the brachial fascia. In the absence of the long head, the short head frequently divides into two terminal tendons that insert either on the radial rudiment, medial epicondyle, or into the joint capsule. The biceps, as well as all the muscles of the anterior compartment, is frequently innervated by the median nerve.

The extensor digitorum communis is rarely absent but is frequently fused with neighboring extensors, primarily with the extensor carpi radialis longus or the extensor digiti minimi or both.

The finger flexors usually show more variations in their superficial component. The flexor digitorum superficialis is commonly present but often incomplete, atrophied, or fused with the deep flexors. The radial head of origin is almost always absent, and the tendon to the index is most frequently missing. Insertions are normal but thin and tight or absent to the index or small fingers or both. The flexor digitorum profundus is more frequently normal than the superficialis, but like the superficialis, the tendon to the index is missing more often than any other tendon. Abnormal insertion sites for the profundus have been described on the bases of the proximal or middle phalanges of the digits.

Muscles rudimentary or fused with other muscles. The coracobrachialis frequently originates with the short head of the biceps as a single fused muscle mass, and it is often innervated by the median nerve.

The brachialis can be normal, rudimentary, fused, or absent. Most frequently it is fused with the biceps at its origin and has no specific site of insertion but becomes continuous with the muscles originating from the common flexor site. It can also insert into the tuberosity of the radius, into the joint capsule, or into the lateral epicondyle.

The pronator teres is frequently fused with the biceps-brachialis mass, with the palmaris longus, or with the flexor carpi radialis and usually inserts into the rudiment of the radius or into the intermuscular septum.

The brachioradialis is described by some as usually absent and by others as missing only in total absence of the radius. If the radius is present, the brachioradialis inserts on it; otherwise, it inserts on the carpus. If it is fused to another muscle, it usually inserts with that muscle.

The extensor carpi radialis longus is often missing but is more often rudimentary or fused with adjacent extensors, primarily the extensor carpi radialis brevis, extensor digitorum communis, or brachioradialis. Its origin is more consistent than its insertion.

Muscles abnormal in origin or insertion. The extensor indicis proprius is frequently missing, but more often it is present with an abnormal insertion. It may send a tendon to the long and index fingers or to the long finger only. It can insert into any phalanx or metacarpal of the index, long, or ring fingers, or it may insert into the carpal bones.

Accessory muscles. Accessory muscles are common. Most authors attempt to identify an abnormal muscle, but the attachments of these muscles vary so drastically that identification is impossible. For this review I have grouped them according to their location.

ACCESSORY MUSCLES ORIGINATING ON THE HUMERUS. Accessory muscles originating on the humerus (1) arise from the anterior humerus just distal to the insertion of the deltoid and cross the elbow joint to insert into the lateral intermuscular septum or join the brachioradialis and (2) originate on the lateral epicondyle and insert onto the carpus, the bases of the metacarpals, or the phalanges. Although the latter are not part of the extensor digitorum communis, in several instances they supply an extensor tendon to a finger otherwise lacking one. Similar muscles have also been described on the flexor surface; they originate from the medial epicondyle, but their insertions usually do not extend beyond the carpus.

ACCESSORY MUSCLES ORIGINATING ON THE ULNA. Most of the accessory muscles that originate on the ulna are extensors that insert into the dorsal radial carpus or into the metacarpals. Some have been vaguely discussed in the literature, but those adequately described lie deep to the superficial wrist and finger extensors.

ACCESSORY MUSCLES ORIGINATING FROM SOFT TISSUE. Abnormal muscles do arise from the soft tissues, primarily the flexor digitorum superficialis and profundus. These accessory muscles take origin from either the belly of a muscle or from its tendon. In cases where they originate from a profundus tendon, it is not uncommon for these accessory muscles to produce one or more lumbrical muscles.

Muscles absent in 50% or more of cases. The anconeus is usually absent but is only rudimentary when it is present.

The supinator is absent except in some cases in which the proximal radius is present. When the muscle is present, it inserts on the connective tissue extending between the radial rudiment and the carpus or into the intermuscular septum.

The extensor carpi radialis brevis is more frequently absent than the extensor carpi radialis longus. When the two muscles cannot be defined, the common muscle belly divides into two tendons, which may insert normally or may insert on the carpus. The common muscle origin is usually from the middle or distal ulna. Bora and associates report that the extensor carpi radialis longus and brevis were absent or ineffective in all 24 limbs in their series.

If the flexor carpi radialis is present, it is usually so highly anomalous that it is hard to identify with certainty.

When the palmaris longus is present, it is frequently fused with the flexor digitorum superficialis or the other flexors and takes a more ulnarward insertion than usual, attaching to the pisiform, to the fifth metacarpal, or into the soft tissue over the third and fourth metacarpals.

The pronator quadratus is usually totally absent. Occasionally it may be represented by a mass of muscle surrounding the distal extremity of the ulna or may be found inserting into the radial side of the carpus or into the second metacarpal.

In partial and total absence of the radius the extrinsic and intrinsic muscles of the thumb are rarely normal whether or not a thumb is present. In more than 50% of the cases of radial hypoplasia the muscles are abnormal or absent with the intrinsic muscles especially affected. When the intrinsics are present, they are usually grossly abnormal, having neither a normal origin nor a normal insertion. A common insertion site is into the metacarpal of the index or long fingers. This aberrant insertion occurs even in cases with a rudimentary or hypoplastic thumb.

The status of thumb muscles in 22 cases reported in the literature is listed in Table 15-2.

Surgical implications. The many aberrations of the muscles make textbook surgical approaches to the area around the wrist impossible. The abnormal

Table 15-2. Status of thumb muscles in 22 cases

Muscle	*Absent*	*Abnormal*	*Normal*
Flexor pollicis longus	14	5	3
Abductor pollicis longus	14	5	3
Extensor pollicis longus	13	4	5
Extensor pollicis brevis	18	2	2
Flexor pollicis brevis	12	8	2
Abductor pollicis brevis	13	7	2
Opponens pollicis	16	4	2
Adductor pollicis	16	5	1

muscle bellies and the aberrant insertions utterly confuse the tyro seeking normal muscle planes. There are normal fascial planes between the abnormal muscles, but fusions between muscle bellies frequently prevent the usual, proximal soft tissue spreading and separation.

Multiple insertions into both sides of the carpus, but particularly on the palmar aspect, prevent correction of the carpal radial deviation and have to be released. I have found no advantage in leaving these insertions and attempting to lengthen their tendons. Branches of the median nerve usually pass to both the dorsal and palmar aspects of the preaxial border and must be carefully retracted while the attachments are released.

The fibrous anlage of the radius, occasionally present in hypoplasia and partial aplasia, may be found deep in the wound and until it is released will prevent free and easy connection of the carpus on the ulna (Fig. 15-4). I believe it important to excise a portion of this fibrous band and advise a very careful proximal and distal definition of the structure to ensure that it is in fact the radial anlage and not an aberrant median nerve.

The extrinsic flexor tendons are frequently undifferentiated at the wrist, and occasionally tendons will branch off to carpal or metacarpal insertions while the main tendons pass on into the fingers. The main tendons to each finger should be identified and their free motion should be demonstrated against passive resistance of the fingers. Free-running tendons are vital if the

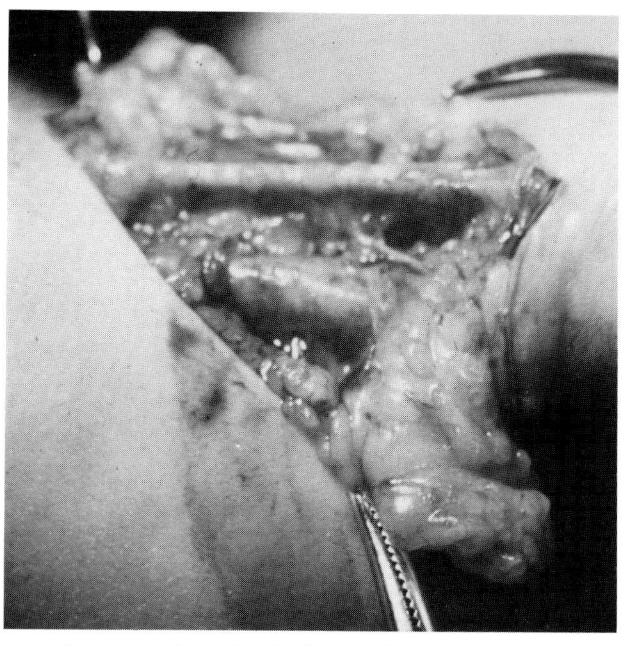

Fig. 15-4. *Fibrous anlage of radius.* The thick, rounded object in the center of the field is the anlage, and the upper, thinner, longer, and more superficial structure is the median nerve. (From Skerik, S. K., and Flatt, A. E.: The anatomy of congenital radial dysplasia, Clin. Orthop. **66**:125-143, 1969.)

full power of the muscles is subsequently to be used in mobilization of the fingers.

Nerves

Disturbances of the normal pattern of nerve distribution may be profound and can be found as far proximal as the root of the brachial plexus. Although generally considered normal, some of the main branches of the plexus contain fibers from a higher cervical segment than usual. Figs. 15-5 and 15-6 illustrate the nerve supply to the upper limb in several cases of radial dysplasia.

Normal nerves. The axillary nerve and the ulnar nerve are usually present and normal.

Absent nerves. The musculocutaneous nerve is most frequently missing. If present, it is usually anomalous, either joined with or substituted by the median nerve. The radial nerve usually ends just above the lateral epicondyle after innervating the triceps. The median nerve provides sensation to the radial side of the hand and anastomoses with the sensory branch of the ulnar nerve on the dorsum of the hand.

Abnormal nerve

Median. The vitally important median nerve is always present, but its distribution may be altered depending upon the status of the other nerves. It supplies the muscles of the anterior compartment of the arm in the absence of the

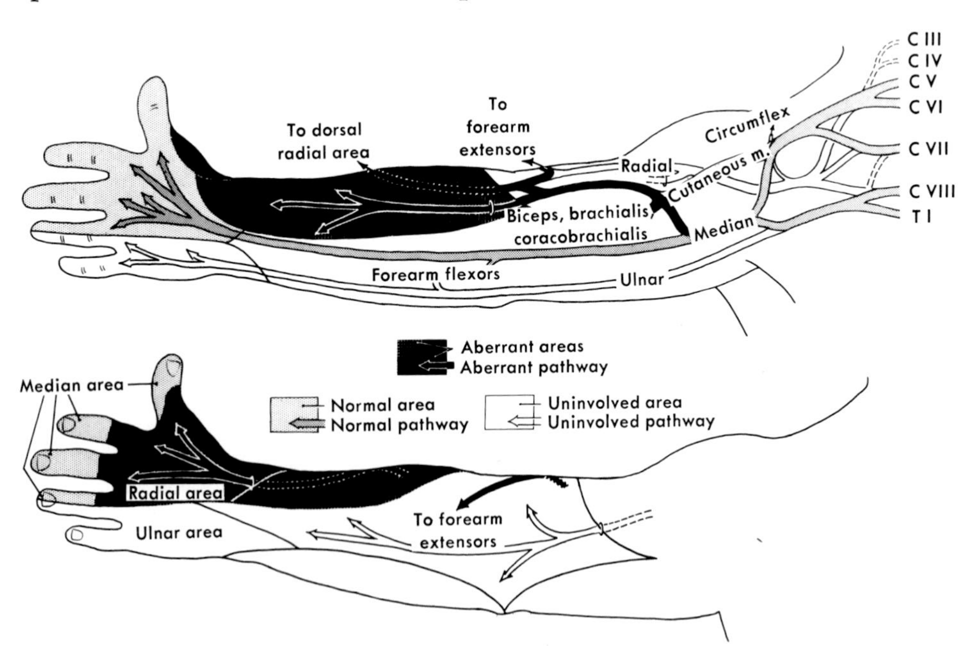

Fig. 15-5. *Aberrant nerve supply—thumb present.* When the thumb is present but the superficial radial nerve is absent, the thumb's sensory supply comes solely from the median nerve through its normal and supplementary branches. Note absence of the musculocutaneous nerve. (From Skerik, S. K., and Flatt, A. E.: The anatomy of congenital radial dysplasia, Clin. Orthop. **66**:125-143, 1969.)

musculocutaneous nerve and usually substitutes for the terminal distribution of the radial nerve. Its course is subject to many strange variations.

The most detailed description in the literature is that of Stoffel and Stempel, who traced and described the course of the nerve in 16 cases. Its route varied in eight ways.

1. It traveled very superficially along the radial edge of the brachioradialis until it entered the palm.
2. It coursed through the forearm on the underside of the palmaris longus and flexor digitorum superficialis to the carpal ligament.
3. It traveled through the forearm along the lateral edge of the flexor digitorum superficialis to the hand.
4. At the elbow it dipped beneath the flexors and supplied them, then reappeared superficially in the middle of the forearm between the brachioradialis and the extensor digitorum communis.
5. After sending strong branches to the flexors of the arm, it went under the pronator teres to the undersurface of the flexors, supplied this area, then came to the surface at the flexor aspect of the wrist, and gave off a dorsal branch to the hand.
6. In two cases it supplied the flexors of the arm and came to the surface between the flexor digitorum superficialis and the brachioradialis, traveling to the radial side of the palm.

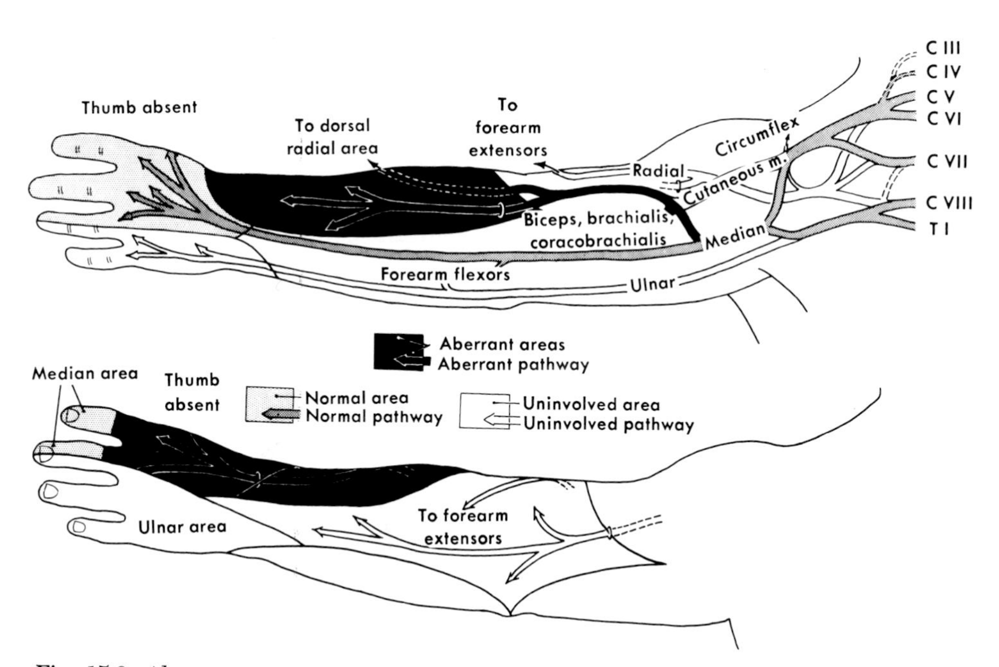

Fig. 15-6. *Aberrant nerve supply—thumb absent.* If the thumb is absent, the median nerve still supplies supplementary cutaneous branches along the preaxial border. Note absence of the musculocutaneous nerve. (From Skerik, S. K., and Flatt, A. E.: The anatomy of congenital radial dysplasia, Clin. Orthop. 66:125-143, 1969.)

7. After branching to the flexors of the arm, it pierced and supplied the brachioradialis. It then followed a normal course in the forearm except for piercing the pronator teres and locating itself between the flexor digitorum superficialis and the flexor digitorum profundus.

8. It supplied the arm flexors and then sent a sensory branch to the brachioradialis, which it pierced to supply the area normally innervated by the absent lateral antebrachial cutaneous nerve.

Surgical comments. Despite the fact that disturbances of the standard neurological pattern occur as far proximal as the roots of the plexus, the distal peripheral course of the nerve is of great surgical significance.

There is great variation in the sensory pattern, but invariably there is good coverage; areas of anesthetic skin are not known to occur. Variations in standard motor patterns also occur but do not have clinical significance.

It is the median nerve above all others that presents a significant surgical challenge. The nerve is thicker than is usually seen in a normal arm of the same age because it carries additional sensory fibers normally distributed by the radial nerve. The course is almost invariably aberrant. It is consistently preaxial and usually lies immediately beneath the fibers of the deep fascial cylinder. The nerve represents a strong and unyielding bowstring of the radially bowed forearm and hand.

If the common pattern of the two-stage corrective operation is performed, the nerve will cause great trouble in the first or soft tissue corrective stage. After the Z-plasty flaps of the skin approach have been mobilized, the deep fascia will have to be split widely up the forearm and the peripheral distribution of the nerve identified (Fig. 15-7).

Release of abnormal muscle insertions and other soft tissue attachments will usually provide a much greater degree of correction than can be tolerated by the nerve. It is hard to judge how much stretching the nerve can withstand and yet provide sensory and motor conduction. I usually err on the side of greater tension

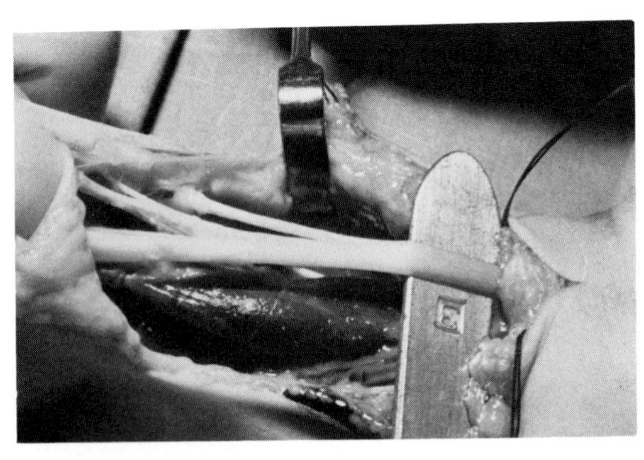

Fig. 15-7. *Aberrant median nerve.* The median nerve lies immediately beneath the deep fascia, and its dorsal sensory branches can be clearly seen. (From Pardini, A. G.: Radial dysplasia, Clin. Orthop. **57**:153-177, 1968.)

and have not yet seen any persistent evidence of anesthesia or paralysis although temporary minor degrees of disturbances have been detected.

I do not repair the deep fascia because to do so would only restore the original restraint on the nerve. Even the skin lengthening supplied by the Z-plasty is usually not adequate, and the nerve bowstrings beneath the flaps as the hand is corrected on the ulna at the end of the operation. Serial plaster casts usually provide full correction without much difficulty, and only very rarely have I had to do a second soft tissue release operation to obtain a satisfactory positioning of the hand on the single forearm bone.

Occasionally, the nerve breaks up into its final small branches well proximal in the forearm, making it both difficult and tedious to dissect out all these important branches (Fig. 15-8). More often the nerve retains its astonishingly large size almost to the level of the wrist crease, and its large size and subcutaneous course has frequently led to its resection by a surgeon under the mistaken impression that it represents the fibrous anlage of the radius (Fig. 15-4).

Vessels

The brachial artery is usually normal. However, it may divide into two branches high in the upper arm or may not divide at all at the elbow. The deep brachial artery may come off the posterior humeral circumflex. Blauth and Schmidt have described 11 arteriograms done on nine children with radial dysplasia and have shown that the vascular anomalies correspond closely to the

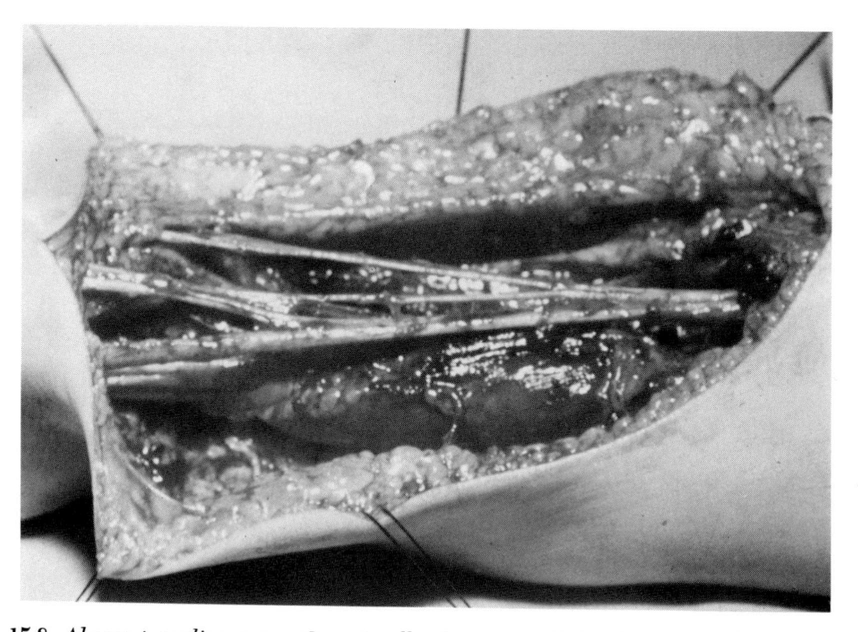

Fig. 15-8. *Aberrant median nerve.* Occasionally the nerve will break up into its final branches well proximal in the limb. These branches must be carefully sought at surgery. (From Skerik, S. K., and Flatt, A. E.: The anatomy of congenital radial dysplasia, Clin. Orthop. **66**:125-143, 1969.)

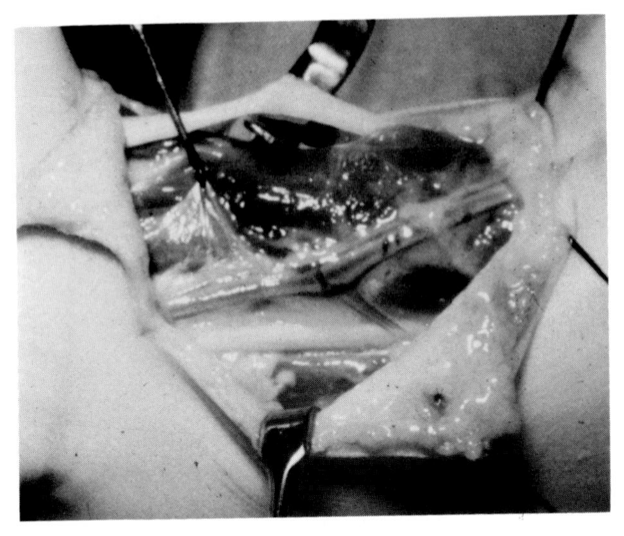

Fig. 15-9. *Displacement of radial artery.* Occasionally the radial artery and its venae comitans are deviated to the midline and lie in close relationship to the ulnar nerve. (From Skerik, S. K., and Flatt, A. E.: The anatomy of congenital radial dysplasia, Clin. Orthop. **66**:125-143, 1969.)

extent of radial dysplasia. The greater the involvement, the more likely the radial artery and palmar arterial arches are to be involved. The ulnar artery is usually present and normal, but it may be anomalous concomitant with the absence of the radial artery. Heikel points out that the vessels, as well as the nerves and tendons in the forearm, course radially at the distal end of the ulna (Fig. 15-9).

The interosseus arteries are usually well developed and may replace the radial or ulnar arteries or both. In severe cases the palmar digital arteries may be of small caliber and even absent on the radial side of the index finger.

Surgical comments. Vascular anomalies cause little or no problem for the surgeon. The tendency for a radialward deviation of the vessels does mean that the normal relative positions of the radial and ulnar arteries and median and ulnar nerves is disturbed. During the first stage soft tissue release operation, it is usually easy to find the ulnar artery, which is frequently enlarged. I deliberately look for and protect this artery because it is frequently the sole—or at least the major—arterial supply for the limb. The second stage operation is done through a dorsal approach and must place at hazard any branches of the interosseus arteries that enter the hand over the dorsum of the metacarpals. If the ulnar artery is damaged during the first stage operation and the radial artery is rudimentary or absent, the blood supply of the hand can be seriously compromised by the second stage dorsal approach.

FUNCTIONAL IMPLICATIONS

A radial clubhand is not a normal hand set on an abnormal wrist; rather, it is a profoundly abnormal hand joined to a poor limb by a bad wrist. The fully

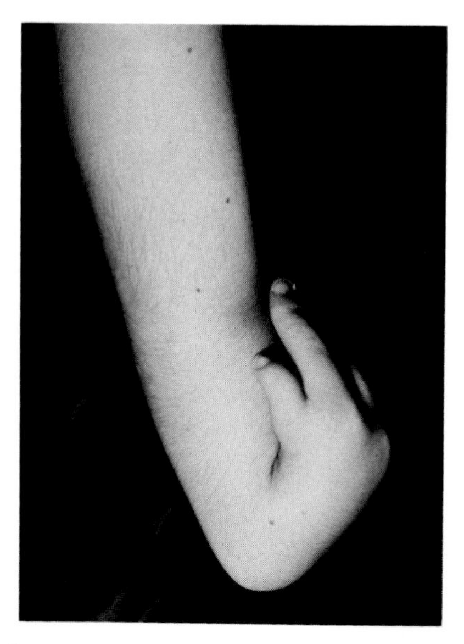

Fig. 15-10. *Untreated radial dysplasia in a child.* In untreated cases the hand assumes a position of radial deviation and palmar flexion, resulting in a hideous deformity and a functional liability. (From Pardini, A. G.: Radial dysplasia, Clin. Orthop. **57**:153-177, 1968.)

developed radial clubhand is a hideous deformity and a functional liability (Fig. 15-10). Lesser degrees of involvement imply a smaller loss of function, but in fact dexterity and skilled activities are severely inhibited. Most children with unilateral involvement become functionally independent, but in those with bilateral deformity functional ability is often less than might be anticipated.

Lamb has carefully studied over 40 patients and stresses that in the unilateral case the other normal arm dominates function and the affected limb is simply used as an aid.

Children with bilateral deformities often have significant difficulties washing, dressing, and feeding themselves. These everyday tasks become virtually impossible when there is associated stiffness of the elbows. Lamb has recorded a high incidence of elbow stiffness in the first and second years of life. Fortunately, as the child grows, motion usually returns to the joint. Full elbow range is virtually never achieved, but eventually nearly 90 degrees of flexion have been obtained in many patients.

The ulnar two digits in these patients are nearly always normal, and the children tend to grip with these fingers (Fig. 15-11). This ulnar prehension is common because the two radial fingers invariably show some degree of stiffness. The index is particularly involved and is often held flexed, and many indexes even show hypoplasia and significant flexion contracture.

We have carefully studied the functional activities of nine individuals with radial dysplasia chosen to represent the spectrum of digital involvement. Three patients had bilateral involvement: one with normal thumbs, one with hypoplastic thumbs, and one with rudimentary thumbs. Three unilateral

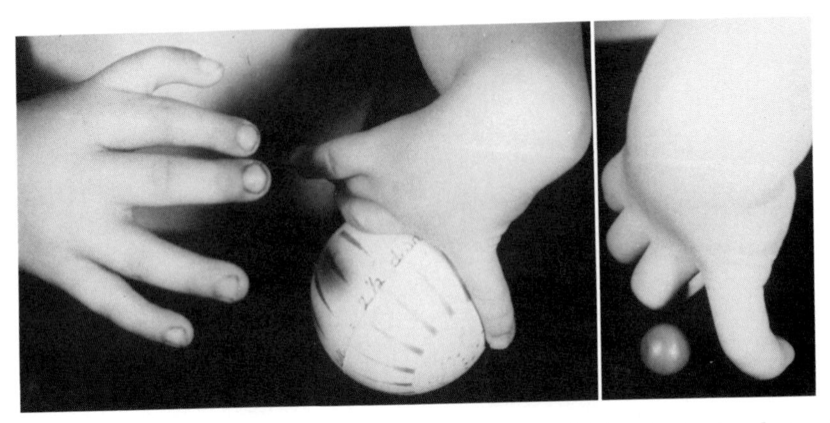

Fig. 15-11. *Small finger substituting for the thumb.* This child with total aplasia of the radius and thumb used the small finger exclusively as a thumb substitute, choosing only a lateral type pinch in the performance of all activities. (From Skerik, S. K., and Flatt, A. E.: The anatomy of congenital radial dysplasia, Clin. Orthop. 66:125-143, 1969.)

cases had no thumbs, two had hypoplastic thumbs, and one had a rudimentary thumb.

In defining maldevelopment of the thumb we have used Heikel's definition of rudimentary as a thumb with no active motion and attached by a soft tissue pedicle. Hypoplastic thumbs are those that are not rudimentary but do deviate from the normal in size, shape, and position. Hypoplastic thumbs were present on half of the hands studied, but they did not in general significantly increase the functional level of the hand over those hands without thumbs. Factors that contribute to this are:

1. Active finger flexion was limited, and opposition between the thumb and the fingers was impaired.

2. Thumb motion frequently was limited. Motion ranged from indifferent carpometacarpal "wiggles" to a fair representation of most motions. Complete active range of motion at each joint was not present in any thumb. Interphalangeal flexion and extension was severely limited in all thumbs. Abduction and extension motions were in most cases far more limited than flexion and adduction motions.

The only significant difference between hands with and without thumbs was the ability of the patients whose hands had thumbs to touch their thumb to the ring or small fingers, which in most cases displayed a greater range of interphalangeal motion than was present in the index and long fingers. Since the thumb was not able to abduct and rotate to a great degree at the carpometacarpal joint, a lateral type pinch was used (Fig. 15-12). This usually worked for fine pinch, but when strength was needed, the index was substituted for the thumb.

Although the hands could in some way perform most activities, they did not demonstrate the generally accepted normal functional patterns. Patterns of use could be identified, and it was found that patients with thumbs used a lateral

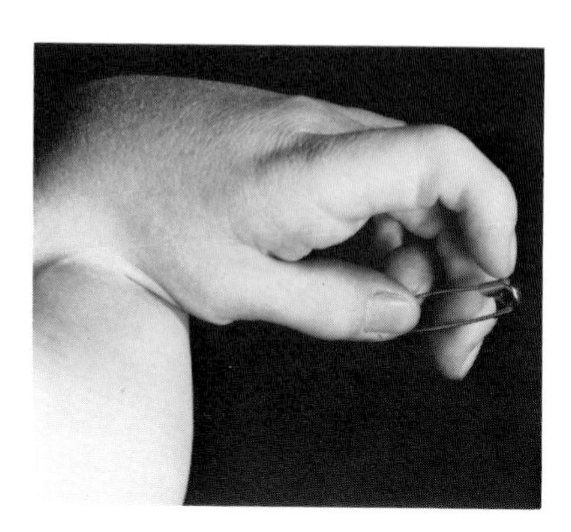

Fig. 15-12

Fig. 15-13

Fig. 15-12. *Lateral pinch.* Whether or not a thumb is present, a lateral pinch is the type of pinch preferred by all patients with radial dysplasia. (From Skerik, S. K., and Flatt, A. E.: The anatomy of congenital radial dysplasia, Clin. Orthop. **66**:125-143, 1969.)

Fig. 15-13. *Spherical grip.* Note the incomplete abduction at the carpometacarpal joints of the thumbs when spherical grip is performed by the hands with total aplasia of the radius. (From Skerik, S. K., and Flatt, A. E.: The anatomy of congenital radial dysplasia, Clin. Orthop. **66**:125-143, 1969.)

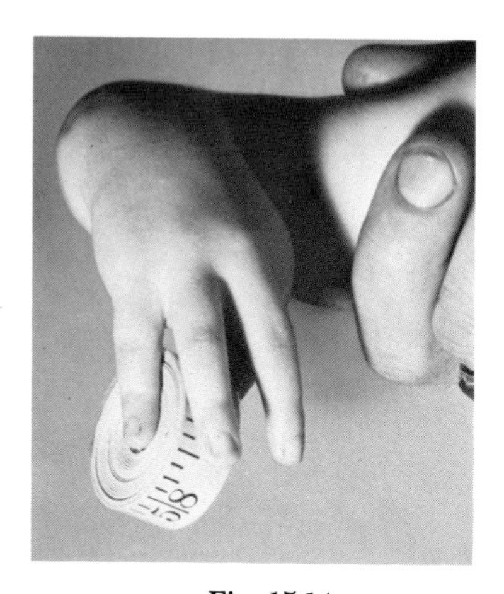

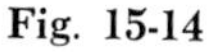

Fig. 15-14

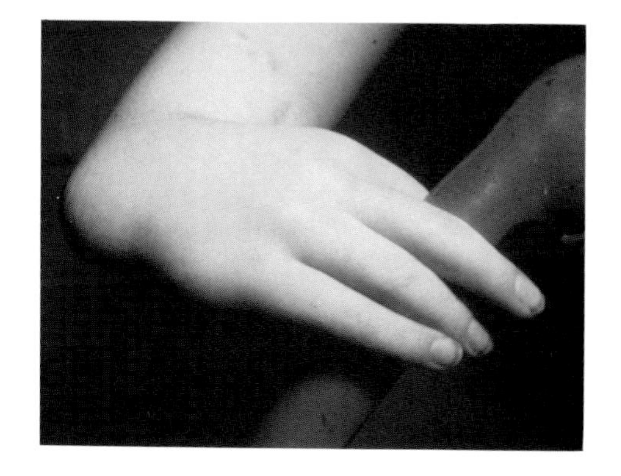

Fig. 15-15

Fig. 15-14. *Tripod configuration of the fingers.* In congenital radial dysplasia with total absence of the thumb, the index passes beneath the long and ring fingers to oppose the pulp of the small finger. (From Skerik, S. K., and Flatt, A. E.: The anatomy of congenital radial dysplasia, Clin. Orthop. **66**:125-143, 1969.)

Fig. 15-15. *Lateral pinch between two fingers.* An attempt to perform a cylindrical grip results in the most prevalent form of pinch used by patients with radial dysplasia. (From Skerik, S. K., and Flatt, A. E.: The anatomy of congenital radial dysplasia, Clin. Orthop. **66**:125-143, 1969.)

pinch. A true tip or palmar type pinch is rarely possible for a hand with congenital radial dysplasia.

Spherical grip was preferred by all patients (Fig. 15-13). Adduction and abduction between the index and long fingers was developed to a greater range of motion than in the normal hand, allowing the index to substitute nicely for a thumb (Fig. 15-14). The index also frequently demonstrated a significant degree of pronation so that its palmar surface faced ulnarward, further enhancing its ability to substitute for the thumb. With a spherical grip, the fingers are positioned in a tripod configuration, allowing a relatively strong and dexterous grip.

Cylindrical grip was less used, probably because the thumb could not abduct sufficiently and the index was not in a position to substitute for a thumb (Fig. 15-15). Power grip was usually impossible. A good fist could not be made because of lack of full flexion of the fingers. To our surprise hook grip was the least preferred even though it is the only form of prehension that does not actively use the thumb. Lack of full finger flexion and weakness of the finger flexors must be the reasons for the general disfavor toward hook grip.

When the thumb is absent, the index tends to deviate ulnarward and the other fingers radialward. Thus in flexion the central two fingers tend to overlap the converging border fingers (Fig. 15-16).

Two types of prehension are seen in these hands that are uncommon in the normal hand: a lateral pinch between any two fingers and what appears to be a variation of spherical grip—the index and small fingers are placed on the sides of an object to be picked up and the long and ring fingers on top of the object. A combination of adduction and flexion of the fingers is then used to pick up the object.

Thus although these patients do not use the more formal or standardized types of prehension, they can do almost anything they wish. Their hands are

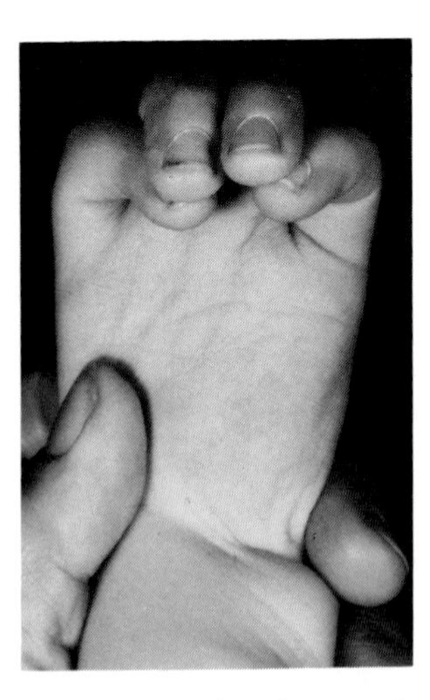

Fig. 15-16. *Rotation of border fingers.* In the absence of the thumb, the rotated border fingers tend to lie beneath the central digits during flexion. (From Skerik, S. K., and Flatt, A. E.: The anatomy of congenital radial dysplasia, Clin. Orthop. **66**:125-143, 1969.)

clumsy and lack skill, but they also lack too many of the necessary anatomical components for surgery to produce dramatic functional improvement.

TREATMENT

Classification of types

Discussion of treatment for a condition carrying such a variety of anatomical and functional possibilities must be based on some form of classification. Heikel's remarkable long-term study of 47 patients provides a very useful and practical classification. He differentiates between the degrees of dysplasia as follows:

1. *Hypoplasia of the radius.* Congenitally inhibited development of the distal epiphysis of the radius with or without shortening of the diaphysis but with the growth zone preserved and further hypoplasia or absence of one or several carpal bones and the bones of the thumb (Fig. 15-17)
2. *Partial aplasia of the radius.* Congenital absence of the distal parts of the radius comprising the distal epiphysis and growth zone together with a part of the diaphysis or, in exceptional cases, of the proximal part of the radius or of the diaphysis only; further hypoplasia or absence of one or several radial carpal bones and the bones of the thumb (Fig. 15-18)
3. *Total aplasia of the radius.* Complete congenital absence of the radius, accompanied by hypoplasia or absence of one or several radial carpal bones and the bones of the thumb (Fig. 15-19)

The care of these three types of involvement is very different and includes both conservative and surgical measures.

Principles of care

Treatment of this condition is difficult because of the balance that has to be made between function and appearance. The fact that adults with fully developed bilateral deformities can work full time, marry, and rear children is hardly a justification for surgical nihilism in the growing child. All the adults I have known with this condition have perforce adjusted to it, but none was happy with his or her lot. Commonly, adults with this deformity demonstrate adequate function and adaptation of their limb to ordinary activities. They often chose occupations appropriate to their limitations, and appeals to improve their cosmetic appearance should usually be resisted. Virtually any surgery may compromise function, and function must remain the principal aim of treatment (Fig. 15-3).

For the child born with partial or total aplasia of the radius, the obvious aim of treatment is to position the hand on the single forearm bone in such a manner that motion, growth, and good appearance are obtained. When one limb is normal, the function of the affected side becomes less important and surgical correction can tend toward appearance more than ability. Lamb has pointed out that in bilateral cases both sides can be operated upon, particularly when the planned procedure for the less functional limb may provide better appearance even at the expense of some function.

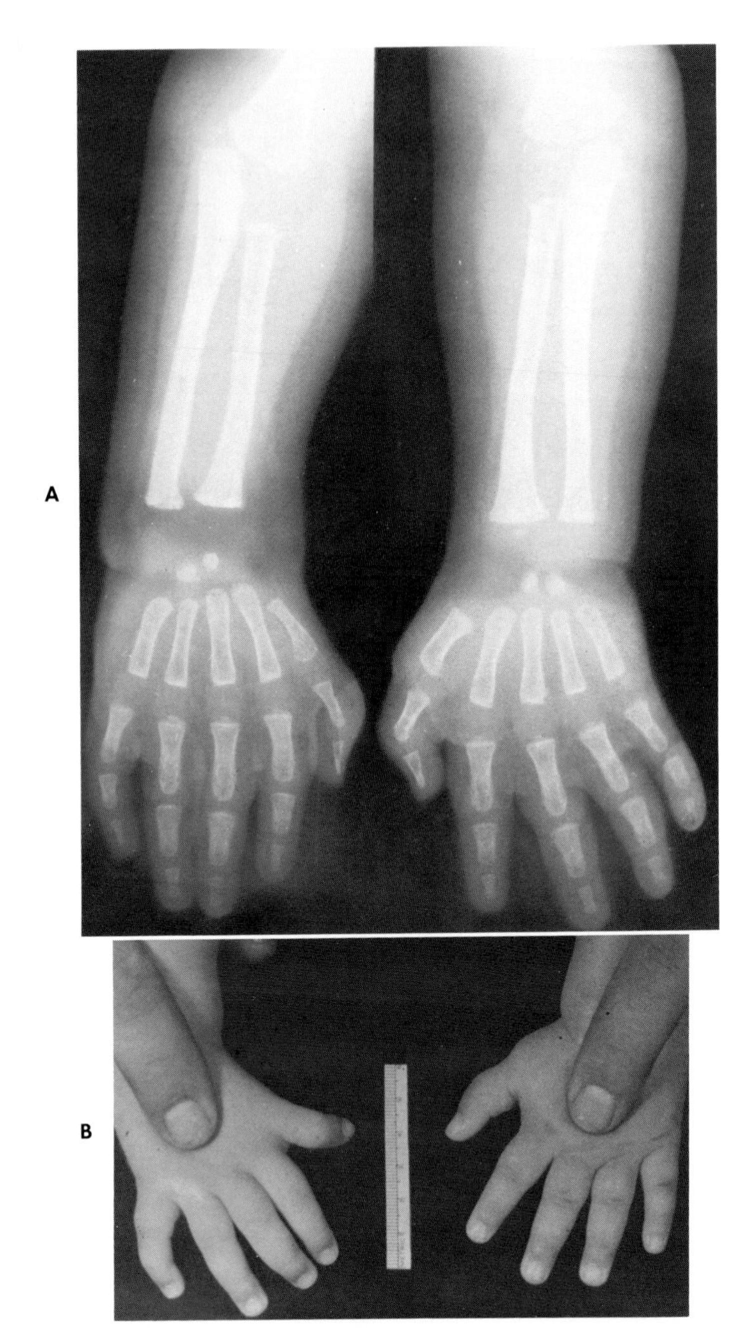

Fig. 15-17. *Hypoplasia of the radius.* **A,** Hypoplasia of the right radius and thumb. There is an inhibition of the development of the distal epiphysis of the radius. Observe the thickness of the right ulna. **B,** Clinically, there is only a discrete degree of radial deviation and a visible hypoplasia of the thumb on the right. (From Pardini, A. E.: Radial dysplasia, Clin. Orthop. **57**:153-177, 1968.)

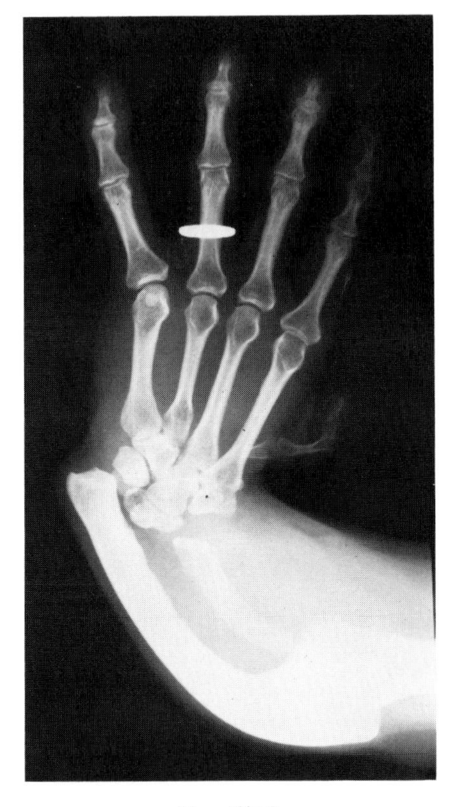

Fig. 15-18

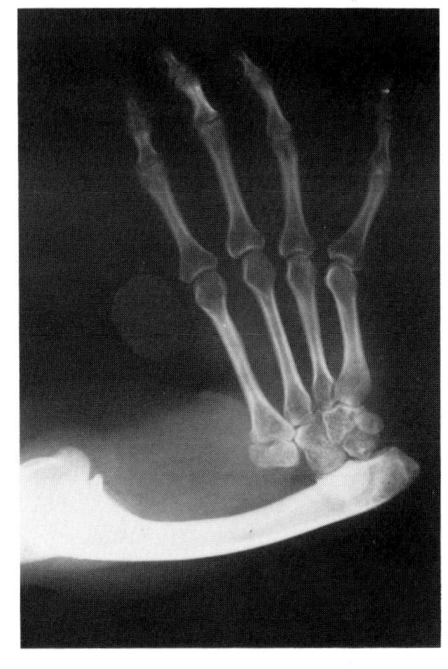

Fig. 15-19

Fig. 15-18. *Partial aplasia of the radius.* Absence of the distal radius, extreme hypoplasia of the thumb, and curvature of the ulna.
Fig. 15-19. *Total aplasia of the radius.* There is hypoplasia of the whole hand, absence of the thumb, and absence of several carpal bones. The hand is dislocated proximally on the radius.

A satisfactory result can only be obtained if treatment is begun early. Because of the lack of broad skeletal support at the wrist, the pull of the extrinsic forearm muscles will topple the hand off the narrow distal end of the ulna. Secondary shortening rapidly occurs, and the deformity will become permanent unless countermeasures are started soon after birth (Fig. 15-20).

Surgery is not needed to reposition the hand of an infant. If manipulations are started soon after birth, the hand can usually be centered on the ulna and the position maintained by some light external fixation.

Many surgical procedures to permanently maintain this centralization have been reported over the last 100 years. These have been extensively reviewed by others, and Bora has recently summarized the history of this period. Centralization was usually accomplished after stretching or division of the deforming soft tissues, osteotomy of the ulna, and excision of the carpal bones. Fixation of the hand in the corrected position was achieved by a variety of means using silk,

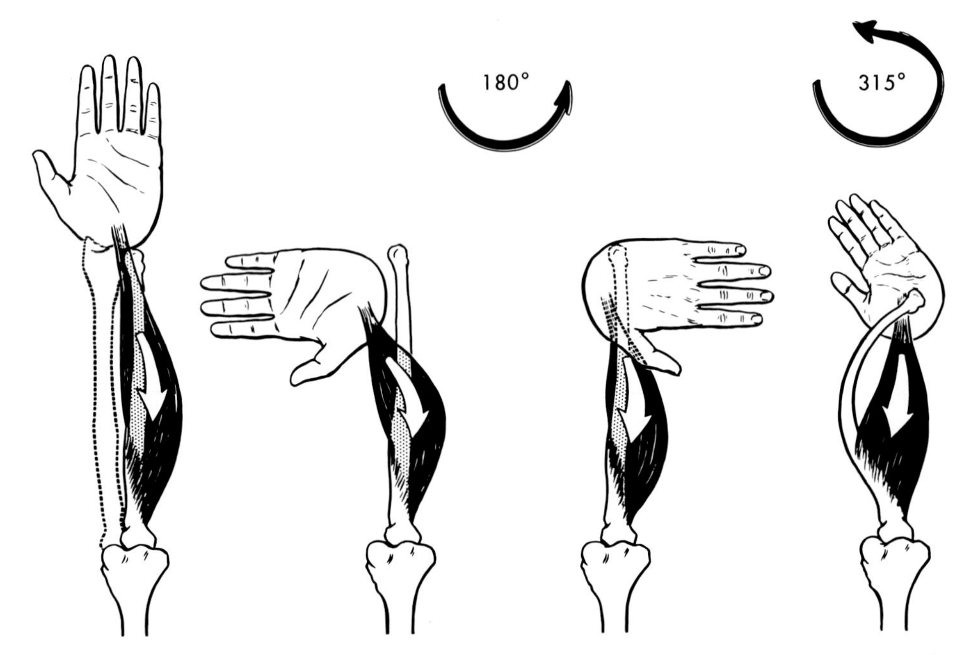

Fig. 15-20. *Absence of the radius.* Lack of skeletal support leads to a sequence of architectural changes in the limb producing curvature of the ulna and gross rotation and malalignment of the hand. (From Entin, M. A.: Reconstruction of congenital absence of the radius, Transactions of the International Society of Plastic Surgeons, Baltimore, 1957, The Williams & Wilkins Co. © 1957, The Williams & Wilkins Co., Baltimore.)

kangaroo tendon, gold wire, chromic catgut, and stainless steel wire and pins. A number of different tendon transfers have been advocated to provide additional support and better balance across the wrist.

Another approach has been to attempt to replace the absent or dysplastic radius by a bone graft. Tibial, ulnar, and fibular grafts have been used to stabilize the hand in proper alignment on the distal end of the forearm. In 1965 Riordan reported his 15 years' experience using the upper end of the fibula. He has now abandoned this procedure and practices early implantation of the distal ulna into the carpus.

Splitting the ulna longitudinally to broaden its distal end is another method that has been described to maintain the hand over the distal end of the ulna. Muscle, silk, and ivory pegs have been placed between the two halves of the split distal end of the ulna in an effort to keep the hand in the correct position. In 1966 Define described a technique in which results were satisfactory after 5 years. The periosteum is elevated from the distal one third of the ulna and the hand displaced ulnarward so that the distal end of the ulna is in contact with the radial side of the carpus. A periosteal tube is then formed on the ulnar side of the ulna with the expectation that a bone peg will be produced. The hand is then supported by the original distal end of the ulna on its radial side and by the newly formed bone peg on its ulnar border. The common theme in all these reports has been the problem of maintaining the stability of the wrist joint.

Wrist stabilization improves function by abolishing the waste of the power of the extrinsic muscles of the digits and by encouraging maximum movement of the fingers. There are promising reports from several surgeons that a combination of early wrist stabilization and appropriate bracing tends to produce a significant and active increase in the range of motion of the elbow and the fingers.

Stabilization of the wrist can only be done in the presence of good controlled elbow motion. Fusing the wrist on an arm with a fused elbow will hold the hand away from the body and cripple the patient. There is an increasing tendency to stabilize the carpus on the ulna at an early age. Generally, the distal ulna is surgically squared so that it can be placed in a slot created in the carpus aligned with the third metacarpal.

There are some surgeons who feel that stabilization by placing the ulna into a carpal slot during early infancy and childhood may cause shortening of the forearm by arresting growth at the epiphysis at the lower end of the ulna. This is true but not significant. The very detailed study by Heikel showed that the ulna was invariably shorter in patients who had not been treated but had been followed to maturity. He also showed that the distal ulna epiphysis developed later and closed earlier than those in normal limbs.

Most descriptions of radial aplasia include curving of the ulna as an integral part of the deformity (Fig. 15-18). This is especially true of untreated cases. Lamb believes, and I agree, that in most cases this curvature is caused by soft tissue tightness on the radial side. In most patients the ulna can be coaxed to grow straight if properly splinted and if soft tissue surgical release is done when appropriate. If the ulna is already bowed at birth, this is probably indicative of the presence of an anlage of the radius creating a tight bowstring (Fig. 15-4). Splinting cannot help these cases, and early surgical release is imperative.

DETAILS OF TREATMENT

Soft tissue correction

Manipulation and splinting. Correction of the angulation caused by soft tissue tightening may be all that is necessary in treatment of the patient with hypoplasia of the radius; in more extensive involvement it is a necessary preliminary to permanent operative correction.

External fixation should be started in the newborn nursery. The only practical method to use is a lightly padded plaster of Paris cast extending proximal to the flexed elbow. The cast must be placed on the infant's limb in three parts in a similar fashion to the application of a cast for clubfoot. The hand itself must first be enclosed in plaster; the thumb should be excluded, but the fingers may be included in the early casts if by this means a good purchase is obtained along the radial border of the hand. The hand is then correctly placed on the forearm, and the wrist and lower forearm are enclosed in plaster. Finally, the cast is extended high up on the arm over the elbow, which is flexed to at least 90 degrees. It is not easy to apply such a cast, and great care must be taken not to make pressure sores from plaster creases. It is impossible to maintain such a cast on the chubby straight arm of an infant, and if the elbow is stiff in extension, a cast cannot be used. A stiff elbow can often be improved by repeated gentle manipulation, and the mother must be taught to do this for short periods many times a day. Parents do not relish the idea of hurting their infant, but they have to be taught that to inflict discomfort is virtuous in this instance. A fussing infant is acceptable, but a bawling and restless baby is in pain, and the degree of correction being obtained is too great for tolerance. This principle applies whenever manipulation is being carried out or after a new splint has been applied.

The cast will need to be changed to gradually improve the position of the hand, but it can usually be left on for several weeks between changes. An alternative to a plaster cast is some form of bracing. I have had poor results from bracing an infant's hand principally because of the difficulty in obtaining a proper fit.

Although bracing will encourage soft tissue relaxation or stretching, it does not always provide the essential ingredient to proper alignment—namely, placement of the hand across the distal end of the ulna. Too often the external device tends to tilt the ulnar border of the carpus on the distal ulna epiphysis. Repeated mild distraction manipulation is usually necessary to separate the carpus from the ulna; the hand should then be displaced ulnarward to line up the long finger metacarpal with the shaft of the ulna. When this has been achieved, a brace can be applied. The carpus can be correctly centered over the distal ulna without there being full correction of the radial deviation. This does not matter since it is this type of forearm that responds readily to the ratchet type splint advocated by Lamb (Fig. 15-21).

When, after treatment in a cast or brace, the hand can be readily centered on the forearm, a bivalved plastic gutter type splint with velcro strapping, which is easily applied by the parents and can be readily cleaned, is used. Cooperation by the parents is essential in the early days of this treatment, and they must be taught how to correctly place the center of the carpus over the end of the ulna every time they reapply the splint. I prefer to leave the digits free to move but usually recommend that the splints be worn continuously in the first few months of life. When full correction can be readily maintained with only the slightest ulnarward pressure, the splints can be left off in the daytime but should be applied during rest periods and at nighttime. Growth will make it increasingly difficult to retain the carpus on the distal end of the ulna. Eventually external splinting can do no more, and surgery will have to be undertaken. I have found that one is forced into a surgical decision well before school entry age and frequently before the age of 2 years.

If a fibrous anlage of the radius is present at birth, if the ulna is curved, or if the soft tissue contracture is so severe that correction cannot be obtained, then an early surgical release of the radial soft tissues is essential.

Surgical release. The approach to unyielding radialward contractures has to be through the concave radial side using a Z-plasty to provide adequate exposure.

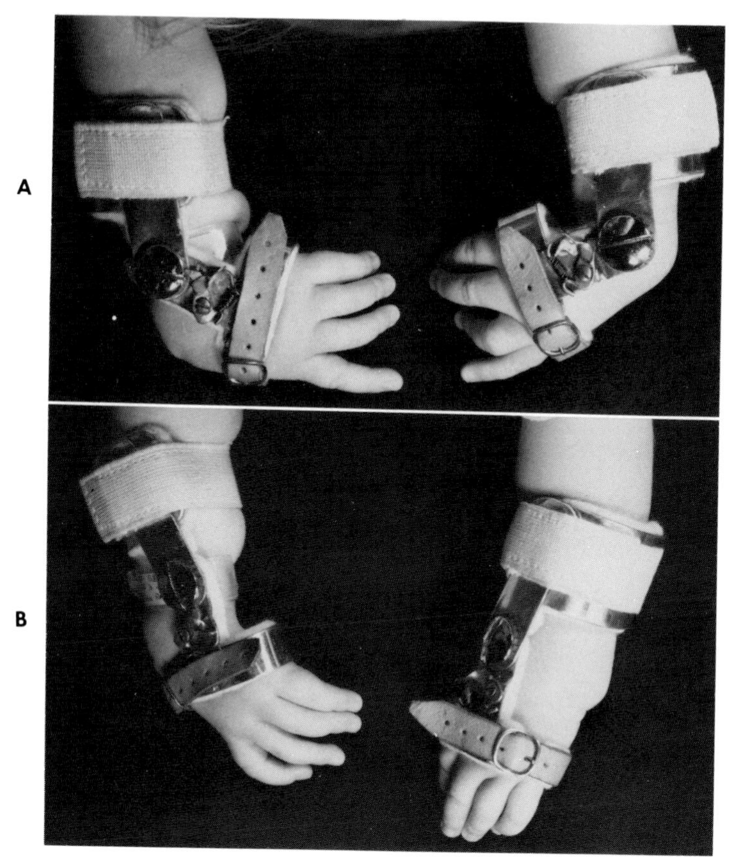

Fig. 15-21. *Radial clubhand brace.* If the carpus can be correctly centered over the distal ulna, then a ratchet type brace will be useful in obtaining full correction of the radial deviation. **A,** Photograph of deformity before correction. **B,** Photograph of same child after relief of fixed deformity by gradual correction. (From Lamb, D. W.: The treatment of radial clubhand, The Hand **4:**22-30, 1972.)

The deep fascia has to be opened, and portions have to be excised; it is not resutured at the end of the operation (Fig. 15-22).

More often than not the median nerve is lying immediately beneath the deep fascia (Fig. 15-7), and on occasion the fibrous anlage of the radius will be found on a deeper plane (Fig. 15-4). Proximal and distal mobilization of the median nerve and release of the anlage may allow full correction.

When the carpus cannot be fully corrected, both normal and abnormal muscle attachments to the carpus must be released. This release must be ruthless; there is no point in inflicting the operation on an infant if full relaxation of the contracture is not obtained. If after release the detached muscles can be reattached to the dorsoulnar portion of the carpus without tension, they should be tacked into place with a few fine absorbable sutures. During later childhood, tendon transfers such as this should be more formally done, but I believe it wrong at

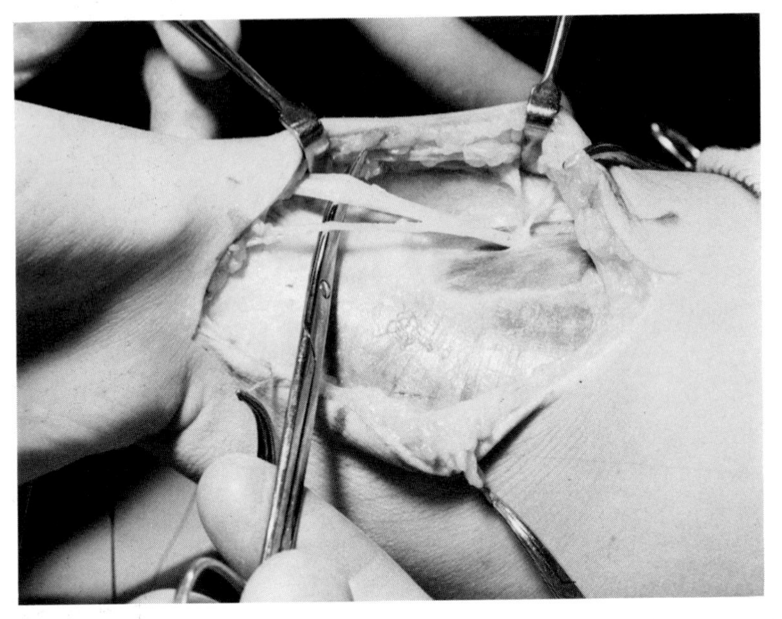

Fig. 15-22. *Deep fascia of the forearm.* The deep fascia is a tough unyielding layer that must be opened and excised to obtain full correction. The median nerve penetrates the fascia proximal to the wrist and is subcutaneous in the rest of its course.

this early age to extend the dissection and expose a lot more tissue in what is frequently a vain attempt to identify the muscles actually being transferred.

Usually the flexor carpi radialis is hard to identify, but if possible it should be detached and reinserted on the carpus as far ulnarward as possible.

Often the median nerve will still be the tightest structure after all soft tissue release has been done. It should be placed under only mild tension, and the appropriate alignment of the hand in relation to the forearm must be determined before the Z-plasty flaps are transposed and the skin is closed with interrupted mattress sutures of fine catgut. The arm and forearm are covered by a lightly padded plaster of Paris cast, which is left on for about 2 weeks. After this the plaster can be removed, the forearm measured for a splint, and a new plaster cast applied while the splint is being made. Since correction of the carpus on the ulna has now been obtained, the future aftercare is similar to that for a patient who does not need operative correction of the soft tissues.

Tendon transfers. Occasionally more formal tendon transfers are needed at a later date but prior to the implantation of the ulna into the carpus. Usually, however, the transfers can be done at the same time as the implantation. Tendon transfers to the ulnar side of the wrist are being done increasingly by surgeons who have cared for significant numbers of children with this abnormality. The concept of establishing a dynamic restraint against recurrence is correct, but the problem is how best to achieve it. This restraint is particularly important for patients in whom the thumb is present to provide an ulnar acting counterforce to the extrinsic and intrinsic thenar muscles. A useful discussion of the choice of

Table 15-3. Incidence of absent finger extrinsic muscles in 22 limbs in the Iowa series

Muscle	*Finger*			
	Index	*Long*	*Ring*	*Small*
Extensor digitorum communis	10	4	0	1
Flexor digitorum superficialis	8	1	0	8
Flexor digitorum profundus	4	0	0	0
Total	22	5	0	9

tendons available has been published by Bora in a paper that advocates the use of the flexor superficialis tendons to the long and ring fingers.

Release of abnormal radial wrist tendons is almost always necessary, but they are so short in relation to normal that they will not comfortably reach to the ulnar side of the hand without prolongation with a tendon graft. To add tendon grafting to the procedure both prolongs and complicates it unnecessarily.

The extensor carpi radialis longus and brevis would normally be the ideal tendons to transfer to the ulnar side of the carpus, but they are usually absent or vestigial. They therefore cannot be relied upon as transfers.

The ulnar wrist tendons are the main components available on the correct side of the limb. They will tighten after the operation and should not be disturbed. Occasionally the extensor carpi ulnaris appears so slack that it can be shortened or reattached more distally on the shaft of the fifth metacarpal. On the palmoulnar side of the hand the hypothenar abductor muscle group can be detached from around the pisiform and reattached more proximally to the underlying soft tissues on the shaft of the ulna.

The only additional tendons available for transfer are those of the extrinsic flexor and extensor digital muscles. The thumb tendons, if present, should not be disturbed, and the extensor tendons are usually absent or maldeveloped in at least the index and long fingers (Table 15-3). Therefore the most suitable tendons available are the finger superficialis group. The long and ring finger superficialis tendons are almost always present, but those of the index and small fingers are frequently absent. Even if the index superficialis is present, it is not wise to consider its use because of its importance as a stabilizer in pinch posture and because the index profundus is often absent.

Bora reported an anomalous flexor superficialis arrangement in six of the 18 extremities in which he performed tendon transfers. In one the superficialis tendon to the long finger was absent, but the superficialis to the small finger was present and was transferred with the ring finger superficialis. In a second the superficialis to the ring finger was unacceptable, but the superficialis to the small finger was present and was used with the superficialis to the long finger. In the remaining four all superficialis tendons were unacceptable for transfer.

Bora stressed the importance of testing for superficialis action before attempting a transfer. This can be done by holding the patient's hand with the fingers held extended at the interphalangeal joints and the metacarpophalangeal and wrist joints hyperextended. When the test finger is released, the patient will be able to flex it if the muscle is present even though the other fingers are held extended. If abnormal flexor tendons are suspected, it is wise to make an exploratory palmar incision to correctly determine which flexor structures are present. This is the only safe way to avoid using a profundus instead of a superficialis tendon as the transfer.

When a superficialis tendon is to be used, it can be released through a transverse incision over the distal palmar portion of the proximal phalanx. The decussating fibers that join the two slips of insertion together must also be divided after the insertions have been severed. If this is not done, the tendon will hang up on the profundus or lumbrical origin when it is pulled proximally.

After release from the finger, each superficialis tendon is passed subcutaneously around the ulnar side of the ulna onto the dorsum of the hand. The long finger superficialis tendon is looped around the index metacarpal shaft and the ring superficialis around the long finger shaft. The tendons lie outside the periosteum and are sutured back on themselves after the ulna has been implanted into a carpal slot. In analyzing his results Bora established two groups, A and B. Group A patients did not have any satisfactory results. All had been operated upon by procedures that did not include tendon transfers. The Group B patients had their tendon transfers done as a secondary procedure following earlier centralization of the hand. Seventy-two percent satisfactory results were obtained in the 11 patients in this group.

Wrist stabilization procedure

I believe that stabilization of the wrist is the best treatment for the young patient with gross displacement of the hand on the ulna or in whom soft tissue correction is no longer possible.

How early should this early surgery be done? Certainly there is general agreement among those who are caring for significant numbers of these children that after 2 years of age is late. Implantation of the ulna should probably be done during the first year of life, and the timing is more dictated by the failure of external splinting than by the calendar. Riordan is now operating around the sixth week of life, but probably most surgeons find the first birthday a reasonable compromise between the size of the structures to be dissected and the tensions on the radial side of the limb.

Details of operation. Several skin approaches can be used depending on the extent of the planned operation. If an extensive amount of soft tissue release and several tendon transfers are anticipated, the incision used by Lamb is appropriate. The approach extends from the dorsal base of the index over the ulnar side of the wrist and across the flexor aspect of the forearm to its radial border. An S-shaped dorsal skin incision gives good access to the dorsum, but one cannot approach the flexor aspect of the wrist. This approach also does not allow excision of the excessive ulnar bulge of skin that is usually present.

If releasing incisions and tendon transfers have already been done, then a good approach is through the protuberant bulge on the ulnar side of the wrist joint. I usually excise this redundant skin and outline my incision in an elliptical fashion but incise only the distal border at the beginning of the operation. The dorsal and palmar apices of the ellipse are placed over the center of the carpus.

The incision is developed down to the level of the tendons on both the dorsal and palmar surfaces. In doing this I like to identify the ulnar nerve, the ulnar artery, and its venae comitans. With these structures localized and protected, I feel more comfortable with the further dissection that is necessary among all the aberrant structures on the radial side of the wrist.

The advantage of this ulnar approach is that the radial dissection only has to be extended until the distal end of the ulna is delivered into the wound and the proximal curved surface of the carpus is mobilized. Despite this advantage, I make it a practice to identify the median nerve and make sure that its branches, which spread onto the dorsal and palmar surfaces of the hand, are free of tension when the carpus is corrected.

The distal end of the ulna is usually covered with a thickened false joint capsule that should be carefully dissected up and left attached on the dorsoulnar side so that it can be sewn back over the top of the ulna after it has been implanted. Often, aberrant insertions of wrist tendons are found joining this capsular tissue. The epiphyseal end of the ulna is a firm oval-shaped mass of tissue in which it may be hard to identify the plane of the epiphysis. This mass has to be trimmed into a squared off end, and this should be done by gentle paring with a scalpel; osteotomes and other impact tools may damage epiphyseal growth. When the size of the distal ulna has been established, the carpus is brought into the corrected position and score marks are made on the carpal bones around the end of the ulna. These marks define the sides of the slot that will have to be cut in the lunate and capitate. Lamb has stressed that to obtain a secure union the sides of the carpal slot must be as long as the width of the ulna. Ideally the carpus should be positioned on the distal ulna in such a fashion that the line of the third metacarpal is perpendicular to the growth plate. If the end of the ulna extends more distally and lies over the whole capitate or even the base of the metacarpal, further soft tissue release will have to be done until the proper amount of distraction is obtained.

The next thing to be decided is the orientation of the hand in relation to the forearm. The single forearm bone precludes pronation and supination, wrist flexion and extension, and even lateral deviation. Assuming that reasonable shoulder and elbow motion are present, the hand should be placed in a position about halfway between neutral and full pronation. Thus the slot in the carpus is not necessarily cut in a plane at right angles to the dorsal surface of the carpus.

After the ulna and the carpal slot have been properly matched, the position is maintained by driving a Kirschner wire through the medullary canal of the long or ring finger metacarpal, across the capitate, into the center of the squared off distal surface of the ulna, and up into the main shaft of the bone (Fig. 15-23).

The size of the Kirschner wire used is subject to some controversy. Delorme recommends using the largest wire or rod that the medullary canal of the metacarpal will accept, but most surgeons use a smaller diameter wire. The diameter of wire selected must vary with the size of the patient's hand, but in children under the age of 1 year I have found a .045 or .062 diameter satisfactory. It is important that both ends of the wire be pointed. Long Kirschner wires that are pointed at both ends may be hard to come by, and I usually sharpen them myself before surgery. Be careful to avoid making a flattened spear end on the wire since the end would then be broader than the diameter of the wire. A spear end

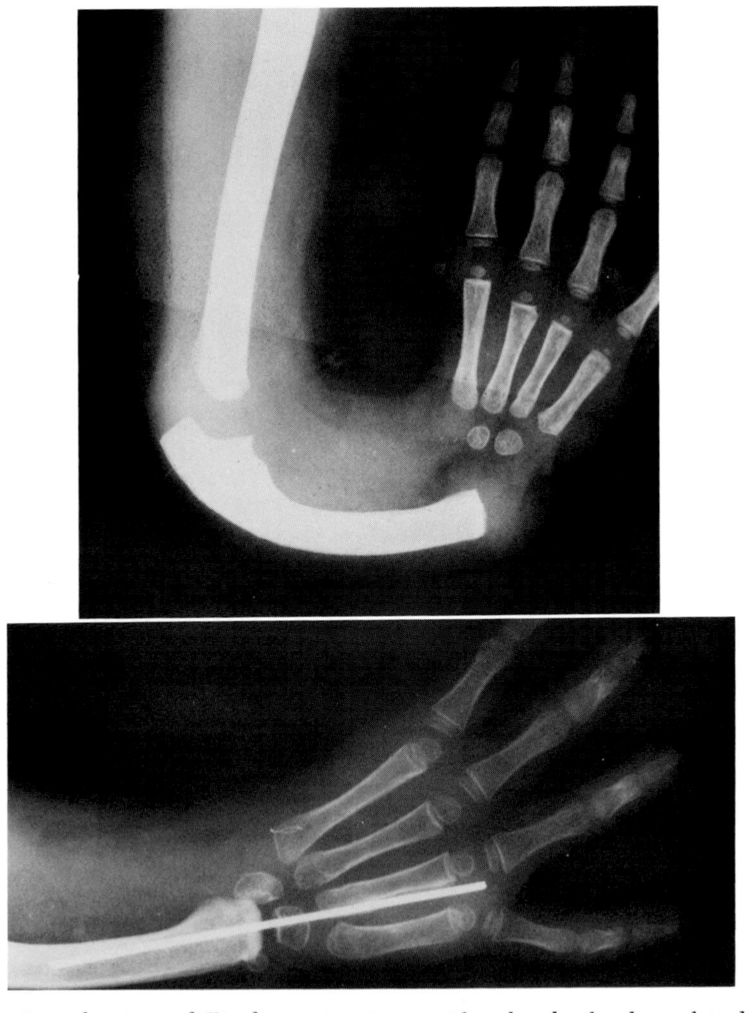

Fig. 15-23. *Centralization and Kirschner wire pinning.* After the ulna has been slotted into the carpus the position is maintained by a Kirschner wire passing through the ulna, the capitate, and the long or ring finger metacarpal. The lower illustration shows that it is not always easy to transfix a metacarpal down its medullary shaft.

would cut a hole with a diameter larger than the wire, which would then be loose in the bone.

The hand should be held with the fingers flexed at the metacarpophalangeal joints while the wire is passed retrograde through the remnants of the capitate, into the base of the metacarpal, through its shaft, and out through the skin. The wire can be felt as it penetrates the metacarpal head, and the overlying extensor tendon is pushed to one side before the wire end is brought out through the skin. The drill chuck is now replaced on the end protruding through the metacarpal and the wire withdrawn until its tip just shows in the proximal surface of the capitate. After the elbow has been flexed, the tip of the wire is then inserted into the exact center of the ulna and passed up into the ulna to eventually exit either along the subcutaneous border of the bone or through the olecranon. The chuck is once again removed and placed on the proximal wire end so that the wire may be withdrawn proximally until the long finger metacarpophalangeal joint can be freely moved passively without grating on the wire still protruding through the metacarpal head.

Once the wrist is stabilized, the hand position must be checked again to ensure that its orientation is the best for the shoulder and elbow motion available in the limb. When the proper position has been obtained, I often place several fine absorbable sutures into the dorsal fascia of the ulna and sides of the carpal slot to get good apposition of the cut cartilaginous surfaces. Sutures can be safely passed through the cartilage if this is necessary to snug up the cut surfaces. The soft tissue flap is then sutured over the carpal slot and the end of the ulna.

Any necessary tendon transfers should now be done before the skin is closed. I usually shorten the effective length of the extensor carpi ulnaris by lifting and reattaching it to the base of the fifth metacarpal. I usually move the origins of the hypothenar muscles more proximally onto the ulna and also bring the pisiform and flexor carpi ulnaris more proximally onto the ulnar, rather than the flexor, aspect of the forearm.

Now that the hand is corrected on the forearm, the extent of resection of the redundant ulnar skin can be readily judged. Some degree of absorption will occur, and the amount excised does not have to be so great as to make tension on the skin edges. Interrupted fine catgut sutures should be used, and I do not usually use a drain.

Postoperative care. There is little motion in these tiny fingers, and postoperative swelling to a marked degree is common, particularly after the ellipse of ulnar skin is excised. Because of this, I often apply a firm compression dressing of Dacron fluff and Kling bandage and elevate the limb. A posterior gutter splint of plaster of Paris is almost always necessary to maintain elbow flexion after traction is applied. The flexed position is more comfortable for the child and is needed to take tension off the radial soft tissue structures, particularly the median nerve. The swelling is usually gone within a week, at which time a complete plaster cast is applied with the elbow held in flexion. Since catgut sutures have been used in the skin, this cast can be safely left on for about 5 or 6 weeks.

A bivalved plastic splint is then made, and it should be worn for a further 3 months. After this the splint can be left off for daytime activities, but it is useful for another 6 months as a rest and nighttime splint.

PROVISION OF A THUMB

All combinations of defects may occur in the thumb in cases of radial dysplasia. Sixty-five percent of those patients with radial agenesis have no thumb, and in another 10% the thumb is so small as to be useless (Fig. 15-24). When it is useless or absent, there is a tendency for the index finger to undergo spon-

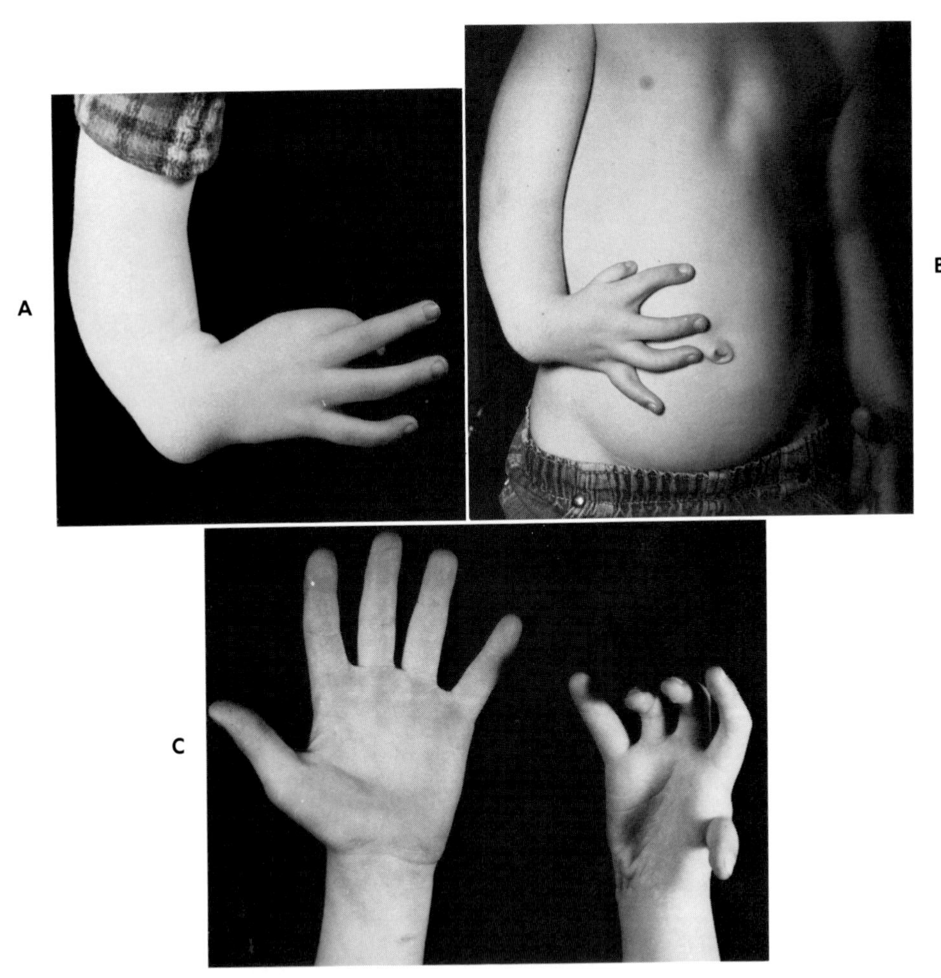

Fig. 15-24. *The thumb in radial aplasia.* **A,** Absence of the thumb and deformity of the index. **B,** Hypoplastic thumb and a well developed space between index and long fingers used for grasp. **C,** The rudimentary thumb is attached by a skin pedicle and has no voluntary motion. Note how the border fingers have rotated in an attempt to provide some degree of opposition. (From Skerik, S. K., and Flatt, A. E.: The anatomy of congenital radial dysplasia, Clin. Orthop. **66**:125-143, 1969.)

taneous pollicization in the sense that it tends to move into abduction away from the rest of the hand and show varying degrees of pronation. It therefore seems logical to complete the process and formally construct a thumb out of the index finger.

The concept is admirable; execution is often difficult. The actual transfer of the index on a neurovascular pedicle is a well established procedure with a high rate of success. The lack of a radial digital artery is not vital. Transfer on the ulnar digital artery alone is possible and satisfactory. The problem in achieving a good functional result is in providing a carpometacarpal joint and in substituting for the absent extrinsic and intrinsic thenar muscles.

Establishing the correct indications for pollicization in these patients is difficult. There is little doubt that in bilateral thumb aplasia pollicization on at least one side can significantly increase functional capacity. Index transfer in unilateral lesions and in the second limb in a bilateral case is less clearly indicated. Eaton has commented that boys, in order to cope with the challenge of occupational competition, have a somewhat greater need when the condition is unilateral. The scars, however minimal they may be, tend to somewhat discourage pollicization in girls with unilateral aplasia. I personally do not consider this a significant contraindication because when the operation is carefully performed, a gratifying functional and cosmetic result can be obtained.

I have found that in bilateral cases most parents request that the second pollicization be performed but that in unilateral cases they are usually not enthusiastic. I do not press the issue in these cases but have on occasion introduced them to patients (and their parents) with bilateral pollicizations so that they can compare functional abilities. The theoretical objections to bilateral pollicization are the risk of loss of a digit and the narrowing of the palm. The former is a real possibility but rare, and the latter is more apparent than significant. Breadth of the hand is always important, but these patients are unlikely to adopt heavy manual occupations, and dexterity is probably more important to them.

Harrison has raised an interesting point in proposing the occasional use of pollicization on the ulnar border of the hand, particularly in cases of complete radial agenesis. These patients frequently show such poor function on the radial side of the hand that ulnar function naturally predominates. In these cases the small finger can be converted to act as a thumb in opposition to the ring finger, which would be a surrogate index finger. He correctly comments that any procedure that will improve function in these severely handicapped children is justified. I agree that it is essential to use to the child's best advantage the anatomical parts available even if it means changing function from the radial to the ulnar side of the hand. Harrison's paper gives the operative details of one case and emphasizes that it is a preliminary report. However, in a recent letter he told me that he has now done the procedure several times and that the results are such that he would not hesitate to recommend it to any suitable patient.

CONCLUSIONS ON TREATMENT OF APLASIA OF THE RADIUS

I have deliberately chosen to end this chapter with a section headed with the title used by Heikel at the end of his excellent monograph. Much of his work was devoted to a critical analysis of the use of a fibular graft to support the hand. I have not reproduced his condemnation of this procedure because it has been generally abandoned. I believe that time has proven the validity of the other main points in his conclusions and that they provide a good basis for a discussion of the present state of the art.

HEIKEL'S CONCLUSIONS

1. Correction of the position of the hand in relation to the ulna without stabilization of the corrected position cannot lead to a permanent improvement.
2. Correction of the position by conservative treatment alone will not succeed.
3. A permanent correction of the position will be achieved if bony ankylosis between the ulna and the carpus or the metacarpals is produced. With incomplete ankylosis or where only the proximal row of carpal bones become fused with the ulna, a partial recurrence may occur through increasing deviation in the intercarpal or carpometacarpal joints.
4. No ankylosis between the ulna and the carpus or metacarpus can be achieved, at any rate as far as my series indicates, after operation carried out on growing infants, without complete or partial destruction of the distal growth zone of the ulna, with resultant arrest or disturbed growth in length of the latter.
5. The function of the fingers is comparatively little, if at all, affected by correction of the wrist. Whether a change in their function can be expected may be ascertained preoperatively by comparison of the digital function in the free and the passively corrected positions of the hand.
6. The function of the extremity as a whole is generally reduced as compared with that of the untreated cases, owing to the fact that loss of mobility of the wrist is not compensated by improved digital function.
7. The appearance of the extremity improves (and may probably be further improved by means of correction osteotomy of the ulna, which was not attempted in my series). But even the cosmetic result is less good in cases operated on at an early age, because the growth in length of the ulna is arrested or reduced, which results in an extreme difference in length between the operated and untreated limbs.
8. As is seen from points 3 to 7, a method of treatment which stabilizes the hand in a corrected position by means of ankylosis between the ulna and the carpus or metacarpus is justified only from a cosmetic and psychological standpoint, while it is to be condemned from a functional point of view. It should be applied only to cases where a reduction in the function of the anomalous hand is of no significance, viz., if the other hand is normal or good enough to enable the patient to manage with this hand alone. In bilateral cases of total or partial aplasia of the radius it should not be attempted. When operation is performed on cosmetic indications, it should be delayed until the longitudinal growth of the forearm has, on the whole, ceased. Psychological factors may possibly indicate a slightly earlier treatment.*

I agree entirely with the conclusions in Heikel's first three points. Conservative treatment does not succeed in stabilizing the carpus on the ulna, and stabilization has to be thorough to be effective. In my own follow-up studies I have found partial recurrence at the level of the intercarpal and particularly the carpometacarpal joints (Fig. 15-25).

*Heikel, H. V. A.: Aplasia and hypoplasia of the radius, Acta Orthop. Scand. (Suppl.) **39:** 1, 1959.

A

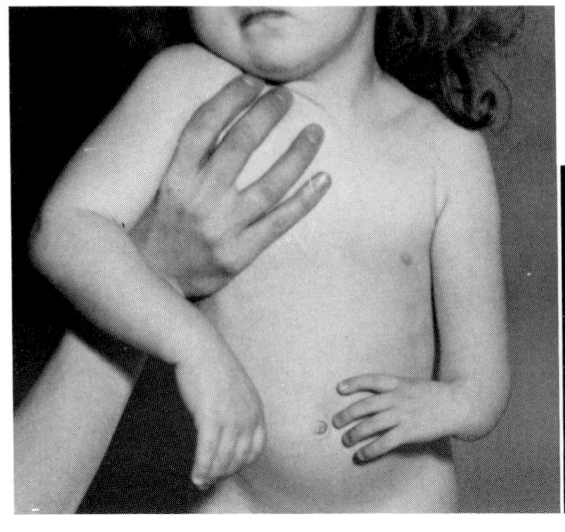

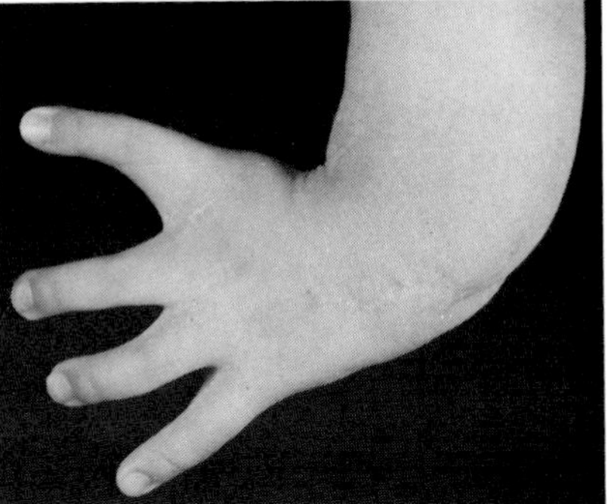

B

Fig. 15-25. *Recurrent radial deviation.* **A,** This child was operated upon at the age of 3 years; the intermetacarpal ligament between index and long finger metacarpals was resected, and the ulna was implanted in the carpus. **B,** Six years later the space between the index and long finger was considerably wider and some radial deviation had recurred at the level of the carpometacarpal joints. No tendon transfers were done in this hand.

On the whole the additional motion caused by the recurrence does not cause loss of function although it does detract from the appearance of the limb. In his fourth point Heikel discusses the loss of length in the limb because of early closure of the distal ulnar epiphysis. Premature arrest has certainly occurred in some of my patients, and I am sure that the same problem has been seen by all surgeons practicing implantation of the ulna. I am not persuaded that occasional early epiphyseal arrest in an area of unpredictable growth potential is too high a price to pay for satisfactory hand stability.

Stability of the hand provides the greatest potential for digital movement, and early stabilization should allow the maximum development of grasp in partially stiff hands. Unfortunately, there are no figures to convince the purist that this does happen. Heikel, in his fifth point, discounts the possibility, and Lamb tends to agree with him. Delorme comments on the improvement in fine prehensile activities, and Bora reports measurable increase in grip after wrist stabilization. Some of our cases at the University of Iowa tend to support Bora, but in others we were unable to show any improvement.

In his sixth point Heikel comments on the reduction of limb function as a whole after the wrist is stabilized. This could be extended into a plea for conservative treatment only, particularly in view of the fact that adaptation to the deformity does result in increased ability as the child grows. Delorme has pointed out that this condition is sufficiently uncommon and sufficiently demanding of surgical treatment that the experience of any one surgeon rarely includes matching lesions in which the results of conservative and surgical management can be compared. An excellent record system devised by Steindler and continued by his

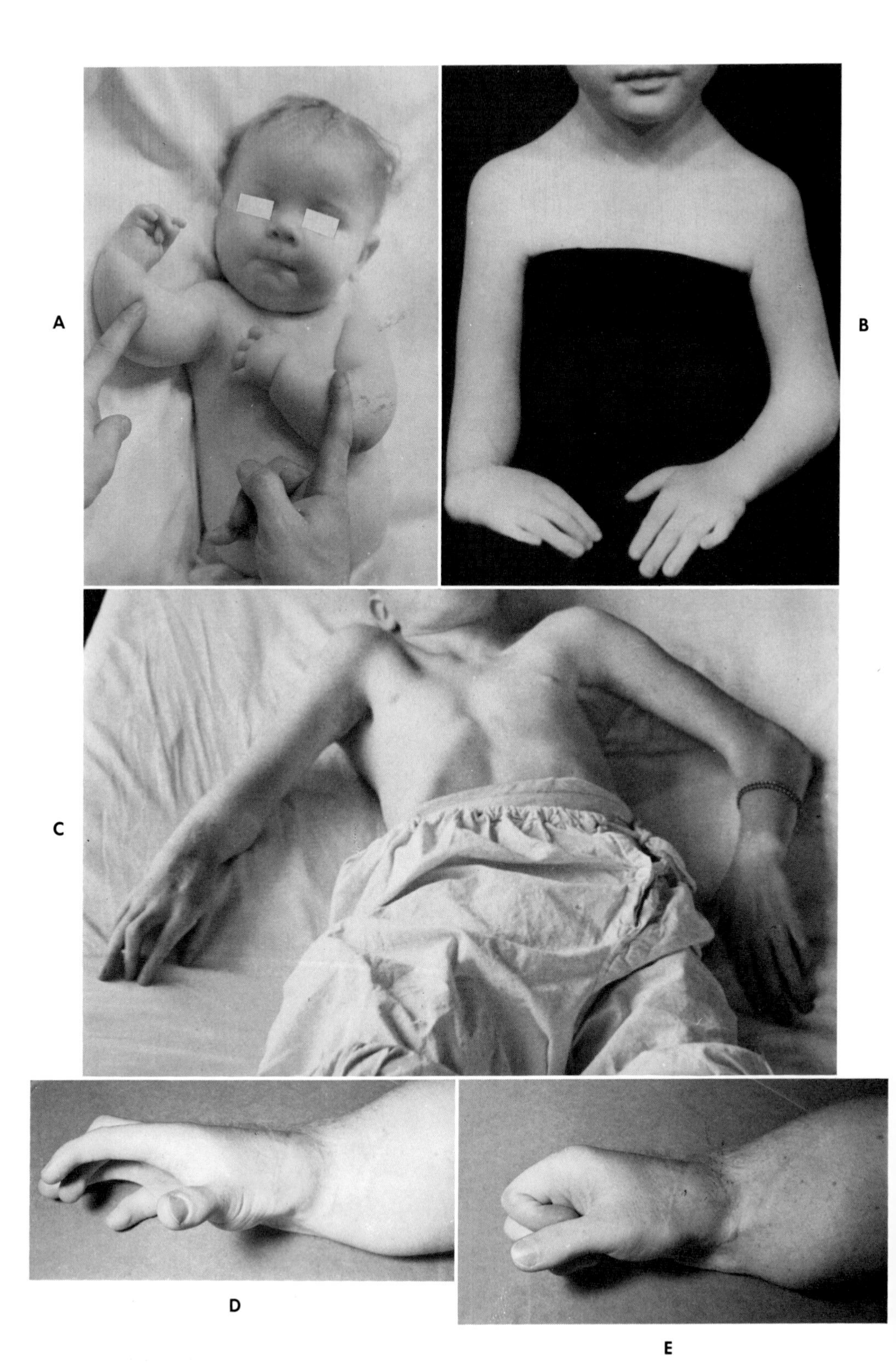

Fig. 15-26. ***Ulna implantation—long-term result.*** **A,** At birth. **B,** Age 6 years, prior to ulna implantation by Dr. A. Steindler. **C,** Age 13 years. **D** and **E,** In adult life full stabilization and excellent grip have developed.

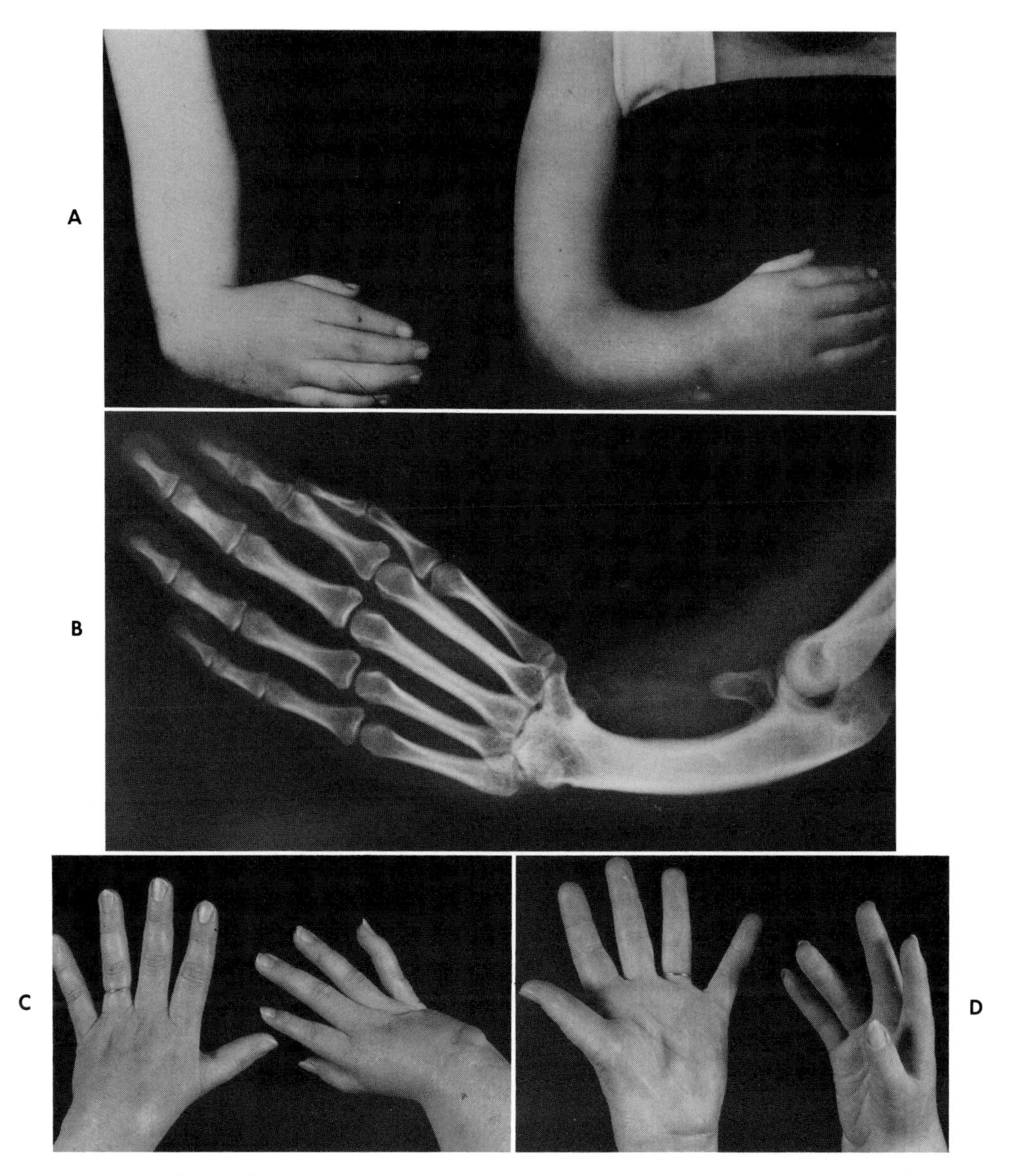

Fig. 15-27. *Ulna implantation—long-term result.* Unilateral implantation patient followed into adulthood. **A,** Before and after implantation at age 5 years (the operation was done 43 years ago). **B,** X-ray film 12 years after implantation showing vestigial proximal radius and solid union of carpus and ulna. **C** and **D,** Recent photographs showing hypoplasia, compared with the normal hand, and slight recurrence of radial deviation.

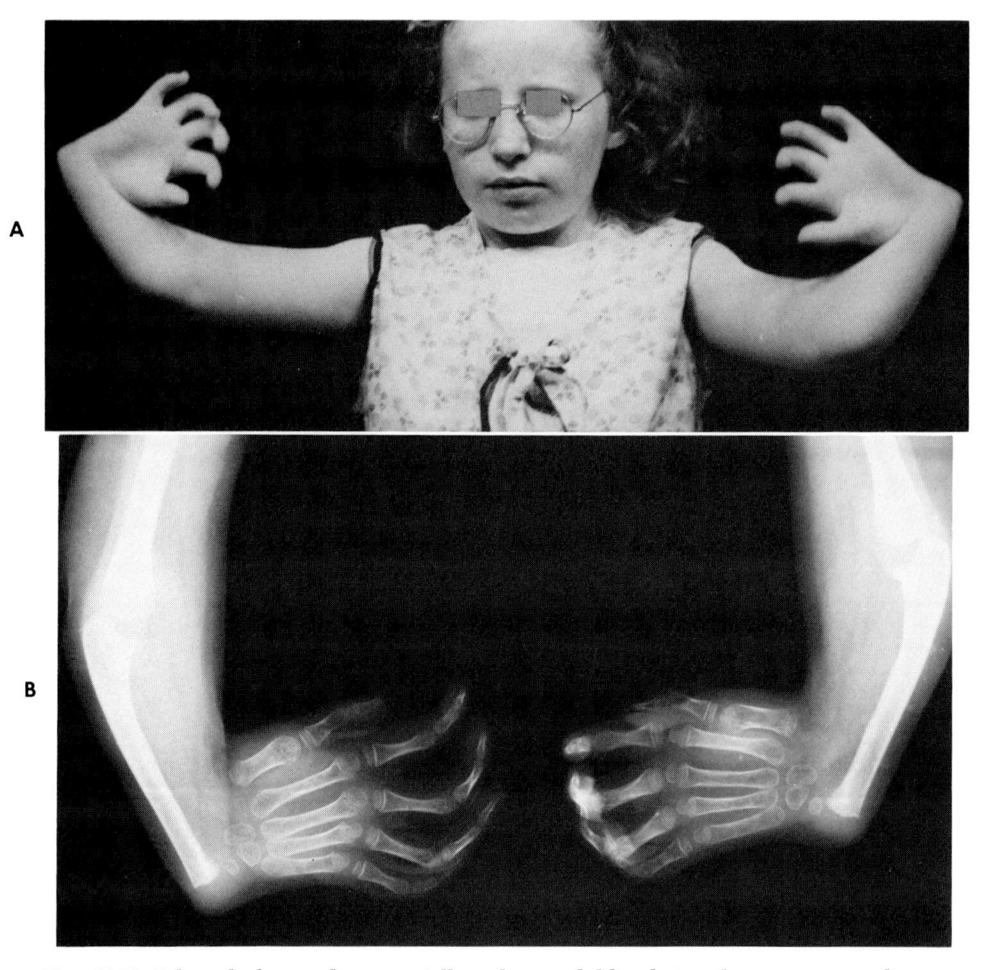

Fig. 15-28. *Bilateral ulna implantation followed into adulthood.* **A** and **B**, Prior to implantation at age 10 years (done 21 years ago). **C** and **D**, Result at age 15 years. This patient was seen recently but declined to be photographed; her stabilization remains sound.

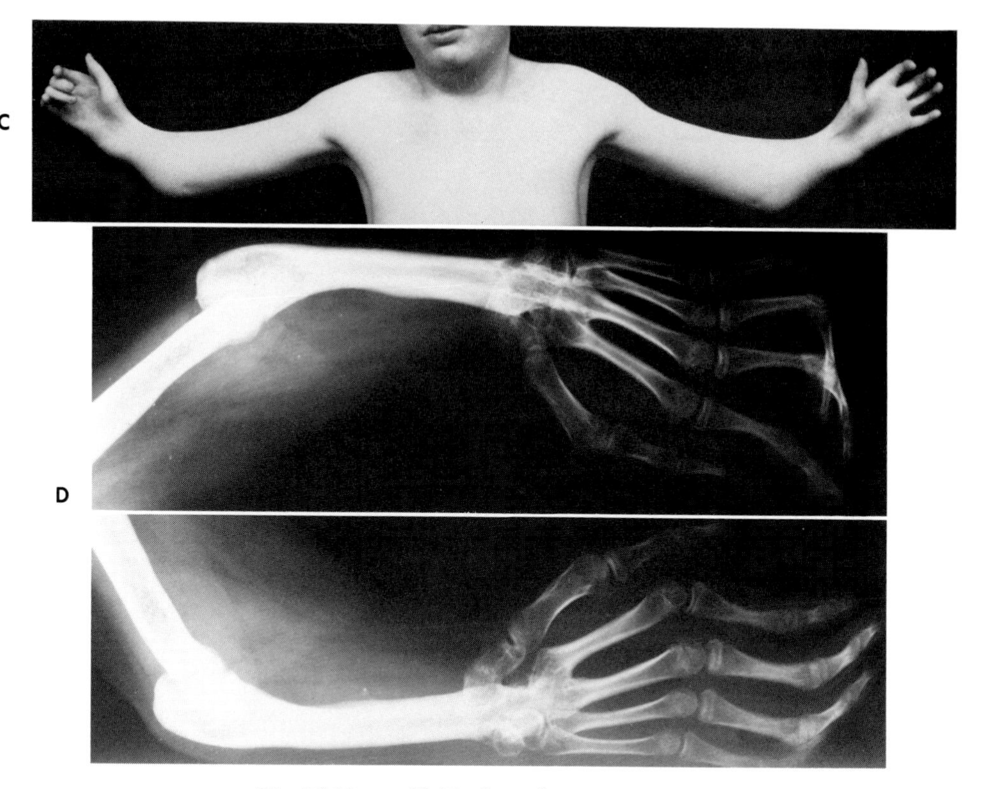

Fig. 15-28, cont'd. For legend see opposite page.

successors has enabled us to follow some long-term results of both conservative and operative treatment. Representative results are shown in Figs. 15-26, 15-27, and 15-28.

I am in no doubt that surgical stabilization of the wrist has produced a happier individual with at least as much dexterity as the unoperated patient. It is impossible to measure attitudes with any accuracy, and I can only report that I believe the brave front put on by the unoperated individual routinely conceals a sad soul who resents his very visible deformity. I will continue to recommend surgical stabilization for all suitable patients.

CHAPTER 16

ULNAR CLUBHAND

Congenital defects of the postaxial border of the forearm and hand are considerably less common than central or radial preaxial deformities. These defects have a low incidence when all congenital deformities are considered. In our University of Iowa series of over 1,475 patients there are only 22 patients with postaxial or ulnar hypoplasia compared with 81 radial clubhands. Thus the proportion of ulnar to radial clubhands is 1 to 3.6, which contrasts with the usually quoted ratio of 1 to 10.

ASSOCIATED ABNORMALITIES

Most cases of ulna dysmelia are sporadic, but some cases have occurred in families in which there are accumulations of joint lesions. Cleft hand has been associated as a mendelian dominant trait, and involvement of the fibula, a homologous structure, has been recorded. These ulnar-fibular defects may be combined with sternal, dental, facial, or renal anomalies. These and other anomalies do not occur as frequently as in radial dysplasia. In our patients we have found clubfoot, absence of the proximal fibula, spina bifida, hypoplasias of the lower extremity, and bilateral patellar absence. Cardiac, pulmonary, and gastrointestinal anomalies do not seem to occur to the same extent as in radial hypoplasia. Associated anomalies such as neurofibromatosis, scoliosis, congenital dislocation of the hip, flexion contracture of the elbow, clinodactyly of the small finger, and Cornelia de Lange syndrome have all been reported.

CLINICAL TYPES

Although the whole spectrum of maldevelopment can occur, from slight hypoplasia of ulnar digits to total absence of the ulna, the common form is incomplete development of the distal portion of the ulna and a varying absence of the ulnar carpal, metacarpal, and digital elements.

In the fully developed case the forearm is shortened and the radius is shorter than normal and has a concavity toward the ulnar side. Ulnar deviation of the hand is associated with slowed growth of the ulnar half of the distal radial epiphysis. The thumb and index finger are usually normal, but the small, ring, and long fingers are frequently absent. Syndactyly is common between those digits that are present. The small and ring finger metacarpals are most commonly missing, and among the carpal bones the frequency of absence starts with the pisiform and progresses through the lunate, the triquetrum, and then the capitate. Carpal fusions occur quite commonly. Internal rotation deformity of the humerus is also associated with ulna dysmelia.

The skeletal changes that can be seen on the x-ray films of all but the very young show a consistent pattern. In all patients the radius shows slight bowing and a tilt of the distal articular surface. The distal portion of the ulna is absent, and very rarely the whole ulna is missing. The carpal bones frequently show anatomical anomalies. The incidence of anomalies in the x-ray films of 11 out of 20 of the patients we studied is shown in Table 16-1.

The main functional disability is associated with the instability of the radial head at the elbow and the ulnar tilt of the hand on the distal radius. Kümmel distinguished three clinical types of ulna hypoplasia on the basis of varying radial-humeral relationships:

Type 1. Normal radial-humeral joint. The radius is normal although slightly curved (Fig. 16-1).

Type 2. Bony ankylosis between radius and humerus. If the proximal segment of the ulna is present, it is fused to the radius or to the humerus or to both. Marked bowing of the radius and a wedge-shaped distal radial epiphysis is usual (Fig. 16-2).

Type 3. Proximal radius dislocated on the humerus. The radius is bowed, and the distal epiphysis is tilted in an ulnar direction (Fig. 16-3).*

Ogden, Kirk Watson, and Bohne have commented that this classification focuses on one area of involvement and does not include the many other defects

*Kümmell, W.: Die Missbildungen der Extremitäten durch Defekt, Verwachsung und Ueberzahl, Heft 3, Cassel, 1895, Bibliotheca Medica.

Table 16-1. Carpal anomalies in ulnar clubhand

	Percentage
Pisiform absent	100
Hamate absent	75
Trapezium, trapezoid, and scaphoid fused	75
Triquetrum absent	50
Capitate and lunate fused	50
Trapezoid absent	25
Capitate, lunate, and triquetrum fused	25
Capitate and hamate fused	25
Trapezium, scaphoid, and lunate fused	25

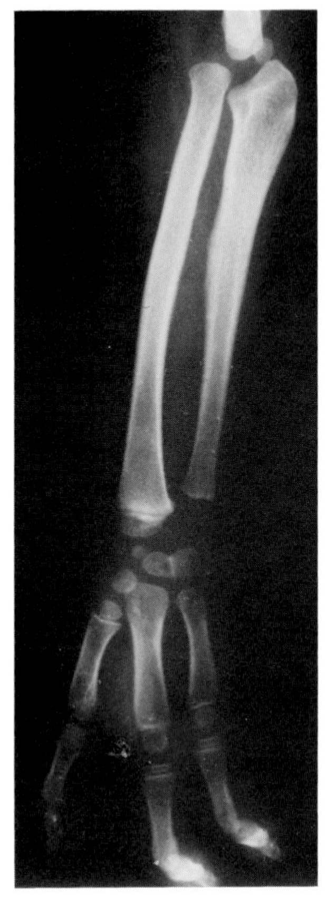

Fig. 16-1

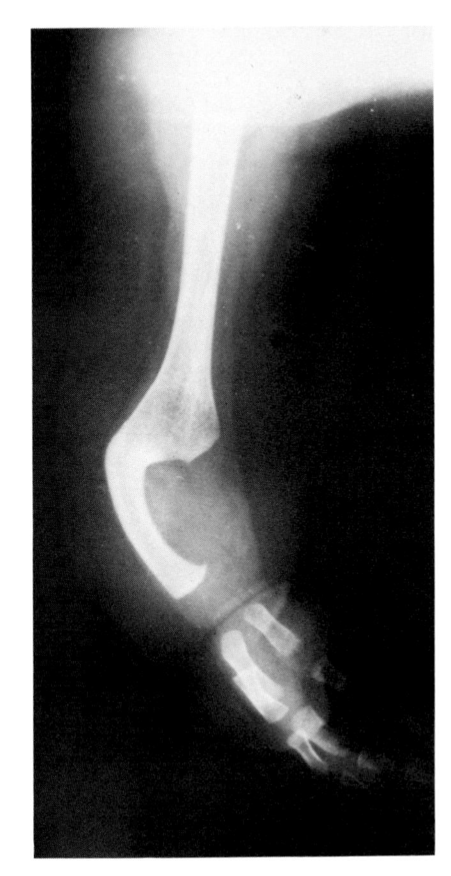

Fig. 16-2

Fig. 16-1. *Ulna hypoplasia—Type 1.* There is a normal radiohumeral joint, and the radius is virtually normal.
Fig. 16-2. *Ulna hypoplasia—Type 2.* There is bony ankylosis between radius and humerus, and marked bowing of the radius has already occurred. (From Pardini, A. G.: Congenital absence of the ulna, J. Iowa Med. Soc. **57**:1106-1112, 1967.)

that are frequently present. They have divided the 11 patients they studied into three groups based on radiological findings: Type I—hypoplasia of an otherwise complete ulna that has a distal epiphysis; Type II—pathological aplasia (absence of distal ulna including the distal epiphysis); and Type III—total aplasia. I prefer this radiological classification because I believe it to be of more value when planning treatment.

FIBROCARTILAGINOUS ANLAGE

At the wrist the distal end of the radius is of normal size and gives stable support to the carpus. Ulnar deviation of the hand does occur but is not as marked as in the radial anomaly, and only rarely does the angulation reach as much as 90 degrees. Straub first reported and Riordan, Mills, and Alldredge later

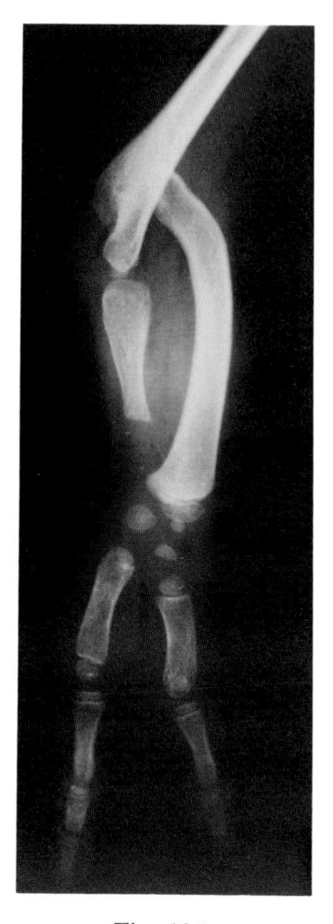

Fig. 16-3

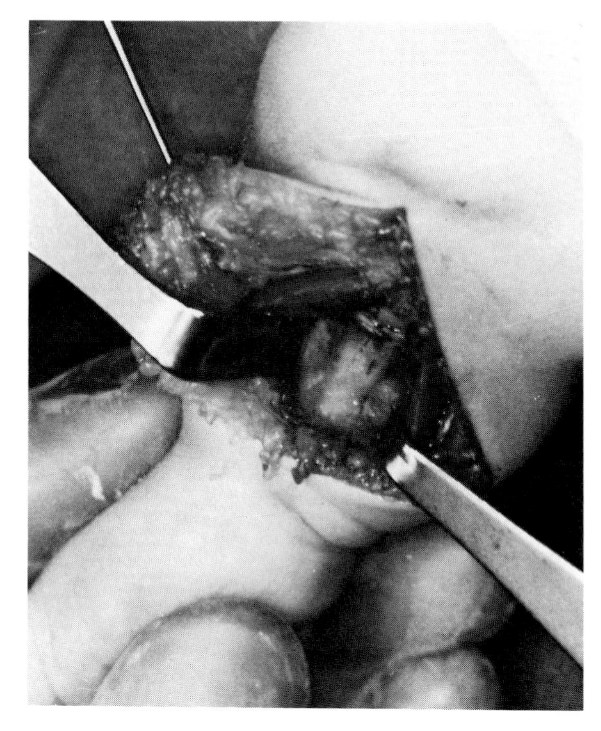

Fig. 16-4

Fig. 16-3. *Ulna hypoplasia—Type 3.* The head of the radius is dislocated, and it is bowed. The proximal ulna is present. (From Pardini, A. G.: Congenital absence of the ulna, J. Iowa Med. Soc. **57**:1106-1112, 1967.)

Fig. 16-4. *Fibrocartilaginous anlage.* The anlage is a hypoplastic, nongrowing fibrocartilaginous band that acts as a fixation point for the ulnar aspect of the carpal bones and thereby inhibits growth of the ulnar half of the distal radial epiphysis. (From Ogden, J. A., Watson, H. K., and Bohue, W.: Ulnar dysmelia, J. Bone Joint Surg. [Am.] **58**:467-475, 1976.)

emphasized the role played by the anlage of the ulna in tilting the distal articular surface of the radius toward the ulnar side. The anlage is a hypoplastic, nongrowing fibrocartilaginous band that acts as a fixation point of the ulnar aspect of the carpal bones and thereby slows down the growth of the ulnar half of the distal radial epiphysis (Fig. 16-4).

This anlage is the cause of the gradual inclination of the hand to the ulnar side. It is also a cause for the gradual bowing of the radius with the convexity to the radial side (Fig. 16-5). The head of the radius may be fused to the distal humerus, but more commonly it articulates with the humerus at birth. As the distal epiphysis grows, the fibrous anlage acts as a tether and the radial head dis-

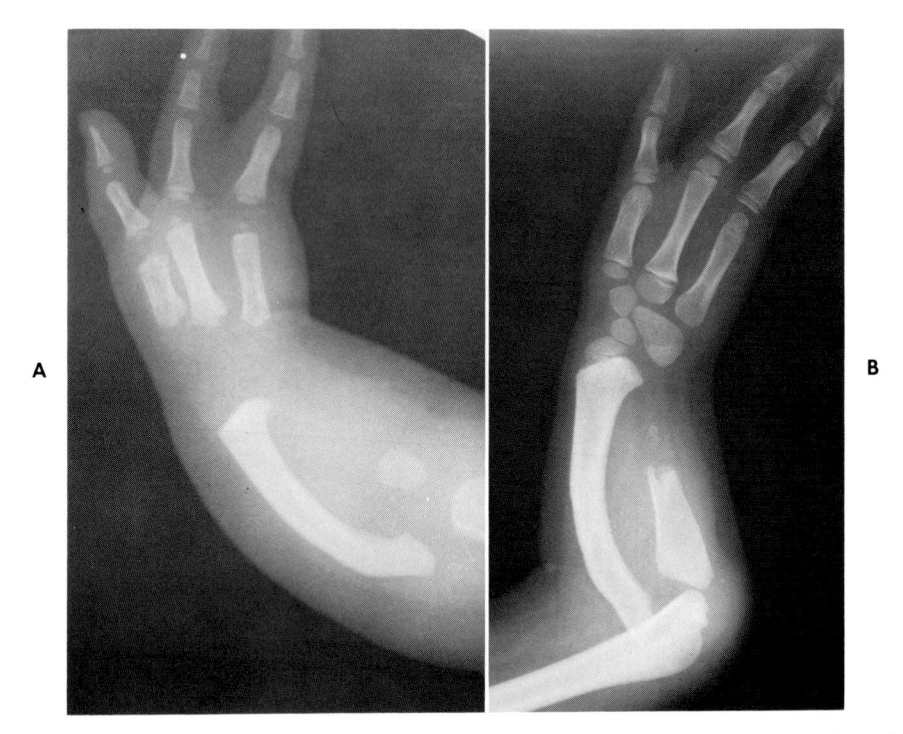

Fig. 16-5. *Curving of the radius.* **A,** Presence of an anlage can be suspected when the radius is curved early in life. **B,** Its presence can be confirmed when the bowing increases and an ossification center appears. A small center can be seen between the distal end of the ulna and the large carpal bone.

locates. Eventually the radius crosses the humerus so that the head lies subcutaneously on the outer side of the arm.

PROBLEMS IN TREATMENT

Two major questions need to be answered in planning treatment for these children. One is whether the fibrocartilaginous anlage is present and is a deforming factor. The other is what would be the functional advantages and disadvantages of fusing the radius to the proximal ulna and creating a single bone forearm.

The anlage

Ossified portions of the ulna may show on an x-ray film at birth, but the fibrocartilaginous anlage will not show although an occasional ossification center may be present (Fig. 16-5, *B*). The anlage links the humerus and the carpus and can readily be palpated beneath the skin. It arises from the humerus whether or not there is an elbow joint and attaches beyond the distal radius to the carpus. Riordan has likened it to a fiberglass fishing rod—it can bend laterally but does not yield in the longitudinal direction. Its presence is shown if there is a resistance to passive full correction of the deviated hand.

In infants, serial plaster casts should be used to try to obtain full correction of the hand on the forearm. Usually one finds good early correction but an almost elastic-like resistance to full correction. Certainly radial deviation of the hand is prevented by the anlage and its attachment to the ulnar aspect of the carpal bones.

It follows therefore that early surgical removal of the fibrocartilaginous tie string would allow the distal radial epiphysis to grow evenly and unhindered. It should lessen the degree of radial bowing and could also prevent or significantly reduce the proximal dislocation of the head of the radius.

I have treated few of these infants but believe this to be the logical and correct method of treatment. I agree with Riordan that it should certainly be done before the infant is 6 months old because by this time most patients will show early dislocation of the head of the radius.

Forearm stability

The proximal portion of the ulna will usually ossify and articulate with the humerus, making a reasonably stable elbow joint. After resection of the ulnar anlage the head of the radius is likely to remain in its normal relationship to the humerus and provide some proximal stability for the radius. Thus the proximal forearm has a stable relationship to the humerus but the distal forearm lacks stability between the two forearm bones.

Many consider that cross union should be established between the two bones to provide in effect a single bone forearm. I have reviewed our patients with ulnar hypoplasia and have been impressed by the versatility of function in those who have not had a single bone forearm made. In the mid 1920s Southwood wrote, "From the functional viewpoint the deformed limb is much more useful than its anatomical condition would have led one to expect."* Fifty years later I feel that this statement still has merit. Certainly I believe that the syndactyly between the digits should be separated, but in almost all our patients the motion in elbow, forearm, and wrist is so good that I do not feel surgery could produce a significant increase in function.

The subcutaneous protrusion of the head of the radius is unsightly, and its removal could be justified on cosmetic grounds; but if the removal is not also accompanied by ulnar-radial fusion, significant instability in the distal forearm can subsequently develop.

It is rash to generalize from one case, but I am particularly impressed by my patient shown in Fig. 16-6. She exhibited such excellent function as a teenager that I did not recommend surgery. Subsequently, she graduated as a physician's assistant and now flies her own plane; if I had made a single bone forearm, her lack of pronation and supination would have been a significant handicap in a small plane cockpit.

Not all people with ulnar hypoplasia and radial head dislocation will attempt

*Southwood, A. R.: Partial absence of ulna and associated structures, J. Anat. **61**:346-351, 1926-1927.

Fig. 16-6. *Functional ability.* This teenage patient showed such excellent function that no surgery was advised.

to fly small planes, but all those who have a single bone forearm created surgically will be forced to provide pronation and supination by a combination of shoulder and spinal movements. I do not believe surgical cross union should be done routinely; particularly I do not think it suitable for infants in whom the surgical excision of the anlage is done in the first 6 months of life because the head of the radius still remains in good position.

If the excision of the distal fibrous ulna is done later in life, the head of the radius is usually already dislocated and the proximal radius can also be resected. In such patients a better case can be made for a single bone forearm since fusion of the proximal ulna to the distal radius provides stability for each bone and therefore the wrist and hand. Many of these patients also have only a limited pronation and supination, and their handicap is not greatly increased by the loss of this limited motion. It is best to fix the hand about midway between pronation and supination.

PLAN OF TREATMENT

For infants

When the child is seen at or shortly after birth, the early treatment should be directed at correcting the ulnar deviation of the hand by use of plaster casts (Fig. 16-7). These casts should be applied in three pieces. First, the hand is en-

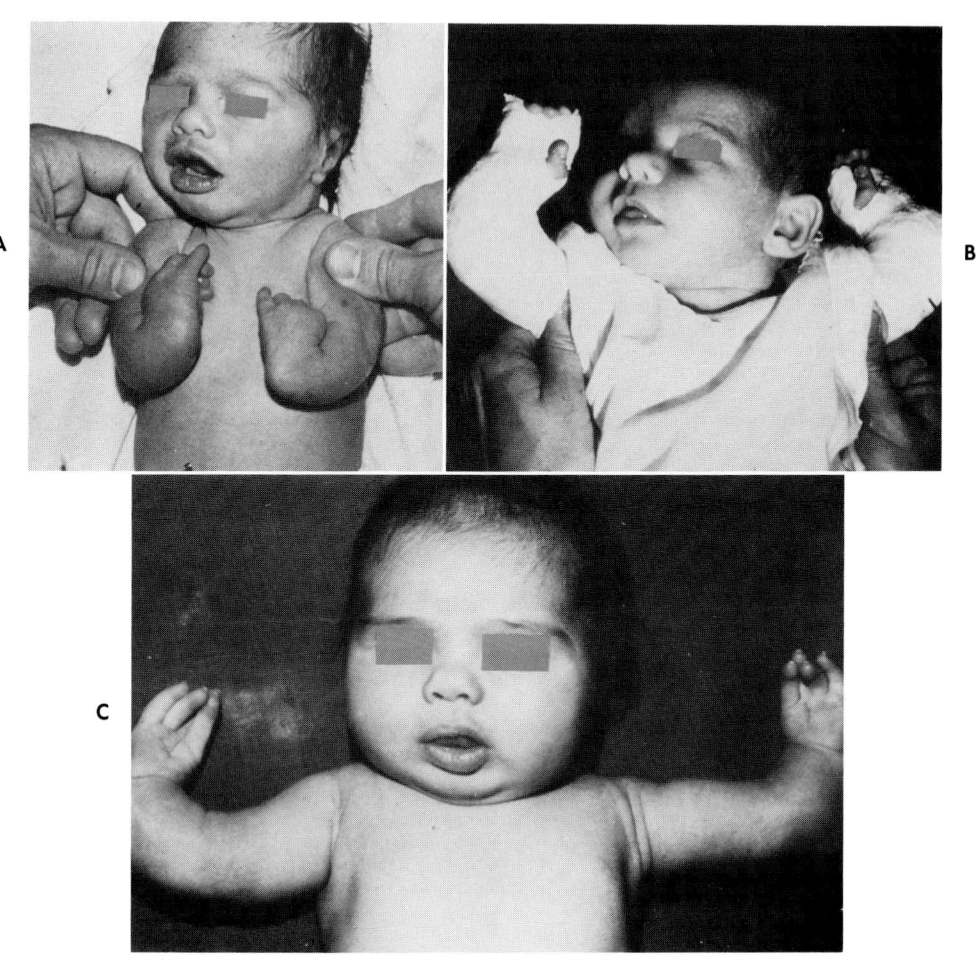

Fig. 16-7. *Plaster cast correction.* **A,** At birth. **B,** A few days after birth, casts correcting the hands on the radius have been applied. **C,** Significant correction can be obtained if casting is started very early in life. (Patient of Dr. Daniel C. Riordan.)

cased; this is then joined to a forearm cast to get the corrected hand-forearm relationship. Finally, the elbow is bent to at least 90 degrees and the cast completed to the axilla. This final portion is necessary to prevent the hand-forearm section from sliding distally on the infant's chubby limb.

These casts will need to be changed repeatedly to accommodate growth and improvements in the correction obtained. Some physicians would use braces, but I have found these considerably more expensive than plaster of Paris and exceedingly difficult to maintain in the correct position on these tiny limbs.

If the child is thriving at about 6 months of age, the ulnar side of the forearm should be explored and the anlage excised. The whole anlage does not need to be removed. The purpose of the procedure is to remove the block to longitudinal radial growth, and this can be done by excision of the central portion or of the distal half of the fibrocartilaginous band (Fig. 16-8). The approach is

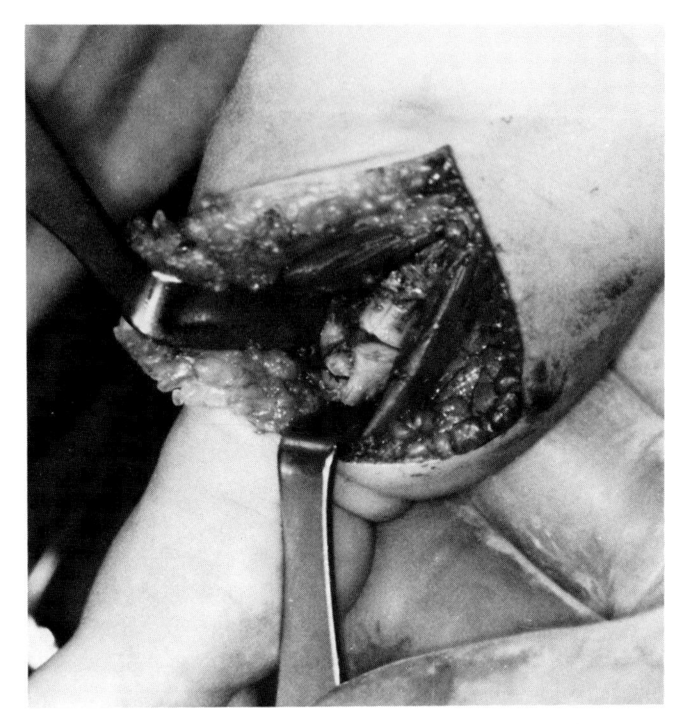

Fig. 16-8. *Resection of anlage.* The whole anlage need not be removed; excision of the central portion, as is shown here, will usually be sufficient to obtain a full correction. (From Ogden, J. A., Watson, H. K., and Bohue, W.: Ulnar dysmelia, J. Bone Joint Surg. [Am.] **58**:467-475, 1976.)

along the postaxial border through a lazy-S type incision, which must be carried across the wrist crease to the midcarpal level. The anatomy is not usually greatly abnormal. The ulnar and median nerves are found in their normal places, as is the ulnar artery. The pisiform is always absent, as is the flexor carpi ulnaris muscle. The hamate and triquetrum are usually absent, and the palmaris longus and the flexor digitorum profundus to the small finger are also often absent. Because of the absence of these extrinsic flexor muscles, the ulnar neurovascular bundle and the anlage lie close together in the subcutaneous tissues. The neurovascular bundle should be freed and protected before the anlage is dissected off its carpal attachment. The proximal level of excision is hard to define—removal of a piece at least one third of the forearm length is desirable, and removal of a greater portion does not significantly improve the longitudinal release.

Provided the head of the radius is still in proper relationship with the humerus, I do not attempt to get cross union between the forearm bones. The soft tissues on the ulnar side of the wrist joint must be incised sufficiently to allow a full correction of the hand on the distal radial articular surface. It is important that the soft tissues be adequately incised; the hand should flop over into neutral or even slight radial deviation. If it has to be pushed over into neutral, further release of soft tissues is necessary.

The wound is closed with No. 6-0 catgut and a well fitting, high above-elbow plaster cast applied. Subsequently, a brace can be made to maintain the correction. If syndactyly is present between the digits, I usually correct it at this operation.

For young children

When the patient is seen for the first time at age 6 to 12 months or even older, the excision of the distal fibrocartilaginous anlage becomes urgent since the earlier it is done the less severe will be the secondary deformities. The surgical technique is the same as for the infant, particularly if the radial head has not dislocated.

If the radial head is already dislocated, a decision has to be made between leaving it untouched and excising it. If there is limited active or passive pronation-supination, then it is easier to justify radial head excision and the creation of a single bone forearm.

When there is good pronation-supination, I find it difficult to justify excision on cosmetic grounds (Fig. 16-9). I usually explain to the parents that our follow-up studies show good function and adaptation in those who have such good forearm motion and suggest a period of observation.

If a single bone forearm is to be made, the incision should be extended in a proximal and posterior direction. The radius should be approached at about its midpoint by lifting the muscles from the interosseus membrane. The level of resection will depend on the length of ulna left after excision of the anlage. Removing the proximal portion of the radius can endanger the posterior interosseus nerve, and it may be wise to extend the incision into the outer side of the arm and positively identify the nerve. Usually the nerve tends to move proximally with the dislocating head of the radius, and it may not be in its normal anatomical site.

Fusion between the two bones is normally easy to obtain if the abutting surfaces are freshed and appropriately shaped to provide good contact. Slotting, end-to-end, or side-to-side junctions are all acceptable; the method most appropriate to the lie of the bones should be used. They should be joined with the hand in neutral pronation-supination. Usually a few small screws are sufficient to hold the hand in position, and occasionally a strong Kirschner wire is adequate (Fig. 16-10).

The fusion site must be protected by an above-elbow plaster cast for a minimum of 2 months, and often a third month of protection is wise.

For those with fixed elbows

In the few patients with Kummel's Type 2 fusion of radius to the humerus, excision of the fibrocartilaginous anlage is still necessary. If it is left as an unyielding tie bar against longitudinal growth, the hand will tilt and the radius bow as growth progresses.

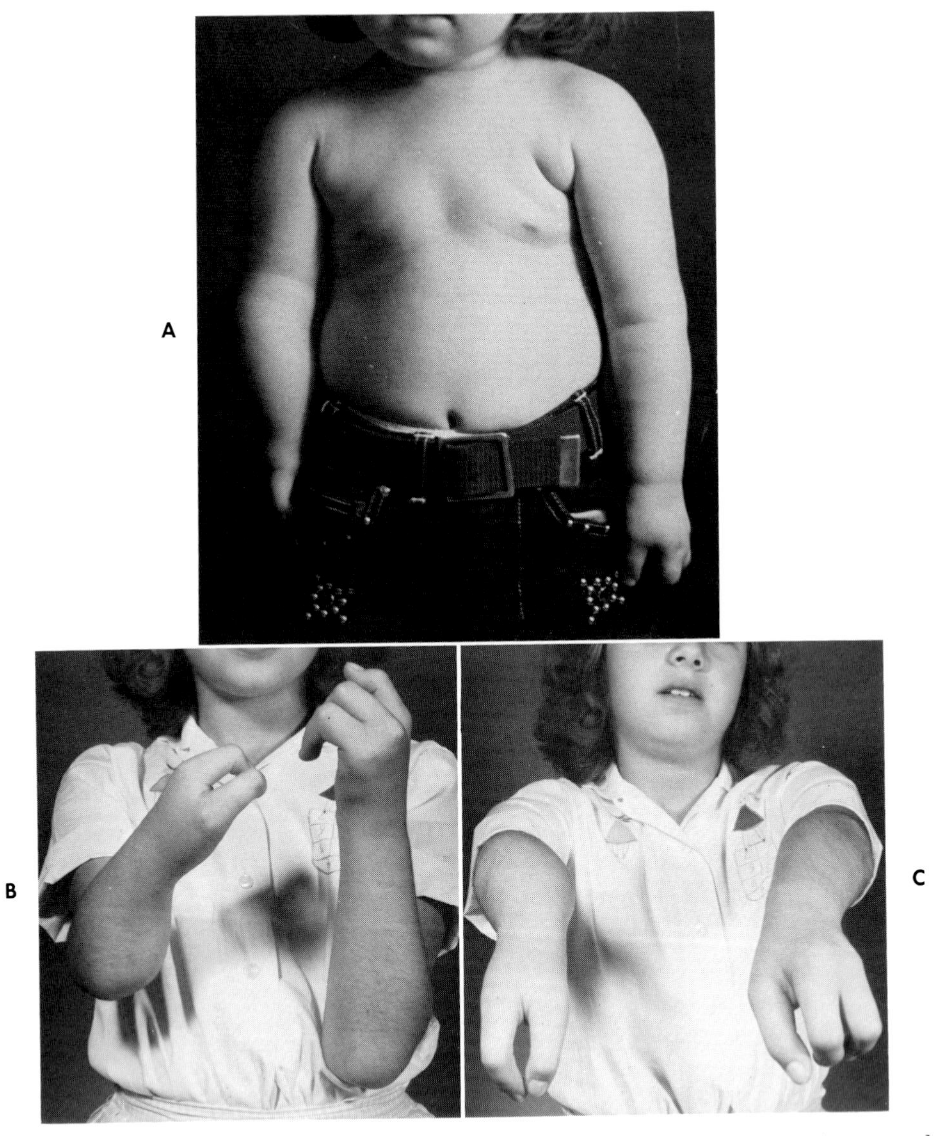

Fig. 16-9. *Bilateral deformity—no treatment.* **A, B,** and **C,** As this child grew up she retained a good range of pronation-supination and developed good hand function. **D** and **E,** X-ray films taken in adult life showing, on the right, the typical dislocated head of radius, curved shaft, and sloping distal articular surface. On the left, the ulna is hypoplastic. Surgery would not have improved her function.

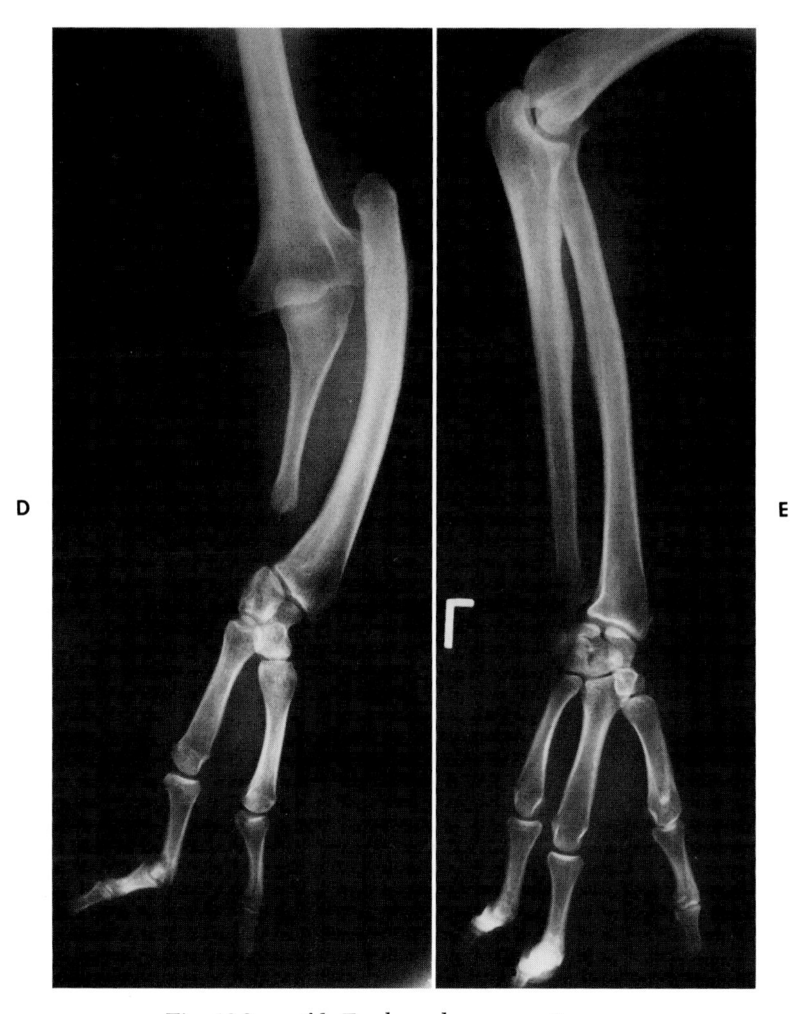

Fig. 16-9, cont'd. For legend see opposite page.

A

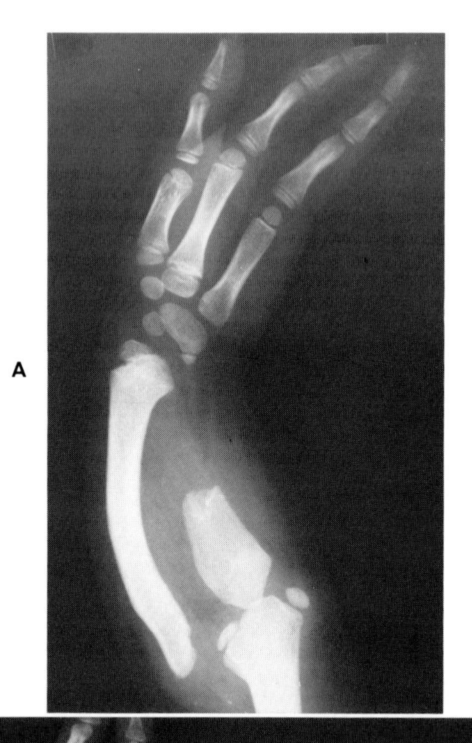

B

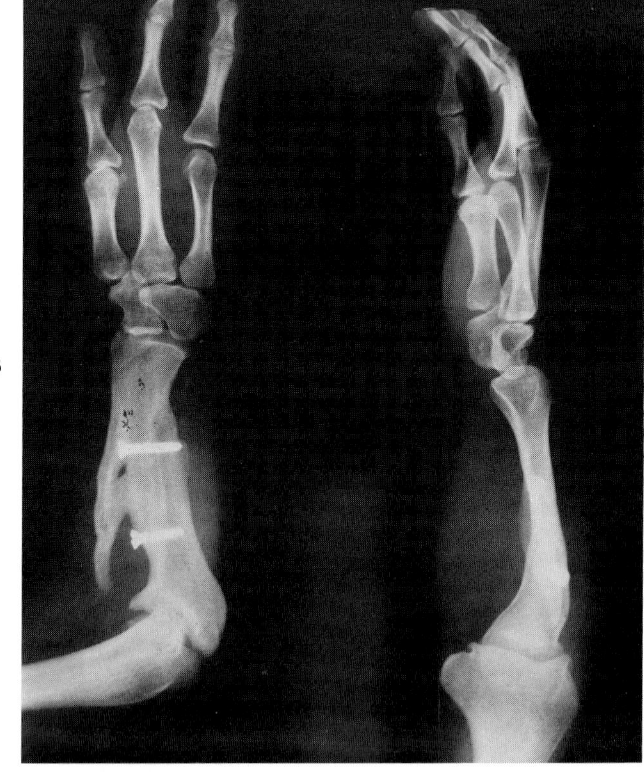

Fig. 16-10. *Single bone forearm.* **A,** In childhood. **B,** A single bone forearm was made when the patient was in late teenage. This girl was distressed by the protuberance of the head of her radius and the relative instability of her wrist. She is pleased with the appearance and function of her forearm after the operation.

For older patients

As the condition progresses with age, three deformities increase. The radial head protrudes, the radius curves, and the hand tilts.

If growth has ceased, a wedge osteotomy to correct the radial bowing and tilt of the hand may be useful. If growth has not ceased, the distal ulnar tie bar must be excised, and this excision can be usefully combined with a wedge osteotomy of the radius.

Unless the protuberance of the radial head is gross, I usually recommend acceptance of the deformity and demonstrate the functional handicap that would be produced by the fusion of the two forearm bones, which would be necessitated by the radial head excision.

In summary I believe the follow-ups of the treatment recommended for infants show that it is worthwhile and helpful. I have not yet convinced myself that a single bone forearm is always a better proposition than pronation-supination coupled with a lumpy elbow. Certainly, I believe few if any adults should have radial head excisions, but they might benefit from wedge osteotomies of their radii.

REFERENCES

I have drawn freely on the writings of many surgeons and have selected in these references work that has been particularly helpful to me. For each chapter I have included papers that are cited or that have direct bearing on the chapter content.

Several books that have been of special help to me and that I believe should be read by all surgeons operating on the child with a congenitally malformed hand are:

Birth Defects, Original Article Series, vol. 5, no. 3. In Bergsma, D., editor: The clinical delineation of birth defects, part III, Limb malformations, New York, 1969, National Foundation-March of Dimes.

Kelikian, H.: Congenital deformities of the hand and forearm, Philadelphia, 1974, W. B. Saunders Co.

Poznanski, A. K.: The hand in radiologic diagnosis, Philadelphia, 1974, W. B. Saunders Co.

Wynne-Davies, R.: Heritable disorders in orthopaedic practice, Oxford, 1973, Blackwell Scientific Publications, Ltd.

Chapter 1/The role of reconstructive surgery

Edgerton, M. T., Snyder, G. B., and Webb, W. L.: Surgical treatment of congenital thumb deformities (including psychological impact of correction), J. Bone Joint Surg. (Am.) **47**:1453-1474, 1965.

Chapter 2/Genetics and inheritance

Bell, J.: On brachydactyly and symphalangism. In Penrose, L. S., editor: The treasury of human inheritance, vol. 5, part 1, Cambridge, 1951, Cambridge University Press, pp. 1-31.

Edwards, J. H.: Familial predisposition in man, Br. Med. Bull. **25**:58, 1969.

Falconer, D. S.: The inheritance of liability to certain diseases estimated from the incidence among relatives, Ann. Human Genet. **29**:51, 1967.

McKusick, V. A.: Mendelian inheritance to man: catalogs of autosomal dominant, autosomal recessive and x-linked phenotypes, ed. 3, Baltimore, 1971, The Johns Hopkins University Press.

Millen, J. W.: Timing of human congenital malformations with a timetable of human development, Dev. Med. Child Neurol. **5**:343, 1963.

Neel, J. V.: Some genetic aspects of congenital defects in congenital malformations. Fishbein, M., editor: First International Conference on Congenital Malformations, Philadelphia, 1961, J. B. Lippincott Co., pp. 63-69.

Poznanski, A. K.: The hand in radiologic diagnosis, Philadelphia, 1974, W. B. Saunders Co.

Temtamy, S., and McKusick, V. A.: Synopsis of hand malformations with particular emphasis on genetic factors. In Bergsma, D., editor: The clinical delineation of birth defects. III. Limb malformations, New York, 1969, National Foundation-March of Dimes, pp. 125-164.

Tjio, J. H., and Levan, A.: The chromosome number of man, Hereditas **41**:1, 1956.

Wynne-Davies, R.: Genetics and malformations of the hand, The Hand **3**:184-192, 1971.

Zwilling, E.: Limb morphogenesis. In Abercrombi, M., and Brachet, J., editors: Advances in morphogenesis, vol. 1, New York, 1961, Academic Press, Inc., pp. 301-330.

Chapter 3/Growth, size, and function

Feingold, M.: Clinical evaluation of the child with skeletal dysplasia, Orthop. Clin. North Am. **7**:291-301, 1976.

Flatt, A. E.: Kinesiology of the hand. In the American Academy of Orthopaedic Surgeons: Instructional course lectures, vol. 18, St. Louis, 1961, The C. V. Mosby Co., pp. 266-281.

Hajnis, K.: The dynamics of hand growth since the birth till 18 years of age, Panminerva Med. **11**:123-132, 1969.

Joseph, M. D., and Meadow, S. R.: The metacarpal index of infants, Arch. Dis. Child. **44**:515-516, 1969.

Keller, A. D., Taylor, C. L., and Zahn, V.: Studies to determine the functional requirements for hand and arm prosthesis, 1947, Department of Engineering, University of California at Los Angeles.

Kuczynski, K.: Development of the hand and some anatomical anomalies, The Hand **4**:1-19, 1972.

Landsmeer, J. M. F.: Studies in the anatomy of articulation. II. Patterns of movement of biomuscular biarticular systems, Acta Morphol. Neerl. Scand. **3**:304-321, 1961.

Landsmeer, J. M. F.: Power grip and precision handling, Ann. Rheum. Dis. **21**:164-170, 1962.

Littler, J. W.: On the adaptability of man's hand, The Hand **5**:187-191, 1973.

Marmor, L.: Hand surgery in rheumatoid arthritis, Arthritis Rheum. **5**:419-424, 1962.

Napier, J. R.: The prehensile movements of the human hand, J. Bone Joint Surg. (Br.) **38**:902-913, 1956.

Poznanski, A. K., Garn, S. M., and Holt, J. F.: The thumb in the congenital malformation syndromes, Pediatr. Radiol. **100**: 115-129, 1971.

Skerik, S. K., Weiss, M. W., and Flatt, A. E.: Functional evaluation of congenital hand anomalies, part I, Am. J. Occup. Ther. **25**:98-104, 1971.

Stetson, R. H., and McDill, J. A.: Mechanism of different types of movements, Psychol. Monogr. **32**(3):18, 1923.

Taylor, C., and Schwartz, R. J.: The anatomy and mechanics of the human hand, Artif. Limbs **2**:22-35, 1955.

Weiss, M. W., and Flatt, A. E.: Functional evaluation of the congenitally anomalous hand, part II, Am. J. Occup. Ther. **25**: 139-143, 1971.

Chapter 4/Classification and incidence

Flatt, A. E.: A test of a classification of congenital anomalies of the upper extremity, Surg. Clin. North Am. **50**:509, 1970.

Frantz, C. H., and O'Rahilly, R.: Congenital skeletal limb deficiencies, J. Bone Joint Surg. (Am.) **43**:1202, 1961.

Hall, C. B., Brooks, M. B., and Dennis, J. F.: Congenital skeletal deficiencies of the extremities: classification and fundamentals of treatment, J.A.M.A. **180**:590, 1962.

International Society for Prosthetics and Orthotics, Working Group: The proposed international terminology of classification of congenital limb deficiencies. London, 1975, Spastics International Medical Publications in association with William Heinemann Medical Books Ltd., and J. B. Lippincott Co.

Kay, H. W.: A proposed international terminology for the classification of congenital limb deficiencies, Orthop. Pros. Appl. J. **28**:33-48, 1974.

Patterson, T. J. S.: Congenital deformities of the hand. Hunterian Lecture delivered at Royal College of Surgeons of England, April, 1959.

Saint-Hilaire, I. G.: Histoire générale et particulière des anomalies de l'organisation chez d'homme et les animaux . . . , Avec Atlas, 4 volumes, Paris, 1832-1837, Editions J-B Baillière.

Swanson, A. B.: A classification for congenital limb malformations, J. Hand Surg. **1**:8-22, 1976.

Swanson, A. B., Barsky, A. J., and Entin, M. A.: Classification of limb malformations on the basis of embryological failures, Surg. Clin. North Am. **48**:1169, 1968.

Chapter 5/The inadequate thumb

Birch-Jensen, A.: Congenital deformities of the upper extremities, Odense, Denmark, 1949, Andelsbogtrykkeriet.

Breitenbecher, J. K.: Hereditary shortness of thumbs, J. Hered. **14**:15-22, 1923.

Davison, E. P.: Congenital hypoplasia of the carpal scaphoid bone, J. Bone Joint Surg. (Br.) **44**:816-827, 1962.

Durham, J. M., and Meggitt, B. F.: Trigger

thumbs in children, J. Bone Joint Surg. (Br.) **56**:153, 1974.

Edgerton, M. T., Snyder, G. B., and Webb, W. L.: Surgical treatment of congenital thumb deformities (including psychological impact of correction), J. Bone Joint Surg. (Am.) **47**:1453-1474, 1965.

Fahey, J. J., and Bollinger, J. A.: Trigger-finger in adults and children, J. Bone Joint Surg. (Am.) **36**:1200, 1954.

Flatt, A. E., and Wood, V.: Multiple dorsal rotation flaps from the hand for thumb web contractures, Plast. Reconstr. Surg. **45**:258-262, 1970.

Harris, H., and Joseph, J.: Variation in extension of the metacarpo-phalangeal and interphalangeal joints of the thumb, J. Bone Joint Surg. (Br.) **31**:547-559, 1949.

Holt, M., and Oram, S.: Familial heart disease with skeletal malformations, Br. Heart J. **22**:236-242, 1960.

Joseph, J.: Further studies of the metacarpophalangeal and interphalangeal joints of the thumb, J. Anat. **85**:221-229, 1951.

Lewis, K. B., Bruce, R. A., Baum, D., and Motulsky, A. G.: The upper limb-cardiovascular syndrome, J.A.M.A. **193**:1084-1086, 1965.

Littler, J. W.: The prevention and the correction of adduction contracture of the thumb, In DePalma, A. F., editor: Clinical orthopaedics, vol. 13, Philadelphia, 1959, J. B. Lippincott Co., pp. 182-192.

Littler, J. W., and Cooley, S. G. E.: Opposition of the thumb and its restoration by abductor digiti quinti transfer, J. Bone Joint Surg. (Am.) **45**:1389-1396, 1963.

Margolis, E., and Hasson, E.: Hereditary malformations of the upper extremities in three generations, J. Hered. **46**:255-262, 1955.

Pfeiffer, R. A.: Associated deformities of the head and hands, Birth defects **5**:18-34, 1969.

Poznanski, A. K., Garn, S. M., and Jolt, J. F.: The thumb in the congenital malformation syndromes, Radiology **100**:115-129, 1971.

Pruzanski, W.: Familial congenital malformations of the heart and upper limbs: a syndrome of Holt Oram, Cardiologia **45**:21-38, 1964.

Stecher, R. M.: The physical characteristics and heredity of short thumbs, Acta Genet. **7**:217-222, 1957.

Strauch, B.: Dorsal thumb flap for release of adduction contracture of the first web space, Bull. Hosp. Joint Dis. **36**:34-39, 1975.

Strauch, B., and Spinner, M.: Congenital anomaly of the thumb: absent intrinsics and flexor pollicis longus, J. Bone Joint Surg. (Am.) **58**:115-118, 1976.

Tupper, J. W.: Pollex abductus due to congenital malposition of the flexor pollicis longus, J. Bone Joint Surg. (Am.) **51**:1285-1290, 1969.

Weckesser, E. C.: Congenital flexion-adduction deformity of the thumb (congenital "clasped thumb"), J. Bone Joint Surg. (Am.) **37**:977, 1955.

Weckesser, E. C., Reed, J. R., and Heiple, G.: Congenital clasped thumb (congenital flexion-adduction deformity of the thumb), J. Bone Joint Surg. (Am.) **50**:1417-1428, 1968.

White, J. W., and Jensen, W. E.: The infant's persistent thumb clutched hand, J. Bone Joint Surg. (Am.) **34**:680, 1952.

White, J. W., and Jensen, W. E.: Trigger thumb in infants, Am. J. Dis. Child. **85**:141, 1953.

Zadek, I.: Congenital absence of extensor pollicis longus of both thumbs: operation and cure, J. Bone Joint Surg. **16**:432, 1934.

Chapter 6/The absent thumb

Ahstrom, J. P., Jr.: Pollicization in congenital absence of the thumb. In Ahstrom, J. P., editor: Current practice in orthopaedic surgery, vol. 5, St. Louis, 1973, The C. V. Mosby Co., chapter 1.

Buck-Gramcko, D.: Pollicization of the index finger: method and results in aplasia and hypoplasia of the thumb, J. Bone Joint Surg. (Am.) **53**:1605-1617, 1971.

Bunnell, S.: Physiological reconstruction of a thumb after total loss, Surg. Gynecol. Obstet. **52**:245-248, 1931.

Cuthbert, J. B.: Pollicization of the index finger, Br. J. Plast. Surg. **1**:56-59, 1948-1949.

Dunlop, J.: The use of the index finger for the thumb: some interesting points in hand surgery, J. Bone Joint Surg. **5**:99-103, 1923.

Edgerton, M. T., Snyder, G. B., and Webb, W. L.: Surgical treatment of congenital thumb deformities (including psychological impact of correction), J. Bone Joint Surg. (Am.) **47**:1453-1474, 1965.

Iselin, M.: Chirurgie de la main, ed. 10, Paris, 1955, Masson & Cie Editeurs.

Littler, J. W.: The neurovascular pedicle method of digital transposition for reconstruction of the thumb, Plast. Reconstr. Surg. **12**:303-319, 1953.

Moore, F. T.: The technique of pollicization of the index finger, Br. J. Plast. Surg. **1**:60-68, 1948-1949.

Riordan, D. C.: Congenital absence of the radius, J. Bone Joint Surg. (Am.) **37**:1129-1140, 1955.

White, W. F.: Fundamental priorities in pollicization, J. Bone Joint Surg. (Br.) **52**: 438-443, 1970.

Chapter 7/Extra thumbs

Aase, J. M., and Smith, D. W.: Congenital anemia and triphalangeal thumbs, J. Pediatr. **74**:471-474, 1969.

Abramowitz, I.: Triphalangeal thumb, S. Afr. Med. J. **41**:104, 1967.

Bilhaut, M.: Guerison d'un pouce bifide par un nouveau procede operatoire, Congres Francais de Chir. (4 session, 1889), **4**: 576-580, 1890.

Burman, M.: An historical perspective of double hands and double feet, Bull. Hosp. Joint Dis. **29**:241, 1968.

Karchinov, K.: The treatment of polydactyly of the hand, Br. J. Plast. Surg. **15**(4): 362-376, 1962.

Lapidus, P. W., Guidotti, F. P., and Coletti, C. J.: Triphalangeal thumb, Surg. Gynecol. Obstet. **77**:178, 1943.

Lenz, W., Theopold, W., and Thomas, J.: Triphalangia of the thumbs as a result of thalidomide damage, Munch. Med. Wochenschr. **106**(45):2033-2041, 1964.

Miura, T.: Triphalangeal thumb, Plast. Reconstr. Surg. **58**:587-594, 1976.

Palmieri, T. J.: Polydactyly of the thumb: incidence, etiology, classification and treatment, Bull. Hosp. Joint Dis. **34**(2):200-221, 1973.

Phillips, R. S.: Congenital split foot (lobster claw) and triphalangeal thumb, J. Bone Joint Surg. (Br.) **53**:247-257, 1971.

Swanson, A. B., and Brown, K. S.: Hereditary triphalangeal thumb, J. Hered. **53**:259-265, 1962.

Wassel, H. D.: The results of surgery for polydactyly of the thumb, Clin. Orthop. **64**:175, 1969.

Wood, V. E.: Treatment of the triphalangeal thumb, Clin. Orthop. **120**:188-199, 1976.

Woolf, R. M., Broadbent, T. R., and Woolf, C. M.: Practical genetics of congenital hand abnormalities. In Littler, J. W., Cramer, L. M., and Smith, J. W., editors: Symposium on reconstructive hand surgery, vol. 9, St. Louis, 1974, The C. V. Mosby Co., pp. 141-143.

Chapter 8/Small and absent fingers

Barsky, A. J.: Congenital anomalies of the hand and their surgical treatment, Springfield, Ill., 1958, Charles C Thomas, Publisher.

Bass, H. N.: Familial absence of middle phalanges with nail dysplasia: a new syndrome, Pediatrics **42**:318-323, 1968.

Bell, J.: On brachydactyly and symphalangism. In Penrose, L. S.: The treasury of human inheritance, vol. 5, part 1, Cambridge, 1951, Cambridge University Press, pp. 1-31.

Dellon, A. L., and Gaylor, R.: Bilateral symphalangism of the index finger, J. Bone Joint Surg. (Am.) **58**:270-271, 1976.

Drachman, D. B.: Normal development and congenital malformation of joints, Bull. Rheum. Dis. **19**(5):536-540, 1969.

Drinkwater, H.: Phalangeal anarthrosis (synostosis, ankylosis) transmitted through 14 generations, Proc. R. Soc. Med. **10**:60-68, 1917.

Elkington, S. G., and Huntsman, R. G.: The Talbot fingers: a study in symphalangism, Br. Med. J. **1**:407-413, 1967.

Flatt, A. E., and Wood, V. E.: Rigid digits or symphalangism, The Hand **7**:197-214, 1975.

Garn, S. M., Fels, S. L., and Israel, H.: Brachymesophalangia of digit five in ten populations, Am. J. Phys. Anthropol. **27**: 205-209, 1967.

Garn, S. M., Hertzog, K. P., Poznanski, A. K., and Nagy, J. M.: Metacarpophalangeal length in the evaluation of skeletal malformation, Radiology **105**:375-381, 1972.

Holmes, L. B., and Remensnyder, J. B.: Hypoplasia of the second metacarpal in mother and daughter, J. Pediatr. **81**:1165-1167, 1972.

Littler, J. W.: Introduction to surgery of the hand, Reconstr. Plast. Surg. **4**:1543-1546, 1964.

McKusick, V. A.: Mendelian inheritance in man: catalogs of autosomal dominant, autosomal recessive and x-linked phenotypes, ed. 3, Baltimore, 1971, The Johns Hopkins University Press.

Strasburger, A. K., Hawkins, M. R., Eldridge, R., Hargrave, R. L., and McKusick, V. A.: Symphalangism: genetic and clinical aspects, Bull. Johns Hopkins Hosp. **117**:108-127, 1965.

Tajima, T.: Congenital anomalies of the

hand, part 3, Rinsho Seikei-geka (Clin. Orthop. Surg.) **11**:475, 1976.

Temtamy, S. A.: Genetic factors in hand malformations. Thesis. Baltimore, 1966, The Johns Hopkins University Press.

Warkany, J.: Congenital malformations, Chicago, 1971, Year Book Medical Publishers, Inc.

Chapter 9/Crooked fingers

Blacker, G. J., Lister, G. D., and Kleinert, H. E.: The abducted little finger in low ulnar nerve palsy, J. Hand Surg. **1**:190-196, 1976.

Blank, E., and Girdany, B. R.: Symmetric bowing of the terminal phalanges of the fifth fingers in a family (Kirner's deformity), Am. J. Roentgenol. Radium Ther. Nucl. Med. **93**:367-373, 1965.

Boix, E.: Deviation des doigts en coup de vent et insuffisance de popene vrose polmaire d'origine congenitale, Nouv. Iconogr. Salp. vol. 10, 1897.

Carstam, N., and Eiken, O.: Kirner's deformity of the little finger, J. Bone Joint Surg. (Am.) **52**:1663-1665, 1970.

Carstam, N., and Theander, G.: Surgical treatment of clinodactyly caused by longitudinally bracketed diaphysis, Scand. J. Plast. Reconstr. Surg. **9**:199-202, 1975.

Courtemanche, A. D.: Campylodactyly: etiology and management, Plast. Reconstr. Surg. **44**:451-454, 1969.

Crawford, H. H., Horton, C. E., and Adamson, J. E.: Congenital aplasia or hypoplasia of the thumb and finger extensor tendons, J. Bone Joint Surg. (Am.) **48**:82-91, 1966.

Inokuchi, S.: Congenital aplasia of the extensor digitorum communis, Rinsho Seikei-geka (Clin. Orthop. Surg.) **8**:877-880, 1973.

Jaeger, M., and Refior, H. J.: Congenital triangular deformity of tubular bones in hand and foot, Clin. Orthop. **81**:139-150, 1971.

Katz, G.: A pedigree with anomalies of the little finger in five generations and seventeen individuals, J. Bone Joint Surg. (Am.) **52**:717-720, 1970.

Kaufmann, H. J., and Taillard, W. F.: Bilateral incurving of the terminal phalanges of the fifth fingers, Am. J. Roentgenol. Radium Ther. Nucl. Med. **86**(3):490, 1961.

Kelikian, H.: Congenital deformities of the hand and forearm, Philadelphia, 1974, W. B. Saunders Co., p. 577.

Kirner, J.: Doppelseitige verkrummungen des kleinfingerendgliedes als selbstandiges, Fortschr. Gen. Rontgen. **36**:804-806, 1927.

Kirner, J.: Doppelseitige verkrummung der endglieder beider kleinfinger, Fortschr. Gen. Rontgen. **78**:745, 1953.

Lundblom, A.: On congenital ulnar deviation of the fingers of familial occurrence, Acta Orthop. Scand. **3**:393, 1932.

McMurtry, R. Y., and Jochims, J. L.: Congenital deficiencies of the extrinsic extensor mechanism of the hand, Clin. Orthop., in press.

Millesi, H.: The pathogenesis and operative correction of camptodactyly, Chir. Plastica Reconstr. **5**:55-61, 1968.

Millesi, H.: Camptodactyly. In Littler, J. W., Cramer, L. M., and Smith, J. W., editors: Symposium on reconstructive hand surgery, vol. 9, St. Louis, 1974, The C. V. Mosby Co., pp. 175-177.

Moberg, E.: Three useful ways to avoid amputation in advanced Dupuytren's contracture, Orthop. Clin. North Am. **4**:1001-1005, 1973.

Oldfield, M. C.: Campylodactyly: flexor contracture of the fingers in young girls, Br. J. Plast. Surg. **8**:312-317, 1956.

Powers, C. R., and Ledbetter, R. H.: Congenital flexion and ulnar deviation of the metacarpo-phalangeal joints of the hand, Clin. Orthop. **116**:173-175, 1976.

Poznanski, A. K., Pratt, G. B., Manson, G., and Weiss, L.: Clinodactyly, camptodactyly, Kirner's deformity, and other crooked fingers, Radiology **93**:573-582, 1969.

Roche, A. F.: Clinodactyly and brachymesophalangia of the fifth finger, Acta Paediatr. Scand. **50**:387-391, 1961.

Smith, R. J., and Kaplan, E. B.: Camptodactyly and similar atraumatic flexion deformities of the proximal interphalangeal joints of the fingers, J. Bone Joint Surg. (Am.) **50**:1187-1204, 1968.

Stoddard, E. E.: Nomenclature of hereditary crooked fingers: streblomicrodactyly and camptodactyly—are they synonyms? J. Hered. **30**:511-512, 1939.

Theander, G., and Carstam, N.: Longitudinally bracketed diaphysis, Ann. Radiol. **17**:355-360, 1974.

Tsuge, K.: Congenital aplasia or hypoplasia of the finger extensors, The Hand **7**:15-21, 1975.

Welch, J. P., and Temtamy, S. A.: Hereditary contractures of the fingers (camptodactyly), J. Med. Genet. **3**:104-113, 1966.

White, J. W., and Jensen, W. E.: The infant's persistent thumb-clutched hand, J. Bone Joint Surg. (Am.) **34**:680-688, 1952.

Wilhelm, A., and Kleinschmidt, W.: New etiologic and therapeutic viewpoints on camptodactyly and tendovaginitis, Chir. Plastica Reconstr. **5**:62-67, 1968.

Wood, V. E., and Flatt, A. E.: Congenital triangular bones in the hand, J. Hand Surg. **2**(3):179-193, 1977.

Woolf, C. M., and Woolf, R. M.: A genetic study of polydactyly in Utah, Am. J. Hum. Genet. **22**:75-87, 1970.

Chapter 10/Webbed fingers

Apert, E.: De l'acrocephalosyndactylie. Bull. Mem. Soc. Med. Hop. Paris **23**:1310-1330, 1906.

Bauer, T. B., Tondra, J. M., and Trusler, H. M.: Technical modification in repair of syndactylism, Plast. Reconstr. Surg. **17**: 385-392, 1956.

Beals, R. K., and Crawford, S.: Congenital absence of the pectoral muscles, Clin. Orthop. **119**:166-171, 1976.

Bing, R.: Weber angeborene muskeldefekta, Virchows Arch. Path. **170**:175, 1902.

Blank, C. E.: Apert's syndrome (a case of acrocephalosyndactyly)—observations on a British series of thirty-nine cases, Ann. Hum. Genet. **24**:151-164, 1960.

Brown, J. B., and McDowell, F.: Syndactylism with absence of the pectoralis major, Surgery **7**:599-601, 1940.

Carpenter, G.: Two sisters showing malformations of the skull and other congenital abnormalities, Rep. Soc. Study Dis. Child. **1**:110-118, 1900-1901.

Carpenter, G.: A case of acrocephaly, with other congenital malformations, Proc. R. Soc. Med. II, part I, pp. 45-53, 1909.

Christopher, F.: Congenital absence of the pectoral muscles, J. Bone Joint Surg. **10**: 350, 1928.

Clarkson, P.: Poland's syndactyly, Guys Hosp. Rep. **111**:335-346, 1962.

Ebskov, B., and Zachariae, L.: Surgical methods in syndactylism, Acta Chir. Scand. **131**:258-268, 1966.

Flatt, A. E.: Practical factors in the treatment of syndactyly. In Littler, J. W., Cramer, L. M., and Smith, J. W., editors: Symposium on reconstructive hand surgery, St. Louis, 1974, The C. V. Mosby Co., pp. 144-156.

Hoover, G. H., Flatt, A. E., and Weiss, M. W.: The hand and Apert's syndrome, J. Bone Joint Surg. (Am.) **52**:878-895, 1970.

Ireland, D. C. R., Takayama, N., and Flatt, A. E.: Poland's syndrome: a review of forty-three cases, J. Bone Joint Surg. (Am.) **58**:52-58, 1976.

Losch, G. M., and Duncker, H. R.: Acrosyndactylism, Transactions of the International Society of Plastic and Reconstructive Surgery, 5th Congress, Australia, 1971, Butterworth Pty. Ltd., pp. 671-676.

Maisels, D. O.: Acrosyndactyly, Br. J. Plast. Surg. **15**:166, 1962.

McKusick, V. A.: Mendelian inheritance in man: catalogs of autosomal dominant, autosomal recessive, and x-linked phenotypes, ed. 3, Baltimore, 1971, The Johns Hopkins University Press.

Pers, M.: Aplasia of the anterior thoracic wall, the pectoral muscle and the breast, Scand. J. Plast. Reconstr. Surg. **2**:125, 1968.

Poland, A.: Deficiency of the pectoralis muscle, Guys Hosp. Rep. **6**:191, 1841.

Poznanski, A. K.: The hand in radiologic diagnosis, Philadelphia, 1974, W. B. Saunders Co.

Saldino, R. M., Steinbach, H. L., and Epstein, C. J.: Familial acrocephalosyndactyly (Pfeiffer syndrome), J. Roentgen. **116**:609-622, 1972.

Sugiura, Y.: Poland's syndrome. Clinicoroentgenographic study on 45 cases, Cong. Anom. **16**:17-28, 1976.

Temtamy, S. A.: Carpenter's syndrome: acrocephalopolysyndactyly, an autosomal recessive syndrome, J. Pediatr. **69**:111-120, 1966.

Walsh, R. J.: Acrosyndactyly: a study of 27 patients, Clin. Orthop. **71**:99-111, 1970.

Warkeny, J.: The upper extremities, Chicago, 1971, Year Book Medical Publishers, Inc., pp. 258-261.

Woolf, C. M., and Woolf, R. M.: A genetic study of syndactyly in Utah, Soc. Biol. **20**:335-346, 1973.

Chapter 11/Constriction ring syndrome

Artz, T. D., and Posch, J. L.: Use of cross-finger flap for treatment of congenital broad constricting bands of the fingers, Plast. Reconstr. Surg. **52**:645-647, 1973.

Barenberg, L. H., and Greenberg, B.: Intrauterine amputations and constriction bands: report of a case with anesthesia below the constriction, Am. J. Dis. Child. **64**:87, 1942.

Browne, D.: The pathology of congenital ring constrictions, Arch. Dis. Child. **32**:517-519, 1957.

Fischl, R. A.: Ring constriction syndrome. Transactions of the International Society of Plastic and Reconstructive Surgeons, 5th Congress, Australia, 1971, Butterworth Pty. Ltd., pp. 657-670.

Kino, Y.: Clinical and experimental studies of the congenital constriction band syndrome, with an emphasis on its etiology, J. Bone Joint Surg. (Am.) **57**:636-643, 1975.

Moses, J. M., Flatt, A. E., and Cooper, R. R.: Congenital constriction ring: a clinical follow up of forty-five cases, J. Bone Joint Surg. (Am.) in press, 1977.

Patterson, T. J. S.: Congenital ring-constrictions, Br. J. Plast. Surg. **14**:1-31, 1961.

Ramakrishnan, M. S., and Nayak, V. S.: Congenital constriction bands of lower extremities, Indian J. Pediatr. **30**:191, 1963.

Torpin, R.: Amniochorionic mesoblastic fibrous strings and amniotic bands; associated constricting fetal malformations or fetal death, Am. J. Obstet. Gynecol. **91**:65-75, 1965.

Chapter 12/Extra fingers

Barsky, A. J.: Congenital anomalies of the hand, J. Bone Joint Surg. (Am.) **33**:35-64, 1951.

Barsky, A. J.: Congenital anomalies of the hand and their surgical treatment, Springfield, Ill. 1958, Charles C Thomas, Publisher, pp. 48-64.

The Holy Bible, King James Version. II Samuel 21:20.

Burman, M.: Note on duplication of the index finger, J. Bone Joint Surg. (Am.) **54**:884, 1972.

Carroll, R.: Congenital anomalies of the hand. Presented at the American Academy of Orthopaedic Surgeons Instructional Course Lectures, Chicago, 1970.

Entin, M. A.: Reconstruction of congenital abnormalities of the upper extremities, J. Bone Joint Surg. (Am.) **41**:681-700, 1959.

Flatt, A. E.: Problems in polydactyly. In Cramer, L. M., and Chase, R. A.: Symposium on the hand, vol. 3, St. Louis, 1971,. The C. V. Mosby Co., pp. 150-167.

Handforth, J. R.: Polydactylism of the hand in southern Chinese, Anat. Rec. **106**:119, 1950.

Harrison, R. G., Pearson, M., and Roaf, R.: Ulnar dimelia, J. Bone Joint Surg. (Br.) **42**:549-555, 1960.

James, J. I.: Congenital abnormalities of the limbs, Practitioner **191**:159, 1963.

Kanavel, A. B.: Congenital malformations of the hands, Arch. Surg. **25**:308-316, 1932.

Kelikian, H., and Doumanian, A.: Congenital anomalies of the hand, J. Bone Joint Surg. (Am.) **39**:1002-1019, 1957.

Millesi, H.: Deformations of the fingers following operations for polydactylia, Klin. Med. (Wien) **22**:266-272, 1967.

Steindler, A.: Congenital malformations and deformities of the hand, J. Orthop. Surg. **2**:639, 1920.

Stelling, F.: The upper extremity. In Ferguson, A. B., editor: Orthopedic surgery in infancy and childhood, vol. 2, Baltimore, 1963, The Williams & Wilkins Co., pp. 304-308.

Strickland, A.: Lives of the queens of England, vol. II, London, 1840-1848, H. Colburn, pp. 589-590.

Temtamy, S., and McKusick, V. A.: Synopsis of hand malformations with particular emphasis on genetic factors, Birth Defects **5**:125-184, 1969.

Turek, S. L.: Orthopaedics principles and their application, Philadelphia, 1967, J. B. Lippincott Co., p. 123.

Vervais, P. W.: The three-dimensional recording of disease and deformity, Clin. Orthop. **28**:216, 1963.

Wassel, H. D.: The results of surgery for polydactyly of the thumb, Clin. Orthop. **64**:175-193, 1969.

Wood, V. E.: Duplication of the index finger, J. Bone Joint Surg. (Am.) **52**:569-573, 1970.

Wood, V. E.: Treatment of central polydactyly, Clin. Orthop. **74**:196-205, 1971.

Chapter 13/Large fingers

Allende, B. T.: Macrodactyly with enlarged median nerve associated with carpal tunnel syndrome, Plast. Reconstr. Surg. **39**:578-582, 1967.

Barsky, A.: Macrodactyly, J. Bone Joint Surg. (Am.) **49**:1255-1266, 1967.

Bean, W. B., and Peterson, P. K.: Note on a monstrous finger, Arch. Intern. Med. **104**:433-438, 1959.

Edgerton, M. T., and Tuerk, D. B.: Macrodactyly (digital gigantism): its nature and treatment. In Littler, J. W., Cramer, L. M., and Smith, J. W., editors: Symposium on reconstructive hand surgery, vol. 9, St. Louis, 1974, The C. V. Mosby Co., p. 157.

El-Shami, I. N.: Congenital partial gigantism: case report and review of literature, Surgery **65**:683-688, 1969.

Hueston, J. T., and Millroy, P.: Macrodactyly associated with hamartoma of major peripheral nerves, Aust. N.Z. J. Surg. **37**:394-397, 1968.

Kaplan, E. B.: Congenital giant thumb, Bull. Hosp. Joint Dis. **8**:38, 1947.

Millesi, H.: Macrodactyly: a case study, In Littler, J. W., Cramer, L. M., and Smith, J. W., editors: Symposium on reconstructive hand surgery, vol. 9, St. Louis, 1974, The C. V. Mosby Co., p. 173.

Streeter, G. L.: Focal deficiencies in fetal tissues and their relation to intra-uterine amputation, Contrib. Embryol. **22**:1-44, 1930.

Thorne, F. L., Posch, J. L., and Mladick, R. A.: Megalodactyly, Plast. Reconstr. Surg. **41**:232-239, 1968.

Tsuge, K.: Treatment of macrodactyly, Plast. Reconstr. Surg. **39**:590-599, 1967.

Tuli, S. M., Khanna, N. N., and Sinha, G. P.: Congenital macrodactyly, Br. J. Plast. Surg. **22**:237-243, 1969.

Wood, V. E.: Macrodactyly, J. Iowa Med. Soc. **59**:922-928, 1969.

Chapter 14/Cleft hand and central defects

Barsky, A. J.: Cleft hand: classification, incidence, and treatment, J. Bone Joint Surg. (Am.) **46**:1707-1719, 1964.

Godunova, G. S.: Operative treatment of congenital split hand. In Godunova, G. S.: Orthopaedics, traumatology and the application of prostheses, no. 6, 1973. Summary reprinted in English I.C.I.B. **13**:14-16, 1974.

Maisels, D. O.: Lobster claw deformities of the hand, The Hand **2**:79-82, 1970.

Maisels, D. O.: Lobster-claw deformities of the hands and feet, Br. J. Plast. Surg. **23**(3):269-281, 1970.

Poznanski, A. K.: The hand in radiologic diagnosis, Philadelphia, 1974, W. B. Saunders Co., pp. 183-187.

Snow, J. W., and Littler, J. W.: Surgical treatment of cleft hand. Transactions of the International Society of Plastic and Reconstructive Surgery, 4th Congress, Rome, 1967, Excerpta Medica Foundation, pp. 888-893.

Walker, J. C., and Clodius, L.: The syndromes of cleft lip, cleft palate and lobster claw deformities of hands and feet, Plast. Reconstr. Surg. **32**:627-636, 1963.

Warkeny, J.: The upper extremities, Chicago, 1971, Year Book Medical Publishers, Inc., pp. 963-965.

Chapter 15/Radial clubhand

Blauth, W., and Schmidt, H.: The implication of arteriographic diagnosis in malformation of the radial marginal ray, Z. Orthop. pp. 102-110, 1969.

Bora, F. W., Jr., Nicholson, J. T., and Cheema, H. M.: Radial meromelia, J. Bone Joint Surg. (Am.) **52**:966-978, 1970.

Define, D.: A aplicacao em cirurgia ortopedica do poder osteogenetico do periosteo na infancia, R. Brasil. Orthop. **1**:42-52, 1966.

Delorme, T. L.: Treatment of congenital absence of the radius by transepiphyseal fixation, J. Bone Joint Surg. (Am.) **51**:117-129, 1969.

Eaton, R. G.: Hand problems in children: a timetable for management, Pediatr. Clin. North Am. **14**:643-658, 1967.

Goldberg, M. J., and Meyn, M.: The radial clubhand, Orthop. Clin. North Am. **7**:341-358, 1976.

Harrison, S. H.: Pollicization in cases of radial club hand, Br. J. Plast. Surg. **23**:192-200, 1970.

Heikel, H. V. A.: Aplasia and hypoplasia of the radius, Acta Orthop. Scand. (Suppl.) **391**:1, 1959.

Kato, K.: Congenital absence of the radius with review of the literature and report of three cases, J. Bone Joint Surg. **22**:589, 1924.

Lamb, D. W.: The treatment of radial club hand, The Hand **4**:22-30, 1972.

O'Rahilly, R.: Morphologic patterns in limb deficiencies and duplications, Am. J. Anat. **89**:135, 1956.

Pardini, A. G., Jr.: Radial dysplasia, Clin. Orthop. **57**:153, 1968.

Riordan, D. C.: Congenital absence of the radius: a 15 year follow-up, J. Bone Joint Surg. (Am.) **45**:1783, 1965.

Sayre, R. H.: A contribution to the study of club-hand, Trans. Am. Orthop. Assoc. **6**:208, 1893.

Skerik, S. K., and Flatt, A. E.: The anatomy of congenital radial dysplasia, Clin. Orthop. **66**:125-143, 1969.

Steindler, A.: Livre jubilaire offert au docteur Albin Lambotte, Brussels, 1936, Vromant S. A.

Stoffel, A., and Stempel, E.: Anatomische studien uber die klumphand, Z. Orthop. Chir. **23**:1, 1909.

Chapter 16/Ulnar clubhand

Conway, H., and Wagner, K. J.: Congenital anomalies reported on birth certificates in New York City (1952-1962 inclusive), N.Y. State J. Med. **65**:1087-1090, 1965.

Goddu, L. A. O.: Reconstruction of elbow and bone graft of rudimentary ulna, N. Engl. J. Med. **202**:1142-1144, 1930.

Kümmel, W.: Die Missbildungen der Extremitäten durch Defekt, Verwachsung und Ueberzahl, Heft 3, Cassel, 1895, Bibliotheca Medica.

Laurin, C. A., and Farmer, A. W.: Congenital absence of ulna, Can. J. Surg. **2**:204-207, 1959.

Lloyd Roberts, G. C.: Treatment of defects of the ulna in children by establishing cross-union with the radius, J. Bone Joint Surg. (Br.) **55**:327-330, 1973.

Ogden, J. A., Kirk Watson, H., and Bohne, W.: Ulnar dysmelia, J. Bone Joint Surg. (Am.) **58**:467-475, 1976.

O'Rahilly, R.: Morphological patterns in limb deficiencies and duplications, Am. J. Anat. **89**:135-193, 1951.

Pardini, A. G., Jr.: Congenital absence of the ulna, J. Iowa Med. Soc. **57**:1106-1112, 1967.

Riordan, D. C., Mills, E. H., and Alldredge, R. H.: Congenital absence of the ulna. Presidential address, American Society for Hand Surgery, New Orleans, 1961.

Southwood, A. R.: Partial absence of ulna and associated structures, J. Anat. **61**:346-351, 1926-1927.

Spinner, M., Freundlich, B. D., and Abeles, E. D.: Management of moderate longitudinal arrest of development of the ulna, Clin. Orthop. **69**:199-202, 1970.

Stoffel, A., and Stampel, E.: Anatomische studien uber die klimphand, Z. Orthop. Chir. **23**:1-57, 1909.

Straub, L. R.: Congenital absence of ulna, Am. J. Surg. **109**:300-305, 1965.

Vitale, C. C.: Reconstructive surgery for defects in shaft of ulna in children, J. Bone Joint Surg. (Am.) **34**:804-810, 1952.

Watt, J. C.: Anatomy of 7 months' foetus exhibiting bilateral absence of ulna accompanied by monodactyly (and also diaphragmatic hernia), Am. J. Anat. **22**: 385-427, 1917.

Zimmerman, W. V.: Incidence of deformities of extremities in Hamburg (1960-1962), Z. Menochl. Vererb. Konstitutions **37**:26-44, 1963.

GLOSSARY

acheiria (achiria) absence of the hand
acheiropodia absence of the hands and feet
acro peak or end
acrosyndactyly fusion of the terminal portion of two or more digits with proximal epithelial lined clefts or sinuses between the digits
adactyly absence of all fingers
agenesis absence or no development
amelia absence of limbs
amelia totalis absence of all four limbs
amputation absence of a distal part of a limb
aplasia absence of a specific bone or bones
apodia absence of foot
arachnodactyly long, slender digits
ateliosis incomplete or imperfect development
baso basal or proximal
brachy shortening
brachybasophalangia short proximal phalanx
brachydactyly short fingers (usually refers to short phalanges)
brachymegalodactyly short broad digit (stub thumb or type D brachydactyly)
brachymelia short limb without absence of bony elements
brachymesophalangia short middle phalanx
brachymetacarpia (brachymetacarpalia) short metacarpals
brachymetapodia short metacarpals and metatarsals
brachymetatarsia short metatarsals
brachyphalangia short phalanx
brachytelephalangia short distal phalanx
camptodactyly curvature of a finger in the plane of flexion of the digit
central defect (cleft or lobster claw hand) absence of one or more of the central rays of the hand, either the second, third, or fourth or any combination thereof
cheiria (chiria) with reference to hand
cleft hand central ray deficiency
clinarthrosis oblique or lateral angular deviation in alignment of joints
clinodactyly deviation of a finger in the plane of the hand
dactylia with reference to digit
delta phalanx triangular ossicle
di double
dicheiria (dichiria) double or mirror hand
dimelia double or mirror limb
dys deformed, ill, bad
dystelephalangia deformed terminal phalanx (Kirner deformity)
ectro absence (from Greek *ektroma*—abortion)
ectrocheiria total or partial absence of hand
ectrodactyly total or partial absence of fingers or hand
ectromelia total or partial absence of a limb
ectrophalangia terminal absence of one or more phalanges
ectropodia total or partial absence of a foot
hemimelia absence of part of a limb
hyper above or increased
hyperphalangia the presence of more than a normal number of phalanges in transverse direction
hypo below or decreased
hypodactyly decreased number of fingers
hypogenesis incomplete development
hypophalangia decreased number of phalanges
hypoplasia incomplete development of specific part
intercalary deficiency absence of a middle portion of a limb while the proximal and distal portions are present
longitudinal deficiency absence of the limb extending parallel to the long axis; it may be preaxial, postaxial, or central
macro excessive size
macrodactyly hyperplasia of the digit
manus referring to hand

megalodactyly hyperplasia of the digit
melia referring to the limb
mero part or partial
meromelia partial absence of a limb
meso middle
micro diminished size
microcheiria (microchira) hypoplasia of all parts of the hand
micromelia short limbs without absence of bone elements
oligo few or little
oligodactyly absence of some of the fingers
parastremma distorted limb
pero deformed or defective
perodactyly deformed fingers
peromelia hemimelia especially for those cases of hands ending in a stump
phoco short (from Greek *phoke*—seal)
phocomelia partial absence of any proximal region of the limb. May be proximal (arm) or distal (forearm) absence
podia with reference to the foot
poikilodactyly irregular or varied digit
poly many or increased number of digits. May be central, radial (preaxial) or ulnar (postaxial)
polyphalangia presence of more than normal number of phalanges in transverse direction
postaxial pertaining to the ulnar side of the limb
preaxial pertaining to the radial side of the limb or thumb
streblo twisted
syn or sym fusion or together
symbrachydactyly short digits with syndactyly
symphalangia bony fusion of phalanges (end-to-end) or clinically rigid digits
syndactyly fusion of adjacent digits. May be complete or incomplete with reference to cutaneous involvement. May be simple or complex with reference to bony involvement
synonychia fusion of fingernail common to two or more digits
synostosis bony fusion
tele distant or end
terminal deficiency absence of the bones which are distal to the proximal limit of the deficiency. This defect may be either transverse or longitudinal
transverse deficiency absence of a portion of the limb gonig across the width of the limb

SYNDROMES INDEX

AARSKOG-SCOTT SYNDROME
(Faciodigitogenital syndrome)

Inheritance: Autosomal dominant
Hands: Mild syndactyly, clinodactyly, camptodactyly, simian crease, swan-neck deformities, and short thumbs
Other: Facial anomalies and hypertelorism, short stature, unusual scrotal skin, foot abnormalities, and cervical spine anomalies

ACROCEPHALOPOLYSYNDACTYLY
(Carpenter syndrome)

Inheritance: Autosomal recessive
Hands: Mild syndactyly of hands and feet, preaxial polydactyly, brachyphalangia (especially middle), clinodactyly, and broad thumb
Other: Acrocephaly, peculiar facies, hypogonadism, obesity, mental retardation, abdominal hernia, and congenital heart disease

(Noack syndrome)

Inheritance: Autosomal dominant
Hands: Mild syndactyly of hands and feet, preaxial polydactyly, clinodactyly, and broad thumbs
Other: Mild acrocephaly, peculiar facies, and duplication of great toe

ACROCEPHALOSYNDACTYLY—TYPICAL
(Type I—Apert syndrome) (Type II—Apert-Crouzon syndrome or Vogt cephalodactyly)

Inheritance: Autosomal dominant with most cases fresh mutant
Hands: Brachysyndactyly (complex), symphalangia, thumb with short broad distal phalanx, synonychia, and carpal and tarsal fusions
Other: Craniosynostosis (usually coronal suture), flattened facies, hypertelorism, often retardation, brachysyndactyly affecting feet, and short great toe resulting from short first metatarsal

ACROCEPHALOSYNDACTYLY—ATYPICAL
(Type III—Saethre-Chotzen syndrome) (Type IV—Waardenburg syndrome) (Type V—Pfeiffer syndrome)

Inheritance: Autosomal dominant with most cases fresh mutant
Hands: Mild syndactyly (may be cutaneous only), occasionally bifid distal phalanges, symphalangia, and short thumb with stubby, secondary, small proximal phalanx
Other: Minimal skull changes, usually brachycephaly, and normal intelligence

ACROPECTOROVERTEBRAL DYSPLASIA
(F-syndrome)

Inheritance: Autosomal dominant
Hands: Short, broad thumbs, radial polysyndactyly, and carpal fusions
Other: Prominent sternum with or without pectus excavatum, spina bifida occulta, impaired intelligence, and similar foot changes

APERT SYNDROME

See Acrocephalosyndactyly—typical

APERT-CROUZON SYNDROME

See Acrocephalosyndactyly—typical

ARTHROGRYPOSIS CONGENITA MULTIPLEX

Inheritance: Usually sporadic
Hands: Increase in carpal angle, intercarpal fusions, syndactyly, camptodactyly, clasped thumb, flexion deformity of wrist, ulnar deviation, and hypoplastic distal phalanges
Other: Multiple joint contractures, clubfoot, vertical talus, tarsal fusions, dislocated hips, scoliosis, and brachycephaly

ASPHYXIATING THORACIC DYSPLASIA
(Jeune syndrome)

Inheritance: Autosomal recessive
Hands: Brachydactyly (MC and phalanges), cone-shaped epiphysis, and postaxial polydactyly
Other: Short extremities, long, narrow thorax, respiratory problems, renal disease, and foot changes similar to those of the hand

BIEMOND I SYNDROME

Inheritance: Undetermined
Hands: Brachymetacarpia
Other: Nystagmus and cerebellar ataxia

BIEMOND II SYNDROME

Inheritance: Irregular autosomal dominant
Hands: Postaxial polydactyly
Other: Coloboma of iris, mental retardation, obesity, and hypogonadism

BLACKFAN-DIAMOND ANEMIA

Inheritance: Autosomal recessive
Hands: Preaxial polydactyly and triphalangeal thumbs
Other: Profound anemia

BLOOM SYNDROME
(Chromosomal break)

Inheritance: Autosomal recessive
Hands: Preaxial polydactyly, syndactyly, and clinodactyly
Other: Short stature, malar hypoplasia, and telangiectatic erythema of face

BRACHYDACTYLY—TYPE A-1

Inheritance: Autosomal dominant
Hands: Brachymesophalangia D2-5, symphalangia middle-distal D2-5, accessory carpal bones, and brachybasophalangia of thumb
Other: Short great toes and short stature

BRACHYDACTYLY—TYPE A-2

Inheritance: Undetermined
Hands: Brachymesophalangia D2, clinodactyly D2, and syndactyly

BRACHYDACTYLY—TYPE A-3

Inheritance: Autosomal dominant
Hands: Brachymesophalangy D5 and clinodactyly D5

BRACHYDACTYLY—TYPE B

Inheritance: Autosomal dominant
Hands: Brachymesophalangia D2-5, distal phalanx rudimentary or absent D2-5, symphalangia, syndactyly, and preaxial polydactyly
Other: Syndactyly toes 2-3 and deformity of great toe

BRACHYDACTYLY—TYPE C

Inheritance: Autosomal dominant
Hands: Brachymesophalangia D2,3,5, clinodactyly D5, hyperphalangia D2,D3, symphalangia, and brachymetacarpia D1

BRACHYDACTYLY—TYPE D
(Stub thumbs)

Inheritance: Autosomal dominant
Hands: Short, broad distal phalanx thumb
Other: Short, broad distal phalanx great toe

BRACHYDACTYLY—TYPE E
(Brachymetacarpia)

Inheritance: Autosomal dominant
Hands: Brachymetacarpia and small hands
Other: Mild shortness

CARPENTER SYNDROME

See Acrocephalopolysyndactyly

CEREBROHEPATORENAL SYNDROME
(Chrs-Zellweger syndrome)

Inheritance: Autosomal recessive
Hands: Camptodactyly especially D5 and simian crease
Other: Hypotonia, hepatic fibrosis, cortical renal cysts, lissencephaly, and punctate patellar ossification

CHRS-ZELLWEGER SYNDROME

See Cerebrohepatorenal syndrome

CORNELIA DE LANGE SYNDROME

Inheritance: Unknown
Hands: Proximally inserted thumb, hypoplastic thumb, brachymetacarpia D1 and D5, simian crease, clinodactyly D5 88%, Kirner deformity, acheiria, absence of digits, and radial ray defects
Other: Growth failure, micrognathia, hirsutism, syndactyly toes 2-3, short limbs, and mental retardation with microbrachycephaly

CRANIOCARPOTARSAL DYSTROPHY

See Whistling face syndrome
See Freeman-Sheldon syndrome

CRI-DU-CHAT SYNDROME
(Partial deletion short arm of 5)

Inheritance: Sporadic
Hands: Brachymetacarpia D5-D4, clinodactyly, and simian crease
Other: Peculiar cat-like cry (only in infancy) and mental retardation

DIASTROPHIC DWARFISM

Inheritance: Autosomal recessive
Hands: Hitchhiker's thumb resulting from short first metacarpal, hand very small, bizarrely shaped epiphysis, accessory carpal bones, carpal angle increased, symphalangia, and brachydactyly (metacarpals and phalanges)
Other: Short-limbed dwarfism, clubfeet, scoliosis, deformed pinnae 82%, and cleft palate

DOWN SYNDROME

See Trisomy 21

EDWARDS SYNDROME

See Trisomy 18

F-SYNDROME

See Acropectorovertebral dysplasia

FACIODIGITOGENITAL SYNDROME

See Aarskog-Scott syndrome

FAIRBANK DISEASE

See Multiple epiphyseal dysplasia

FANCONI (ANEMIA) SYNDROME
(Pancytopenia-dysmelia syndrome)

Inheritance: Autosomal recessive
Hands: Radial ray defects, preaxial polydactyly, clinodactyly D5, and syndactyly
Other: Pancytopenia, hypoplastic genitalia, pes planus, syndactyly feet, Klippel-Feil, and congenitally dislocated hip

FIBRODYSPLASIA OSSIFICANS PROGRESSIVA

See Myositis ossificans progressiva

FOCAL DERMAL HYPOPLASIA
(Glotz syndrome)

Inheritance: X-linked dominant
Hands: Syndactyly, polydactyly, cleft hands, adactylia, clinodactyly, and camptodactyly
Other: Dermal hypoplasia, dystrophic nails, hypoplastic teeth, scoliosis, microcephaly, spina bifida, and genu valgus

FRANCESHETTI SYNDROME

See Treacher Collins syndrome

FREEMAN-SHELDON SYNDROME

See Craniocarpotarsal dystrophy
See Whistling face syndrome

GARGOYLISM

See Mucopolysaccharidosis I-H

GLOTZ SYNDROME

See Focal dermal hypoplasia

HAND-FOOT-UTERUS SYNDROME

Inheritance: Autosomal dominant
Hands: Hypoplastic thumb (short first metacarpal), brachymesophalangia D5, clinodactyly D5, pseudoepiphysis, and carpal fusions
Other: Small feet and short great toes, duplication of female genital tract, and tarsal fusions

HOLT-ORAM SYNDROME
(Cardiomelic syndromes)

Inheritance: Autosomal dominant
Hands: Bilateral upper limb defects not necessarily symmetrical, carpal fusions with accessory bones, hypoplastic and/or triphalangeal thumb, brachymesophalangia D5, clinodactyly D5, and Radial ray defects
Other: Cardiac lesions (especially atrial and ventricular defects)

HURLER DISEASE

See Mucopolysaccharidosis I-H

JEUNE SYNDROME

See Asphyxiating thoracic dysplasia

JUBERG-HAYWARD SYNDROME

Inheritance: Autosomal recessive
Hands: Hypoplastic thumb
Other: Cleft lip and/or palate, microcephaly, and syndactyly of feet

KLIPPEL-FEIL SYNDROME

Inheritance: Sporadic
Hands: Not usually affected
Other: Cervical vertebral fusions of varying degree, cervical rib, kyphoscoliosis, and basilar artery insufficiency

KLIPPEL-TRENAUNAY-WEBER SYNDROME

Inheritance: Sporadic
Hands: Macrodactyly and syndactyly
Other: Hypertrophy of skeletal and soft tissue (usually unilateral), angiomas, nevi, varices, and lipomas

LAURENCE-MOON-BARDET-BIEDL SYNDROME

Inheritance: Autosomal recessive
Hands: Polydactyly (postaxial), brachymetacarpia, brachytelephalangia, brachymesophalangia D5, syndactyly, and clinodactyly
Other: Obesity, retinitis pigmentosa, hypogonadism, and mental retardation

MANDIBULOFACIAL DYSOSTOSIS
(Treacher Collins syndrome) (Franceshetti or Franceshetti-Klein syndrome)

Inheritance: Autosomal dominant
Hands: Radial ray defects, central ray defects, and polydactyly
Other: Hypoplastic mandible, malar hypoplasia, and ear abnormalities

MARCHESANI SYNDROME

See Spherophakia brachymorphia

MARFAN SYNDROME

Inheritance: Autosomal dominant
Hands: Arachnodactyly and clinodactyly
Other: Lax joints, lens subluxation, pectus excavatum or carinatum, aortic dilatation, and kyphoscoliosis

MECKEL SYNDROME

Inheritance: Autosomal recessive
Hands: Postaxial polydactyly and syndactyly
Other: Encephalocele, polycystic kidney, cleft lip and palate, microcephaly, and dwarfism

MOBIUS SYNDROME
(Congenital facial diplegia)

Inheritance: Autosomal dominant
Hands: Syndactyly, absence of digits, polydactyly, and brachydactyly
Other: Mental retardation and cranial nerve palsy (especially VI and VII)

MOHR SYNDROME

See Orofaciodigital syndrome—Type II

MONGOLISM

See Trisomy 21

MORQUIO DISEASE

See Mucopolysaccharidosis IV

MUCOPOLYSACCHARIDOSIS I-H
(Gargoylism or Hurler syndrome)

Inheritance: Autosomal recessive
Hands: Brachytelephalangia, camptodactyly (MCP and PIP) "clawing," clinodactyly D5, delayed and irregular carpal ossification, metacarpals taper proximally, decreased carpal angle, and stiff digits
Other: Mental retardation, corneal clouding, skeletal dysplasia, deafness, death before age 10 to 20 years

MUCOPOLYSACCHARIDOSIS I-S—SCHEIE
(Formerly mucopolysaccharidosis V)

Inheritance: Autosomal recessive
Hands: Camptodactyly (MCP and PIP) "clawing," median nerve compression at wrist, impaired wrist motion, and stiff fingers
Other: Skeletal dysplasia, corneal clouding, deafness, and possibility of survival to adulthood, no retardation

MUCOPOLYSACCHARIDOSIS II—HUNTER

Inheritance: X-linked—affecting males only
Hands: Camptodactyly (MCP and PIP), "clawing," impaired wrist motion, and rigid digits
Other: Skeletal dysplasia, mental retardation, and deafness

MUCOPOLYSACCHARIDOSIS III—SANFILIPPO

Inheritance: Autosomal recessive
Hands: Mild camptodactyly and stiff digits
Other: Skeletal dysplasia, mental retardation, deafness, death at puberty

MUCOPOLYSACCHARIDOSIS IV—MORQUIO

Inheritance: Autosomal recessive
Hands: Camptodactyly (MCP and PIP), ulnar deviation of hands, joint laxity, short, thick, tubular bones, tapering of proximal metacarpals, and brachymetacarpia
Other: Skeletal dysplasia, corneal clouding, deafness, dwarfism, platyspondyly, and no retardation

MUCOPOLYSACCHARIDOSIS VI—MAROTEAUX-LAMY

Inheritance: Autosomal recessive
Hands: Limited wrist and elbow motion and camptodactyly (MCP and PIP)
Other: Skeletal dysplasia, corneal clouding, deafness, possibility of survival to adulthood, and no retardation

MUCOPOLYSACCHARIDOSIS VII—WINCHESTER

Inheritance: Autosomal recessive
Hands: Claw hand and destructive arthritis MCP and PIPs
Other: Skeletal dysplasia, mental retardation, and corneal clouding

MULTIPLE EPIPHYSEAL DYSPLASIA
(Fairbank disease)

Inheritance: Autosomal dominant and recessive
Hands: Fragmented carpals, malformed epiphysis, pseudoepiphysis, and brachydactyly (MC and phalanges)
Other: Short stature and malformed epiphysis (especially hips)

MYOSITIS OSSIFICANS PROGRESSIVA
(Fibrodysplasia ossificans progressiva)

Inheritance: Autosomal dominant
Hands: Hypoplastic thumb, brachymesophalangia D5, and clinodactyly
Other: Progressive ossification by striated muscle, pain, decreased range of motion, and hypoplastic great toes with synostosis

NEUROFIBROMATOSIS

Inheritance: Autosomal dominant
Hands: Hypoplasia, hyperplasia, coned epiphysis, syndactyly, and macrodactyly
Other: Multiple neurofibromas, cafe au lait spots, and multiple bone lesions

NOACK SYNDROME

See Acrocephalopolysyndactyly

OCULODENTODIGITAL SYNDROME

Inheritance: Dominant
Hands: Camptodactyly D5, clinodactyly D5, and syndactyly
Other: Ocular anomalies, characteristic facies, and hypoplastic teeth

OROFACIODIGITAL SYNDROME—TYPE I

Inheritance: X-linked dominant
Hands: Brachydactyly, polydactyly, mild syndactyly, clinodactyly, and camptodactyly
Other: Mental retardation, cleft tongue and palate, hypoplastic nasal ala, and polydactyly of feet

OROFACIODIGITAL SYNDROME—TYPE II

Inheritance: Autosomal recessive
Hands: Brachyphalangia, polydactyly, syndactyly, small hands, and clinodactyly
Other: Cleft lip, lobate tongue, broad nasal root, dystropia canthorum, hearing deficit, polysyndactyly of feet, and short first metatarsal

PANCYTOPENIA-DYSMELIA SYNDROME

See Fanconi (anemia) syndrome

PATAU SYNDROME

See Trisomy 13

PECTORAL APLASIA DYSDACTYLY SYNDROME

See Poland syndrome

PFEIFFER SYNDROME

See Acrocephalosyndactyly—atypical

POLAND SYNDROME
(Pectoral aplasia dysdactyly syndrome)

Inheritance: Sporadic
Hands: Syndactyly, hypoplastic hand, and brachyphalangia
Other: Unilateral absence of the sternocostal portion of the pectoralis major, absence of nipple ipsilateral, and ipsilateral hypoplastic forearm and/or arm

POPLITEAL PTERYGIUM SYNDROME

Inheritance: Dominant and recessive
Hands: Syndactyly, symphalangia, camptodactyly, and ectrodactyly
Other: Cleft lip and palate, popliteal webbing, toenail dysplasia, and scoliosis

PSEUDOHYPOPARATHYROIDISM AND PSEUDO-PSEUDOHYPOPARATHYROIDISM

Inheritance: X-linked or autosomal dominant
Hands: Brachydactyly (MC and phalanges), brachymetacarpia D4 and D5 80% to 90%, decreased carpal angle, cone epiphysis, and soft tissue calcification
Other: PHPT—hypocalcemia, hyperphosphatemia, unresponsiveness to Para-thor-mone
PPHPT—Normal blood chemistry
Both—Short stature, obesity, and retardation

PSEUDOTHALIDOMIDE SYNDROME

Inheritance: Autosomal recessive
Hands: Radial aplasia, synostosis metacarpals, and clinodactyly
Other: Reduction defects of limbs, flexion contractures, facial anomalies, long bone bowing, and humeroradial synostosis

RING D SYNDROME
(Deletion long arm 13)

Inheritance: Sporadic
Hands: Hypoplasia or aplasia of thumb, synostosis fourth and fifth metacarpals, brachymesophalangy D5, and clinodactyly
Other: Dislocated hips and spinal anomalies

ROTHMUND-THOMPSON SYNDROME

Inheritance: Autosomal recessive
Hands: Acroosteolysis, soft tissue calcification, and small hands
Other: Poikiloderma, congenital cataracts, and short stature

RUBINSTEIN-TAYBI SYNDROME

Inheritance: Sporadic
Hands: Thumb broad and deviating radially, delta phalanx thumb, clinodactyly D5, polydactyly, and simian crease
Other: Mental retardation, broad great toes, short stature, straight or beaked nose, high arched palate, and vertebral anomalies

SAETHRE-CHOTZEN SYNDROME

See Acrocephalosyndactyly—atypical

SILVER SYNDROME
(Russell-Silver dwarf)

Inheritance: Sporadic
Hands: Clinodactyly D5, asymmetrical hand size, Kirner deformity, syndactlyly
Other: Dwarfism, asymmetry, cafe au lait spots, and mental retardation

SPHEROPHAKIA BRACHYMORPHIA
(Weill-Marchesani syndrome) (Marchesani syndrome)

Inheritance: Autosomal recessive
Hands: Brachydactyly
Other: Short stature, small spherical lens, myopia and glaucoma, and cardiovascular anomalies

THROMBOCYTOPENIA-RADIAL APLASIA SYNDROME

Inheritance: Autosomal recessive
Hands: Bilateral radial ray defects, thumb hypoplastic but present, syndactyly, clinodactyly D5, and brachymesophalangia
Other: Thrombocytopenia, anemia, and leukemoid reactions

TREACHER COLLINS SYNDROME

See Mandibulofacial dysostosis

TRISOMY 13 SYNDROME
(Patau syndrome)

Inheritance: Sporadic
Hands: Polydactyly of hands or feet 76% (usually postaxial), hyperconvex nails, triphalangeal thumb, broad thumb, simian crease 92%, thumb retroflexible, radial ray defects, syndactyly, and clenched hand—index over long, small over ring
Other: Cleft lip and palate, microphthalmia, colobomas, hypotelorism, hypoplastic ear canals, cardiac defects, and delayed maturation

TRISOMY 18 SYNDROME
(Edwards syndrome)

Inheritance: Sporadic
Hands: Clinodactyly caused by brachymesophalangia, ulnar deviation between index and long fingers, radial ray defects (all degrees), syndactyly, simian crease, and clenched hand—index over long, small over ring
Other: Death in first year of life (most patients), skull elongated, abnormal facies (micrognathia and low set ears), congenital heart disease (ventricular septal defect 96%, patent ductus arteriosus 69%), renal anomalies, foot abnormalities (rocker bottom foot, short great toe with triangular distal phalanges), and delayed skeletal maturation

TRISOMY 21 SYNDROME
(Down syndrome—mongolism)

Inheritance: Usually mutant—one in 660 newborns
Hands: Simian crease in 50%, clinodactyly D5 caused by brachymesophalangia 40% to 50%, relatively small hands, increased carpal angle, and syndactyly
Other: Congenital heart disease 40% to 60% (septal defects), hypotonia, mental retardation, slanted palpebral fissures with hypotelorism, small stature, Brushfield spots, lens opacities, small skull with hypoplastic facial bones, atlantoaxial subluxation, and flattening of acetabular slopes

TURNER SYNDROME
(45XO)

Inheritance: Sterile mutant—one in 5000 newborns
Hands: Brachymetacarpia D4, D5, D3, Madelung deformity (decreased carpal angle), intercarpal fusions, osteoporosis, drumstick distal phalanx, clinodactyly, and simian crease
Other: Female phenotype, web neck, short stature, coarctation of aorta, congenital heart defects, gonadal dysgenesis, and renal anomalies (horseshoe kidney)

WAARDENBURG SYNDROME

See Acrocephalosyndactyly—atypical

WHISTLING FACE SYNDROME
(Freeman-Sheldon syndrome) (Craniocarpotarsal dystrophy)

Inheritance: Autosomal dominant—sporadic
Hands: Flexion contractures (especially thumb) and ulnar deviation
Other: Puckered (whistling) lips, short stature, small nose, clubfoot, and pectus deformity

XXXXY SYNDROME
(49XXX)

Inheritance: Sporadic
Hands: Elongation of distal ulna (plus variant) 89%, pseudoepiphysis (thumb metacarpal with distal epiphysis, D2-5 metacarpal with proximal epiphysis) 84%, clinodactyly D5 56%, and simian crease
Other: Hypogonadism, mental deficiency, lack of elbow pronation (proximal radioulnar synostosis) 32%, coxa valgus 50%, scoliosis, kyphosis, and abnormality about the elbow visible on x-ray film 100%

INDEX

Boldface numbers indicate pages with illustrations.